The Right Brain and the Unconscious

*Discovering the
Stranger Within*

The Right Brain and the Unconscious

Discovering the Stranger Within

Dr. R. Joseph

Plenum Press • New York and London

Library of Congress Cataloging-in-Publication Data

Joseph, Rhawn.
    The right brain and the unconscious : discovering the stranger
  within / R. Joseph.
      p.   cm.
    Includes bibliographical references and index.
    ISBN 0-306-44330-9
    1. Cerebral dominance. 2. Subconsciousness.   I. Title.
    [DNLM: 1. Brain--physiology. 2. Defense Mechanisms.
  3. Dominance, Cerebral. 4. Unconscious (Psychology).   WM 460.5.U6
  J83r]
  QP385.5.J67   1992
  150--dc20
  DNLM/DLC
  for Library of Congress   92-49646
                                                                 CIP

ISBN 0-306-44330-9

Printed in the United States of America

Acknowledgments

When I was a young child, I was very fortunate in that my grandmother, Mrs. Regina Becker, would read to me stories from the books of Moses and Job—and try to explain to me the unfathomable and unknown face of God. I cannot say if my grandmother's reading sparked in me an interest in biblical archaeology, philosophy, psychology, evolution, and ancient history, in particular that of ancient Egypt, but these subjects appealed to me greatly, and I increasingly spent a good part of my childhood, teenage, and adult years thinking and reading about these and related topics. It wasn't clear to me at first, but I was trying to solve a puzzle. However, it wasn't the mind of God I was attempting to discern, but our own unconscious origins. Nevertheless, although I studied Freud, Jung, Nietzsche, Sartre, mythology, ancient history, anthropology, and so on, the puzzle remained exactly that, a puzzle with many missing pieces. Fortunately, after beginning college, I met a brilliant graduate student, David Duvall, who awakened in me a new interest, the brain. I realized then that, if I could learn to understand how the brain worked, I might be able to discover the missing pieces of the enigma that had intrigued me for so many years.

Over the course of the last two decades, I pursued that goal and have had the opportunity to conduct a number of research and behavioral studies in neuroanatomy, neurophysiology, and neuropsychology and to treat and study a wide range of patients suffering from any number of symptoms and disturbances. If not for these experiences, in addition to what I have read and learned, this book and these ideas could not have been written. In this regard, I am most grateful to UHS/ The Chicago Medical School and the North Chicago Veterans Administration Medical Center, Departments of Psychology, Neurology, and Speech Pathology; the Yale/VAMC Seizure Unit, Dr. Robert Novelly and the Yale Medical School neurology and neurosurgical staff; Vanderbilt University Medical School, Departments of Anatomy and Pharmacology, and Drs. Oakley Ray and Vivian Casagrande; Dr. Jerome Siegel, the Institute for Neuroscience and the University of Delaware; the California State Department of Disability; California Home Care; the Santa Clara County and Santa Cruz County Superior Courts; and the Palo Alto Veterans Administration Medical Center, Department of Speech Pathology, and its chief, Dr. Arlene Kasprisin.

I first began writing out some of the ideas for this book almost twenty years ago, and over that time, I have been fortunate to have had a number of friends and colleagues who were willing to challenge my thoughts, assist me in my projects, and listen for hours on end to my pontificating and arguing. In this regard, I am truly indebted to Drs. J. Josephine Iwanaga, Elizabeth Birecree, Nancy Forrest, and Roberta Gallager. In producing this book, I have been very fortunate to have the assistance of Shawna Borden, R.N., and especially that of Linda Greenspan Regan, Senior Editor, Plenum Press, both of whom provided me with exceptionally skilled and extremely valuable editorial assistance and critical feedback. Finally, it might well have been impossible these last twenty years to have spent so many late nights studying, working in the lab finishing up some experiment, or reading, writing out, or researching some topic if not for the faithful companionship of three noble and highly intelligent beasts, Nietzscha, Jesse, and Sara, best friends, one and all.

Caveats

Most of the patients and cases presented here are based on case studies published by me or other scientists or patients from my private practice in neuropsychology and psychotherapy. I have taken great liberties and pains to disguise the identities of these individuals so as to maintain confidentiality, and in two instances (one involving a couple), having heard similar stories, I created fictional composites. There are a few examples that are in the public record or are meant to be taken completely in a satirical fashion; these, I believe, are obvious. It is also important for the reader to note that I have simplified many complex issues so as to make them easier to understand. For a highly detailed, fully referenced treatment of many of these topics, I refer the reader to my recent book *Neuropsychology, Neuropsychiatry, and Behavioral Neurology*. The reader should also be forewarned that the ideas expressed here reflect my own biases and that some scientists and scholars may take extreme exception to some of my statements and conclusions. I do not consider that a problem, for how else can science and understanding advance?

Contents

Introduction: The Unconscious Mind 1

I. NEURODYNAMICS

1. Freud, Jung, and the Evolution and Duality of the
   Mind and Brain 11

2. Right Brain--Left Brain and the Conscious
   and Unconscious Mind 29

3. Right-Brain Unconscious Awareness: Socialization,
   Self-Image, Sex, and Emotion 57

4. Right-Brain Limbic Language and Long-Lost
   Childhood Memories 75

5.   The Split Brain: Two Brains, Two Minds, and the Origin of Thought 91

6.   The Limbic System and the Most Primitive Regions of the Unconscious 107

7.   Speculations on the Evolution of Mind, Woman, Man, and Brain 139

II. PSYCHODYNAMICS

8.   The Four Ego Personalities and the Unconscious Child and Parent Within 165

9.   The Unconscious Child Within 185

10. Unconscious Conflicts between Child, Parent, and Self 201

11. Unconscious Parent and Child Repetition Compulsions 227

12. Repetition and Rejection: Dreaming, Self-Fulfilling Prophecies, and the Seeking of Failure 251

13. Love, Criticism, Sex, and Abuse 281

III. THE DEFENSE MECHANISMS

14. Reaction Formation and the Defense Mechanisms 301

15. The Misinterpretation of Needs: Limbic Needs 313

16. Self-Deception and Denial 329

17. Projection and the Modeling of Abuse 343

IV. APPLICATIONS

18. Choice and Responsibility 367

Notes 379

Index 407

Many fear
the unknown self
for the claims it makes
upon them.

Brain Plates

Figure 1. Lateral view of the right side of the brain exposed within the skull. From F. A. Mettler, *Neuroanatomy*, 1948. Courtesy of the C. V. Mosby Co., St. Louis, MO.

Figure 2. Lateral surface of the left half of the brain. From S. J. DeArmond, M. M. Fusco, and M. M. Dewy, *Structure of the Human Brain* (New York: Oxford University Press, 1976). Reprinted by permission.

Figure 3. Superior surface, top view of the brain. From S. J. DeArmond, M. M. Fusco, and M. M. Dewy, *Structure of the Human Brain* (New York: Oxford University Press, 1976). Reprinted by permission.

Figure 4. Medial (split-brain) view of the right half of the brain. From S. J. DeArmond, M. M. Fusco, and M. M. Dewy, *Structure of the Human Brain* (New York: Oxford University Press, 1976). Reprinted by permission.

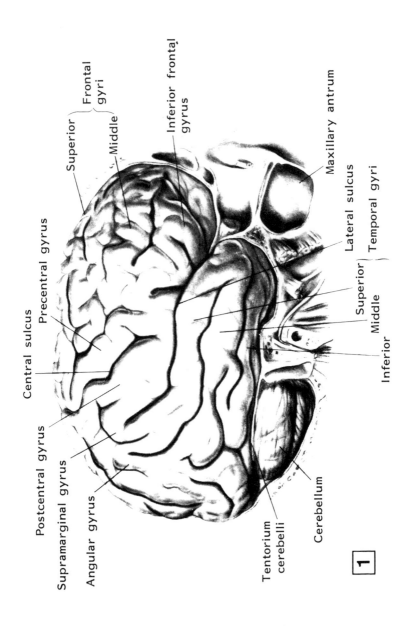

Central sulcus

Precentral gyrus

Postcentral gyrus

Supramarginal gyrus

Angular gyrus

Superior ⎤
          ⎥ Frontal
Middle   ⎦ gyri

Inferior frontal gyrus

Maxillary antrum

Lateral sulcus

Superior ⎤
          ⎥ Temporal gyri
Middle   ⎦

Inferior

Tentorium cerebelli

Cerebellum

1

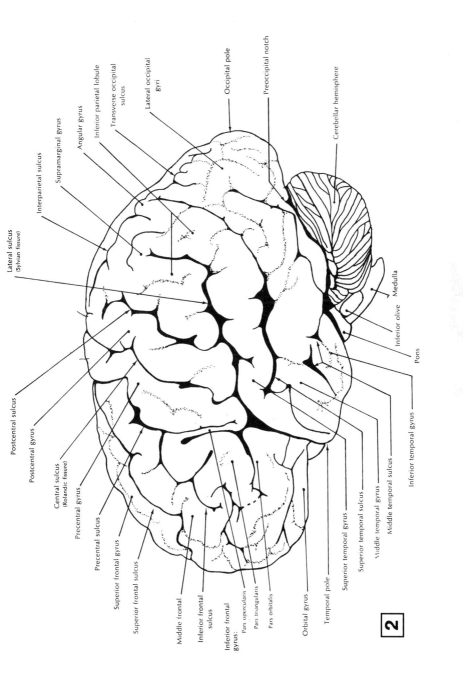

Lateral sulcus (Sylvian fissure)

Interparietal sulcus

Supramarginal gyrus

Angular gyrus

Inferior parietal lobule

Transverse occipital sulcus

Lateral occipital gyri

Occipital pole

Preoccipital notch

Cerebellar hemisphere

Medulla

Inferior olive

Pons

Postcentral sulcus

Postcentral gyrus

Central sulcus (Rolandic fissure)

Precentral gyrus

Precentral sulcus

Superior frontal gyrus

Superior frontal sulcus

Middle frontal

Inferior frontal sulcus

Inferior frontal gyrus:
  Pars opercularis
  Pars triangularis
  Pars orbitalis

Orbital gyrus

Temporal pole

Superior temporal gyrus

Superior temporal sulcus

Middle temporal gyrus

Middle temporal sulcus

Inferior temporal gyrus

2

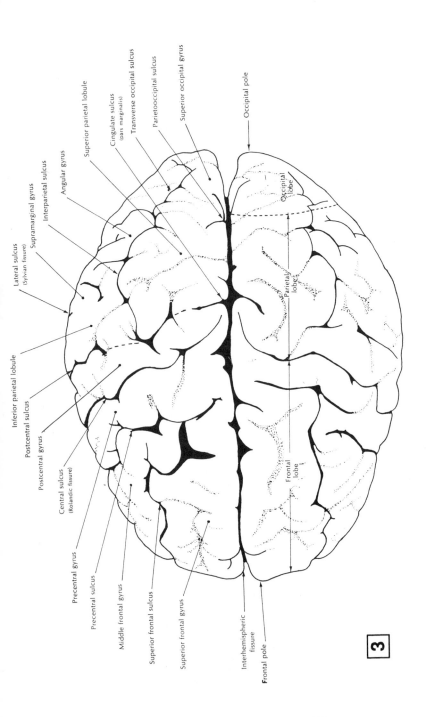

Lateral sulcus (Sylvian fissure)

Supramarginal gyrus

Interparietal sulcus

Angular gyrus

Superior parietal lobule

Cingulate sulcus (pars marginalis)

Transverse occipital sulcus

Parietooccipital sulcus

Superior occipital gyrus

Occipital pole

Inferior parietal lobule

Postcentral sulcus

Postcentral gyrus

Central sulcus (Rolandic fissure)

Precentral gyrus

Precentral sulcus

Middle frontal gyrus

Superior frontal sulcus

Superior frontal gyrus

Interhemispheric fissure

Frontal pole

Occipital lobe

Parietal lobe

Frontal lobe

3

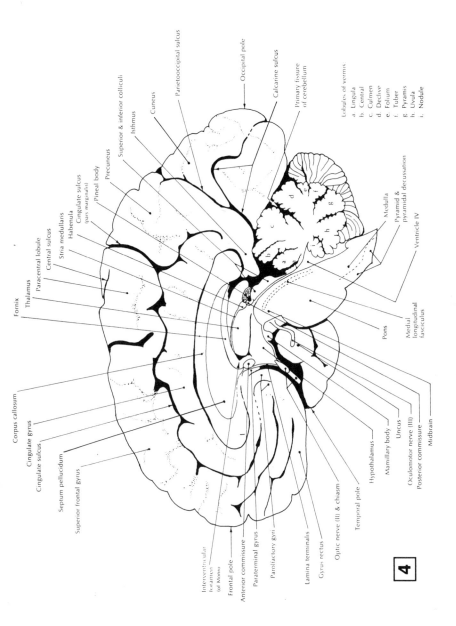

Corpus callosum

Cingulate gyrus

Cingulate sulcus

Septum pellucidum

Superior frontal gyrus

Interventricular
foramen
(of Monro)

Frontal pole

Anterior commissure

Paraterminal gyrus

Parolfactory gyri

Lamina terminalis

Gyrus rectus

Optic nerve (II) & chiasm

Temporal pole

Hypothalamus

Mamillary body

Uncus

Oculomotor nerve (III)

Posterior commissure

Midbrain

Fornix

Thalamus

Paracentral lobule

Central sulcus

Stria medullaris

Habenula

Cingulate sulcus
(pars marginalis)

Precuneus

Superior & inferior colliculi

Isthmus

Cuneus

Parietooccipital sulcus

Occipital pole

Calcarine sulcus

Primary fissure
of cerebellum

Pineal body

Medulla

Pyramid &
pyramidal decussation

Ventricle IV

Medial
longitudinal
fasciculus

Pons

Lobules of vermis

a Lingula
b Central
c Culmen
d Declive
e Folium
f Tuber
g Pyramis
h Uvula
i Nodule

4

Introduction: The Unconscious Mind

Unconscious Influences and Conscious Explanations

Some people claim they do not believe in an unconscious or the possibility that our actions are sometimes influenced by feelings or impulses that originate outside the conscious mind. There are entire schools of psychological and philosophical thought that utterly reject these notions as useless and outdated.

This view is exemplified by the various "behaviorist" philosophies espoused by John B. Watson, B. F. Skinner,[1] and their followers; that is, all behavior is the result of association and conditioning and the reinforcement of certain actions with positive rewards. Someone rewarded for performing a specific action is likely to do it again, and behavior is shaped accordingly. If rewards are withheld or provided immediately following an action, all subsequent behavior can be modified and directed via association or through the pairing of various stimuli (e.g., a bell followed by food) with certain natural responses (e.g., hunger). Even psychotic, criminal, or other types of behavior can be either created or extinguished in this manner. As a form of therapy, this has been referred to as *behavior modification*.

According to the strict behaviorists, even our conscious thoughts and feelings are irrelevant and have no bearing on how we go about navigating through reality, as they, too, are the product of reinforcement. We are thus completely at the mercy of our environment.

The behaviorists are not alone in rejecting the unconscious as a serious construct. In his magnum opus, *Being and Nothingness*, the great existential phenomenologist Jean-Paul Sartre described the notion of an unconscious as "absurd" and as being based on an attempt to escape responsibility for one's freely chosen actions.[2] Sartre also asked how it is possible to have two psychic dimensions—that is, a conscious and an unconscious mind—that are somehow linked yet separate, without resorting to magic and mysticism. Moreover, why should the unconscious hide things and try to deceive us? Isn't this just "bad faith" or, rather, "self-deception"? Others feel similarly.

Many people argue that they are very much in touch with their feelings and thoughts and in full control of their behavior. Even when they "lose control," they have a ready explanation. As far as they are concerned, notions of the unconscious seem rather obsolete, even insipid.

Indeed, many individuals, despite having repeatedly suffered from "bad luck," "bad judgment," or bad relationships, still believe it is well within their capability to reason out what has caused them to behave in a certain manner. Of course, in the process of conjuring up reasons and explanations, some people simply blame "fate," "God," their employer, or their spouse and are thus able to avoid facing their own unconscious motives and responsibility for what has occurred.

It is often this kind of thinking and overreliance on language, thought, and reason that gets them into trouble in the first place. Because we are so smart and have highly developed left-brain language and thinking capabilities, we are often able to justify and develop perfectly logical and eminently believable explanations for our behavior. Nonetheless, often all we have done is to outsmart and fool ourselves. Just because an explanation is logical and makes sense does not mean it is correct.

For example, in a study performed in a supermarket many years ago, two experimenters, R. E. Nisbett and T. D. Wilson, laid out four identical nylon stockings from left to right on a table and invited shoppers to indicate their preference.[3] The majority of shoppers picked the nylon on their far right, even though it was in all respects identical to the other three. Although the experimenters repeatedly changed the order, most shoppers still picked whatever stocking happened to be on their right.

When asked the reasons for their preference, all gave perfectly logical, rational, and sensible explanations: "The color"; "The texture"; "The shading"; "It seems more durable"; "More elastic"; "Softer"; "Smoother." However, not a single shopper came up with the real reason: The stocking was on their right. Nevertheless, all were perfectly satisfied that the explanation they provided was the correct one.

Most people (at least those who are right-handed) have a right-sided response bias. They are more likely to gesture with the right hand rather than the left when talking, and they are more likely to spontaneously activate the right half of the body when asked to make a movement. Many advertisers are aware of this response bias and usually put "Brand X" to the left and give their own product the favored right-side exposure. This is especially evident in television commercials.

A right-sided bias, of course, is not due to emotions or impulses arising from within the unconscious. It is a product of the differential organization of the human brain and the fact that we have two brains, a right brain and a left brain, which control different abilities and functions. It is because of this same differential organization of the brain that people often produce inaccurate explanations that they erroneously believe.

Unconscious Origins and Influences

The conscious mind is admittedly a very highly prized possession. Nevertheless, it is not the be-all and end-all. It is not responsible for or even involved in a good many of our everyday activities. Many actions are produced and mediated by the unconscious half of the mind and brain.

Indeed, we often engage in very complex activities without the aid of conscious reflection or scrutiny. For instance, we may engage in complex and purposeful behaviors such as dancing, throwing or catching a football, or riding a bicycle without the aid of thought, reason, or linguistic conscious assistance. We may even drive a car for miles through heavy traffic or across twisting roads while our conscious mind and our thoughts are "elsewhere."

In these instances, whatever controlled our driving seemed to function independently of the conscious mind. In other cases, such as dancing, the generation of thought may actually hinder the fluidity of movement. That is, once people begin to think about or consciously contemplate their dancing movements, they may begin to stumble or move awkwardly. Of course, while dancing we are not unconscious, for we are certainly completely *aware* of our movements.

We may also engage in other complex behaviors that are self-destructive, insulting, or inappropriate, and that also seemingly originate outside the conscious mind. Thus, we may say the wrong thing at work, use the wrong tone of voice, forget to perform some important task, be overcome with rage, or be blinded by love and "not know what came over us."

However, if the behavior originated in us, how can we "not know what came over us"? Who or what is keeping this information secret? In truth, it is the conscious half of the brain that "does not know," and in this regard, it is telling only a half-truth. If the nonconscious region of the mind had complete access to language and the speech areas of the brain and could talk, it would probably share its well-kept secrets.

When Things Go Wrong

Some people are plagued by misfortune or "bad luck." "Something always seems to go wrong," or they may be "accident-prone," "absent-minded," "too trusting," "too loving," or "too giving" and thus are always getting hurt.

Some become involved in relationships that always turn out badly, and they then have difficulty "letting go." They may be drawn and attracted to people who want to fight and argue, lie, use or abuse them, or are in need of rescuing, or to mates who break promises or dates or even cheat on them.

Others engage in obvious self-destructive behavior, acting on dangerous, ill-conceived impulses; making "thoughtless," embarrassing statements or "slips of the tongue," or offending or disappointing friends, relatives, loved ones or employers over and over again. Some people repeatedly demonstrate lapses of judgment, constantly flirt with disaster, or become engaged in ill-conceived schemes that result in financial loss or even incarceration.

When it is pointed out that they have engaged in some foolish or inappropriate action or have again become involved with someone who is not good for them, some people respond by blaming "bad luck," fate, drugs, the "blindness of love," circumstances beyond their control, or unscrupulous characters who have it in for them. A few claim to "have no idea of what came over" them, or they deny your sanity: "You're crazy"; "I didn't say that"; "My boyfriend wouldn't do that"; "I would never do that"; "It didn't happen"; "It would never happen"; "It's all in your imagination."

Sometimes people who are repeatedly subject to misfortune in love, friendship, or business may begin to suspect that there is some force or power that is influencing their lives in a negative manner, causing them to repeatedly make similar mistakes or to repeatedly experience bad luck and misadventures. They may conclude they were "born under a bad sign," committed some sin that has offended God, or deduce that it is "bad karma" affecting them from a past life. A few just blame others for their difficulties.

The Foundations of Experience

Is there some force or power that creates cycles of misfortune, bad luck in business, or a succession of unhealthy relationships? Indubitably.

However, it is not the astrological sign under which we were born or bad karma from a previous life. Although it may sound obsolete, sometimes these unpleasant patterns and repetitive experiences are partly due to the influences of two all-powerful, godlike beings who ruled our early existence: our parents. Still, there is so much more to it than that.

Some of us purposely place ourselves in self-destructive situations because we have unconsciously learned to seek out or create havoc in our lives and to associate with people who hurt us. Certain new situations provide us with the opportunity to re-create familiar feelings of unpleasantness that we experienced during infancy or childhood. Because these early experiences form the foundations for all future experience, we seek out situations that are similar, familiar, and relatively *normal* and thus tolerable, even if unpleasant.

The Unconscious "Self-Concept"

The manner in which we were treated as children and how our parents treated each other, good or bad, was observed, responded to emotionally, and stored away in memory. If we were neglected, abused, ignored, teased, or ridiculed and made to feel bad, insignificant, or incompetent by our parents or other children, these feelings, hurts, and fears not only define us, they are stored away and become part of us. These experiences become familiar, and to a child who has little with which to compare them, they seem "normal."

Fortunately, good feelings, happy memories, and impressions of warmth and love are also present. Accomplishments that made us proud, the knowledge of what we are best at, and all the little rewards we received contribute to our formation as well.

Sometimes, however, the good and the bad do not commingle and cannot be recalled or felt together. The reason is that positive emotional experiences seem to be more easily stored in verbal memory. Negative, painful, unpleasant emotions and memories are often stored in a completely different fashion. Indeed, there is some evidence to suggest that, although the right half of the brain is clearly dominant in regard to all aspects of emotional processing, the left half is generally more likely to verbally code and store in memory emotions that are positive. Consequently, the conscious mind, with which the left brain is associated, has greater access to the positive than to the negative. However, the negative and the positive are stored within the right brain as well. However, there is also much more to it than that.

Children, like adults, define themselves according to how they are treated, and then they attempt to live up, or down, to the labels and expectations others apply. It is on these emotional building blocks that our entire self-concept is erected.

We never outgrow our foundations; rather, we build on them. And foundations do not disappear or go away; they just tend to get buried. Nevertheless, our entire future rests on these foundations.

Children do not analyze or process their experiences in the manner that an adult would process similar information. Children do not have well-developed language skills and are more greatly ruled by the immediacy of emotion. Hence, a great deal of this early experience, including children's initial self-concept and all attendant joys, triumphs, traumas, and bad feelings, is internalized and stored away in that portion of the psyche that is not controlled by the language-dependent conscious mind, that is, the unconscious. This is why it is so difficult to remember events, be they good or bad, that occurred when we were young.

Infants and children are more emotional in their psychological orientation. Verbal thinking and the maturation of the conscious mind (which is heavily dependent on language and linguistic processing) develop much later. Emotion and nonconscious mental functioning are present from the very beginning.

It is within this emotional, nonconscious psychic realm that our early experiences are stored. It is this same realm that later gives birth to emotional conflicts, intuitive leaps of the imagination, creativity, slips of the tongue, "thoughtless" behavior, daydreams, and even the melodies to which we dance and sing. As we shall see, this nonconscious domain is maintained and mediated by the right half and limbic regions of the brain.

I

NEURODYNAMICS

1

Freud, Jung, and the Evolution and Duality of the Mind and Brain

(From below) cried ID: "If thou be Christ, save thyself and us."
But Superego rebuked ID, saying: "Dost not thou fear God?"

The Duality of Mind

For many thousands of years human beings have speculated on the possible existence of a nonconscious psychic or spiritual domain from which intuitive leaps, inspirations, dreams, and forbidden impulses originate. Some of the ancients believed that the ultimate source of these mysterious forms of experience were the various gods who populated the heavens. For their own unknown reasons and purposes, the gods transmitted these thoughts, dreams, and ideas into human hearts and heads. There was no unconscious mind. There were unconscious gods. For a primitive human being to truly know himself and his destiny, he had to supplicate these gods and study the dreams they sent.

Not all of the ancients, however, looked to impersonal, external gods for self-understanding. Six thousand years ago, the Sumerians, the presumed inventors of writing and civilization, worshiped and believed in a personal god whose source and abode was within themselves, and which served almost as a conscience.[1,2] It was to this personal deity that the individual sufferer bared heart and soul about personal shortcomings, sins, and misdeeds, and to which the individual admitted responsibility for bad behavior. This personal god, in

11

turn, acted as a mediator with the pantheon of gods that ruled the world at that time. However, insofar as the Sumerians (and many other ancient cultures) were concerned, the mind and soul were to be found in the blood-colored liver. Hence, if sufficient blood were lost, life was lost and the soul and spirit departed.

Several centuries later, the ancient Egyptians, concerned about "judgment day" and their well-being after death, conceived of a personal soul and spirit that represented the self in the hereafter. However, human beings have demonstrated a belief in an afterlife for well over fifty thousand years; our *Homo sapiens* cousin, the Neanderthals, during the last ten thousand years of their reign, began to bury their dead in a fetal position, to sprinkle the bodies with flowers, and even to leave meat,[3,4] presumably in case the dead became hungry on their trip to the hereafter. Hence, they, too, appear to have believed in the existence of some sort of soul or spirit.

The ancient Egyptians referred to these psychic entities as the *Ba* and the *Ka*, which they thought existed within oneself, and later, after death, within the realm of the supernatural either as a spiritlike self-image or as a godlike creative life force. Like modern humans, the Egyptians also believed the soul to be affected by sin and to contain all those aspects of the self considered both base and noble. However, it was not the brain, but the bowels and the heart in which these entities were to be found. The brain, in fact, was discarded after death, whereas these other organs were preserved with the mummified remains.

Not long after Egyptian culture and civilization rose to prominence, the people of the Indus Valley of ancient India began to worship Brahma. Brahma was described as an all-pervasive, sustaining, and creative force that exists in oneself as the most essential aspect of one's being, one's soul. In addition to this all-encompassing god within, these people also postulated the existence of a surface layer of the self that engages in the seeking of knowledge, referred to as the *atman*. It was this psychic cleavage and thus the duality of the mind and spirit (atman and Brahman) which prevented as well as made possible the seeking of one's hidden self and essential being, and thus the attainment of unity and enlightenment.

It is this same dualistic philosophical system, first written in Sanskrit and in the ancient Vedic literature, that eventually gave rise to modern Hinduism and that also influenced Plato and Kant and thus the very foundations of Western philosophical thought. Hence, the duality of the mind has been recognized and debated for at least four thousand years, if not longer. Nevertheless, with the advent of Greek philosophical thought, as advocated by Plato, and with the establishment of the Hippocratic school of physicians, around 350 B.C., the brain was finally posited as the abode of the mind and the source from which emotions arose.

Over the ensuing centuries, Western and Middle Eastern religious and spiritual thought continued to be influenced by the ancient Sumerians[5] and the Egyptians, as well as the Greeks. In consequence, a marriage of these different philosophical and religious belief systems resulted, at least in part. Hence, by the nineteenth century, the duality of the mind (including both conscious and nonconscious elements), the presence of a personal soul or spirit, the existence of a hidden unknown self, and the possibility that the seat of the mind was to be found within the brain, as well as other body organs, were widely debated beliefs.

Moreover, the "unconscious" (as well as the pineal gland of the brain) began to be viewed by some as the source within which the soul could be found. In fact, as recently as the beginning of the twentieth century, the distinguished American psychologist William James argued that the unconscious mind, particularly during sleep, acted as a doorway or portal through which God could exert his influences: "If there be higher spiritual agencies that can directly touch us," it is via "our possession of a subconscious region which yields access to them."[6]

The belief in a link between the spiritual and the supernatural and in a hidden, unknown, unconscious self continues unabated even today. Indeed, the very roots of modern-day psychology are completely entangled in such beliefs, for even the term *psychology*, as many people know, originally meant the "study of the soul."

Unfortunately, this same presumed link between the unconscious and the spiritual has so traumatized modern scientists that many of them refuse to consider the unconscious as a serious construct; they fear the taint of association and the ridicule of their colleagues for toying with what they believe to be mysticism. Indeed, this has been a major complaint regarding Sigmund Freud and Carl Jung in particular. Presumably, mystical experiences are not worthy of the attentions of a true scientist.

Nonetheless, it is noteworthy that extremely heightened feelings of the mystical, hyperreligiousness, hypermorality, and spirituality sometimes result when certain areas of the brain (i.e., the limbic system and the temporal lobes) are injured, abnormally activated, or subject to seizure disorders.[7–9] Fantastic verbal and visual hallucinations (often filled with dreamlike or religious imagery) may be experienced, and the erroneous assignment of religious and emotional significance to commonplace, mundane events may occur. In fact, individuals who were never very religious may suddenly "find religion" when this area of the limbic brain has been injured. Indeed, some scholars have even postulated that Moses, who may have been learning-disabled, had a speech disorder, and suffered from murderous rages and tremendous headaches (all signs of possible brain damage and temporal-lobe–limbic seizures), may also have suffered from a temporal-lobe epileptic seizure disorder. This, in turn, would also explain his supposed hallucinations (for example, seeing a burning bush and hearing the voice of God), his hyperreligious fervor, his hypermorality, and his need to write, which was so pronounced that, although he lacked pen and ink, he spent months and maybe even years carving his thoughts and God's laws in stone; a condition resembling hypergraphia. These conditions are all associated with an irritative lesion that abnormally activates the temporal lobes and the limbic system of the brain. However, even if he did suffer from this condition (and I am not saying that he did), does this invalidate his experiences?

The Abode of the Mind

Although the Egyptians recognized that speech disorders may result from left-cerebral injuries and the ancient Greeks acknowledged the importance of the brain in emotions, mind, and even epilepsy, the brain has only recently been seriously thought by mainstream scholars to be an organ of the mind. Indeed, as noted, the soul, spirit, or mind was thought by various ancient and even recent cultures and philosophers to reside in the liver, heart, lungs, or diaphragm, and it was from these organs that inspiration, love, emotion, and desire were believed to have their source and origin; thus the changing in the rhythm of the heart or the intensity of one's breathing when beset by strong feelings and emotions. In fact, God and the Holy Spirit were long associated with the wind and the "breath of life." However, it was also thought that these spiritlike forces might be transmitted from the various internal

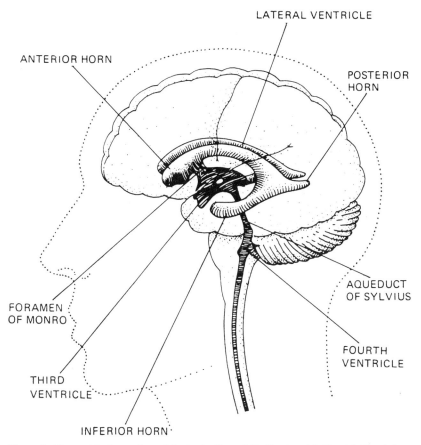

LATERAL VENTRICLE

ANTERIOR HORN

POSTERIOR HORN

AQUEDUCT OF SYLVIUS

FORAMEN OF MONRO

FOURTH VENTRICLE

THIRD VENTRICLE

INFERIOR HORN

Figure 5. The ventricular system of the brain. From J. DeGroot and J. Chusid, *Correlative Neuroanatomy* (20th ed.), 1988. Courtesy of Appleton & Lange, East Norwalk, CT.

organs to the fluid-filled ventricular cavities of the brain, where they could then be expressed.

Such beliefs still influence modern humans, popular literature and psychiatry. For example, the term *schizophrenia* originally and long ago referred to a splitting of the soul (or mind). This was thought to be caused by damage or a splitting of the phrenic nerve, which serves the diaphragm and thus the ability to breathe, and within which was the abode of the mind, hence *schizo-phrenic*.

Figure 6. The ventricular system was thought by medieval scholars to house the soul, the senses, and the mind. From Albertus Magnus (1206–1280), *Philosophia Pauperum*.

By the dawn of the nineteenth century, neither the lungs, the heart, the pineal gland, nor the fluid-filled ventricular cavities of the brain, were thought by most scientists to have much psychic significance and were viewed as exerting little if any influence on mental or emotional functioning except indirectly. Increasingly, the brain itself (and not its cavities) was being viewed as the abode of the mind. Although seriously challenged and debated, the duality of the mind and the presence of a nonconscious psychic domain continued to be recognized by many as exerting significant influences on mental activity, creativity, and emotional functioning.

The Unconscious Mind

Many modern theorists have referred to this nonconscious portion of the mind as the *subconscious*, the *preconscious*, the *unconscious*, or the *collective unconscious*. Indeed, almost ninety years ago, Sigmund Freud and Carl Gustav Jung began to develop elaborate models of the mind and posited the existence of an unconscious psychic realm in which dreams, intuitive understanding, emotional conflicts, and embarrassing and unpleasant impulses have their source and origin.

They also believed that the unconscious contained many ancient and archaic elements and was much more primitive than the more recently evolved conscious mind. Freud and Jung also believed that the unconscious, in fact, consisted of two levels. Freud classified these levels as the unconscious and the preconscious, whereas Jung posited the existence of a personal unconscious and a collective unconscious.

The Collective Unconscious

According to Jung, for a good part of our ancestral history, all mental functioning was governed by the Original Mind. Over time, consciousness appeared and became increasingly differentiated and separated from the original unconscious psyche. Consciousness is thus a relatively recent evolutionary development. This separation was not complete, however, and both remain linked.

According to Jung, the difference between the conscious and unconscious mind is, in part, a matter of distinctness, organization, and differentiation, as both tend to process and respond to information in a different fashion. If we descend from the surface levels of consciousness into the unconscious, the material we encounter becomes increasingly less distinct or logical and more disorganized, and the capacity to specifically focus the mind's eye quickly wanes. Delving deeper into the swirling mists of the unconscious, we encounter forgotten and painful memories, as well as a variety of significant impulses and ideas, some of which are about to emerge into or have just been pushed out of consciousness.

Jung noted, however, that the unconscious contains a very ancient psychic dimension, the contents of which form the core of what he referred to as the "collective unconscious."

Jung argued that, if we continue to descend we would soon enter a level of the mind where the personal layer ends. It is here that the residues of ancestral life and the earliest memories and most profound experiences of our ancient ancestors begin to be encountered. We have thus entered the collective unconscious.

The collective unconscious is an aspect of the mind and brain that functions in accordance with inherited tendencies to respond to certain experiences with emotional, spiritual, mythical, or religious feelings, images, thoughts, and ideas. Collectively, these primordial feelings and images are called *archetypes*. Archetypes are often encountered in dreams (what Freud and others have referred to as the *royal road to the unconscious*) and mythology. However, these archetypes can also ascend into consciousness and are represented in art, film, literature, and architecture.

When archetypes reveal themselves as images, it is usually as emotionally invested symbols or concepts that tend to transcend meaning or precise definition. God is such an archetype, as is the ideal of motherhood, as well as the cross and the Christmas tree, which, one might argue, are presumably symbolically descended from or related to the Egyptian key of life, the Hebrew tree of life, and the tree of good and evil (as described in Genesis), and all of which are probably related in some unknown fashion to the fact that our extremely remote ancestors at one time spent a good deal of their lives in trees for several million years. Hence, even when people from completely different cultures encounter these and other archetypical symbols, they often view them with similar emotional, spiritual, or mystical awe.[10]

For instance, in the sixteenth century, when the Spanish conquistadores first arrived in Mexico, they were astonished not only by the highly advanced culture and grand architecture they found everywhere but by what appeared to them to be the sign of the cross (as well as the swastika), which repeatedly appeared on various buildings and temples. The Aztecs and the many other tribes that inhabited these regions had never heard of Christ, however, nor were they familiar with the ancient Romans, who also used the swastika (as did the Nazis several hundred years later). Although the Aztecs held the cross in reverence, it was in no way associated with a crucified god.[11]

Jung argued that, because large masses of humanity respond similarly to certain symbols, such as the cross, across cultures, continents, and time, and because many people have similar dreams and myths with similar archaic and often mysterious or spiritual imagery, these elements must be the residue of a collective memory of the same ancestral heritage; because these experiences and dreams contain symbols or actions that have no bearing on one's personal experience, they can represent only something that is inherited. However, it is not the image per se that is inherited, it is the *feelings* and *emotions* associated with certain images and ancestral experiences, argued Jung.

Insofar as these ancestral experiences had significantly influenced many generations and large masses of ancient humanity, they eventually became figuratively engraved into the minds, memories, and even the genetic structure of *Homo sapiens sapiens*. These memories have passed down succeeding generations and are forever recalled in the dreams and myths of the masses. Hence, ten thousand years from now, our descendants will probably be dreaming of mushroom clouds and atomic explosions.

Interestingly, Jung was convinced that if we continue our journey into the farthest reaches of the collective unconscious, we will reach a point where all human memory and experience ends, and where the ancestral and spiritual memories of our nonhuman ancestors begin.

Freud's Theory of Mental Functioning

The Conscious, Preconscious, and Unconscious Mind

According to Sigmund Freud, the psyche consists of three mental domains; the unconscious, the preconscious, and the conscious mind. The unconscious, argued Freud, is the deepest, least accessible region of the mind, which contains hidden and seemingly forgotten memories and unacceptable thoughts, feelings, and ideas.

The preconscious contains information and memories that exist just below the surface of consciousness, and in this respect, it is part of the unconscious. Once information reaches the preconscious, it becomes relatively accessible to the conscious mind. However, the preconscious also contains information that is pushed out of consciousness.

Initially, all psychic material—thoughts, feelings, and desires— exists as unconscious impulses that attempt to discharge into consciousness. To reach the conscious mind, however, these unconscious impulses must first pass through the preconscious. The preconscious mind serves in some respects as a psychic corridor that links yet separates these two mental realms.

All impulses must also be approved by two censors that lurk at opposite ends of the preconscious corridor. One censor guards the door that links the unconscious to the preconscious; the other guards the door that links the preconscious with the conscious mind. If an impulse is deemed undesirable by either censor, entry into consciousness is met with resistance and is prevented. The information is either transformed into something less threatening or repressed and remains unconscious. Although repressed, it may continue to influence the conscious psyche, however.

Freud argued that the conscious mind has a very minor role in psychic functioning, being only the "outermost superficial portion of the mental apparatus." Rather, all behavior is determined by the interaction of unconscious mental forces that he termed the *id*, the *superego*, and the *ego* (the latter of which is both conscious and unconscious). Ultimately, all behavior, even slips of the tongue or forgetfulness, is due to impulses that initially exist totally beyond conscious awareness; there is no such thing as an accident, as all behavior is purposeful and serves unconscious desires.

Thus, at a conscious level, we have no control and no choices over our actions and thoughts and are instead pushed about by unconscious forces that literally have a mind of their own.[12–17]

The Ego, Id, and Superego

In developing his theory, Freud posited the existence of an ego, which simultaneously occupies the conscious, preconscious, and unconscious; a superego, which can be found in the preconscious and unconscious mind; and an id, which is wholly unconscious. The id (a term possibly coined by Nietzsche) was posited as the seat of the pleasure principle and instincts. It acts to maximize pleasant feelings and minimize unpleasant tensions.

The superego acts to instill guilt and was thought to represent an internalized parental voice. It serves in an inhibitory fashion so as to promote high standards of behavior while quashing all that which had been taught to be undesirable. At a very basic level, the id states figuratively, "I want it," whereas the parental superego states, "No! You can't have it. Don't do that. That's wrong!"

Essentially, the ego is identified with our conscious sense of self. However, it is also unconscious and acts as a mediator between the id and the superego, so that compromises can be fashioned. The ego has the additional duties of censorship, as it is the responsibility of the unconscious portion of the ego to protect the conscious aspect of the mind by standing guard at either end of the interlinking psychic corridor, the preconscious.

The ego thus serves as both mediator and censor and can enable an impulse which the superego finds objectionable to be resisted and repressed, or to be refashioned and transformed into something more acceptable to the conscious self-image. This process reduces tension by permitting the original impulse to be partially expressed, although in disguise or indirectly.

For example, the id may spy something that catches its fancy and proclaim (nonverbally), "I want it," and then try to obtain it regardless of the consequences. The superego may admonish the id for these desires and in its own unique language proclaim, "You can't have it." The ego, acting as mediator, may then try to meet the wishes of the id while attending to the standards of the superego. Instead of just taking the item (which the id urges but the superego finds objectionable), the ego pays cash and rationalizes the purchase through the generation of numerous believable explanations.

The Mind and the Brain

It is noteworthy that, long before developing his psychoanalytical theories, Sigmund Freud had spent many years studying the nervous system and the effects of brain damage on the mind. In 1895, he even wrote an almost forgotten treatise, "Project for a Scientific Psychology," in which he proposed a neurological theory of the psyche.[18] However, he soon abandoned his mind-brain proposals and, in fact later argued against attempts to draw any parallels between psychoanalytic theory and brain functioning, as he believed that simply not enough was then known about the brain.

Fortunately, much has been learned about the brain in the last one hundred years. Moreover, parallels between certain aspects of brain functioning and specific features of Freud's and even Jung's theory of the mind do, in fact, exist, and some scholars (such as David Galin, J. Jaynes, K. D. Hoppe, Robert Ornstein, and especially Laurence Miller[19–23]) have expended considerable effort exploring these and similar relationships, as limited as they may be. Indeed, a few neuroscientists and researchers who specialize in the study of the brain (such as I[24–26]) have argued that the so-called unconscious mind is in fact a manifestation of right-brain and limbic-system mental activity. Conversely, what we regard as the conscious mind appears to be localized within and to be maintained and supported via the functional integrity of the left half of the brain in most people.

The Two Brains

The human brain is organized so that two potentially independent mental systems coexist, literally side by side: one psychic system within the right, the other within the left, half of the brain. In fact, over the course of evolution, and particularly over the last fifty thousand years, each half of the brain has developed its own strategy for perceiving, processing, and expressing information, as well as specialized neuroanatomical interconnections that assist in mediating these functions.[27]

For instance, it is well known that, in the majority of the population, the left half of the brain controls the ability to speak, write, and comprehend spoken and written language and to perform arithmetical operations. It is responsible for naming, spelling, writing, and counting. The left brain also controls the right hand, as well as the ability to perform skillful sequential movements such as those involved in manual tool making, typing, and the ability to communicate via sign language.

The right half of the brain controls the left hand and is good at manipulating and constructing complex shapes, puzzles, and block designs (as in the skills of carpentry). It also controls the perception of visual, spatial, and geometric relationships, such as depth, distance, location, movement, and motion.[28] It is more concerned with the processing and mimicking of emotional, musical, and environmental sounds (e.g., wind, rain, thunder, and animal cries) than the left brain as well.[29] The right brain, moreover, is responsible for nonlinguistic forms of communicating, such as facial expression, body language, and the expression of sounds that convey emotional nuances. In fact, the right brain is associated with the capacity to sing, swear, and even pray.[30]

Speculations on the Evolution of Mind-Brain Development

Well over fifty thousand years ago, and perhaps as recently as ten thousand years ago, it is quite possible that the relationship between the two halves of the brain was quite different. As there was no math, reading, writing, or spelling in existence until recently, the left half of the brain was perhaps occupied with different functions. Indeed, until about fifty thousand years ago, our *Homo sapiens* cousins (e.g., the Neanderthals) were probably extremely limited in their ability to articulate and produce many of the sounds common among most languages today[31] simply because their voice box was quite rudimentary. As a result, they were unable to create many vowel sounds and could not form certain consonants, such as *g* and *k*. Indeed, their ability to produce speech sounds was only about 10 percent of that of modern humans.[32] Moreover, their mouth was quite big and was used as a tool,[33] reducing further their ability to articulate.

Rather, the language of the Neanderthals (a people who may have lived from 130,000 to about 30,000 years ago), like that of the primitive *Homo sapiens* who came before (who may have first appeared 500,000 years ago), may possibly have consisted of clicks and the rising and falling of tones, as well as occasional screams and yells.[34] Although they

Figure 7. Over forty thousand years after he was buried, a Neanderthal sees the light of day and seemingly glowers from his grave at those who have disturbed his rest. From R. Solecki, *Shanidar: The First Flower People*, 1971. Courtesy of Alfred A. Knopf, New York.

may well have employed a limited vocabulary of words,[35] it is likely that predominantly they communicated via gesture, facial expression, body posture, and the mimicry of animal and environmental sounds,[36] as well as through the melodic, musical, and tonal qualities of the voice. It is not likely that complex sign language played a significant role in communication, as their ability to engage in complex temporal-sequential motoric activities (such as are involved in tool making) was very primitive, their tool kit ranging variably from thirty to sixty items made of stone.[37] Hence, even those aspects of the left brain concerned with language may have been only poorly developed until about forty thousand years ago, when the Cro-Magnon people appeared on the scene.[38]

Moreover, unlike some modern-day apes that have undergone intense one-to-one individual training in sign language with Ph.D.'s and graduate students in highly sophisticated and controlled settings, and that have learned to communicate with some signs, similar opportunities were not available when our ancient ancestors roamed the planet. Nor is it likely that a well-worked-out system of signs was available to learn from.

If language was minimal, what functions did the left brain perform in our remote ancestors? Probably functions quite similar to those performed by the right brain, which then, as now, was extremely well equipped for enabling us to live among the elements as our hunting, scavenging, and gathering ancestors did so well for over a million years. Although the brain may have begun its process of differential organization almost two million years ago,[39] it nevertheless appears that, originally, we may have been equipped with two similarly functioning brains and a more-or-less unified mind that was specialized for maximizing survival among the plains, mountains, forests, and savannas.

However, perhaps because of environmental and evolutionary pressures associated with the drastic climatic changes that were occurring, genetic mutation, and the further development and refinement of preferential and temporal sequential hand use, tool making, and gathering, the right and left halves of the brain appear to have undergone a major functional reorganization about forty thousand years ago, an event that may have coincided with the appearance of our *Homo sapiens sapiens* ancestors, the Cro-Magnon. Indeed, during their reign, the frontal lobes of the brain significantly increased in size, sexual and social bonding became much more complex, and there was literally a

Figure 8. Cro-Magnon art: The head and chest of a bull and the head of a horse, Aurignacian-Périgordian. Picture gallery, Lascaux. Courtesy of Bildarchiv Preubischer Kulturbesitz.

cognitive explosion in the ability to communicate and express ideas, particularly via drawings, paintings, etchings, carvings, and sculpture[40]—talents that some scholars have argued equal those of modern artisans.

Coinciding with this rapid evolution in cerebral size and functioning, there was probably an elaboration in the capacity to speak, so that language skills may have become quite similar to those of modern humans. This process may have begun in earnest at least 100,000 years ago[41] and then possibly underwent a rapid acceleration beginning about 40,000 years ago,[42] followed by yet another possible rapid progression during the last 10,000 years with the invention of reading, writing, and math. Human beings now had the ability not only to speak and think in words but to write them down. It certainly seems plausible to assume that the brain had evolved accordingly.

It could thus be argued that, over the course of evolution, the left brain had become increasingly specialized for language, thus giving

Figure 9. Cro-Magnon art: Two small horses, Aurignacian-Périgordian. Picture gallery, Lascaux. Courtesy of Bildarchiv Preubischer Kulturbesitz.

rise to linguistic consciousness. Correspondingly, one might suppose that because of the increase in language representation, former functions maintained by the left brain probably had to be discarded or minimized in importance and thus became the sole domain of the right hemisphere. Possibly, older functions were thus crowded out of the left brain by language because there is only a limited amount of neocortical space available. However, it is likely that capacities originally associated with the right brain not only were maintained and strengthened but continued to evolve rapidly.

Indeed, these notions are not based merely on speculation, as it has been demonstrated by Robert Novelly and me,[43] as well as by many other neuroscientists (e.g., M. Dennis, H. A. Whitaker, P. S. Gott, S. Krashen, and H. Lansdell), that if the left brain is damaged early in infancy, language functions can be partly acquired by the right brain. However, when this occurs, capacities normally associated with the right half of the cerebrum are diminished because of the effects of

Figure 10. Cro-Magnon art: Hewn from a reindeer antler, a bison turns and licks its flank. This four-inch ornament was thought to be part of a spear thrower. Courtesy of Prehistoire de l'Art Occidental, Editions Citadelles & Mazenod, photos by Jean Vertut.

functional crowding. That is, language takes over cortical space devoted to later appearing right-brain functions, which, in turn, suffer because of lack of representation. Conversely, if the right brain is damaged early in life, many of its capacities may be, in part, acquired by the left brain, in which case language begins to suffer. There is only so much cortical space available, and there is much competition for representation.[44,45]

Eventually, over the course of evolution, it appears that the functioning of the two brains eventually became quite different. Where there was once an original unified mind adapted to scavenging, gathering and the hunting of small prey, now there were two rapidly evolving psychic entities, one in the right and the other in the left half of the brain:

> These semi-animals, happily adapted to the wilderness, to war, free roaming, and adventure, were forced to change their nature. Suddenly they found all their instincts deval-

ued, unhinged. A terrible heaviness weighed upon them, for in this new, unknown world they could no longer count on the guidance of their unconscious drives. They were forced to think, deduce, calculate, weigh cause and effect—unhappy people, reduced to their weakest, most fallible organ, their consciousness.

—NIETZSCHE

2

Right Brain–Left Brain and the Conscious and Unconscious Mind

The Left Brain

Language and Consciousness

Language is often a tool of consciousness. We usually think in words and use language to label, describe, and communicate our experiences. It is through language and thought that we are able to manipulate the world, describe ourselves, make predictions about the future, and symbolize aspects of the past in verbal memory and in written form. Via these modalities, we are able to analyze and describe the world as we view it and to express ourselves in a multimodal, multidimensional fashion.

Not only are consciousness, language, and linguistic thought intimately interrelated, but all are supported and maintained by the left half of the brain. It is the left brain that controls the ability to talk and think in words, and it is the left brain that listens and analyzes spoken and written language. The abilities to produce linguistic knowledge and verbal thought, to engage in mathematical and analytical reasoning, or to process and express information in a temporal-sequential, grammatical, and rhythmical fashion are associated with the functional integrity of the left half of the brain in most of the population. It is the left brain that we associate with linguistic consciousness.

In contrast, the right half of the brain cannot read, spell, or write and cannot understand many aspects of human speech except for a few simple words, particularly those that are emotional. However, it can still express itself vocally through singing, swearing, crying, praying, mimicking, or cooing sounds of love and sorrow. The right brain cannot talk. It is the left half of the brain that does all the talking.

In fact, there is one area within the left brain that controls the ability to speak, and there is yet another region that mediates the ability to understand speech. These regions are referred to as *Broca's expressive speech area*, located in the left frontal side of the brain, and *Wernicke's receptive language area*, which is found along the left temporal side of the brain.[1,2]

Language and the Left Brain

If a person were to suffer a left-brain stroke such that Broca's area were injured, word finding and the ability to speak would be greatly curtailed. This condition has been referred to as an *expressive aphasia*. It has also been called *Broca's aphasia* (named after Dr. Paul Broca, who first localized and described this symptom in detail). Interestingly, although unable to talk, many patients are still able to sing, curse, and even pray.[3–5] Such patients may have little problem understanding the speech of others.

If a patient suffered an injury to the left temporal area of the brain such that Wernicke's receptive language area was damaged, the ability to understand and comprehend human speech or written language would be greatly diminished. The reason is that the left half of the brain and Wernicke's receptive speech area are highly involved in discerning temporal sequences, including the units of speech. That is, Wernicke's area acts to organize and separate incoming sounds into a temporal and interrelated series so as to extract linguistic meaning via the perception of the resulting sequences.

Hence, when this area of the brain is damaged, a spoken sentence such as the "big black dog" might be perceived as "the klabgigdod," or as "thbelickblacdokdg." However, comprehension is improved when the spoken words are separated by long intervals.

The loss of verbal comprehension is referred to as *receptive aphasia*.

This condition has also been called *Wernicke's aphasia* (named after Dr. Carl Wernicke, who first localized and described this condition in detail). These patients can still talk, however. Unfortunately, a lot of what they say sounds like nonsense.

The Language Axis and the Train of Thought

Wernicke's area and adjacent brain tissue not only sequentially decode and organize incoming speech sounds but also act to verbally organize the temporal sequential order and the auditory-verbal associations of everything a person is planning to say. When a person wishes to talk, these verbal associations are then transmitted from Wernicke's to Broca's area, so that they can be expressed. Because of this relationship, patients with receptive (Wernicke's) aphasia often speak nonsense, as

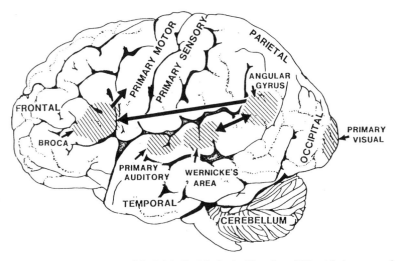

Figure 11. The language axis of the left half of the brain: Broca's and Wernicke's areas and the angular gyrus. Heavy lines and arrows indicate the probable interactive pathways involved in the formulation of language and thought. Also depicted are the primary visual, auditory, motor, and somesthetic (sensory) areas. From R. Joseph, "The Neuropsychology of Development," *Journal of Clinical Psychology* (1982).

the temporal, sequential, and grammatical order and arrangement of their outgoing speech have become abnormal. Phrases such as "the big black dog" may come out as "belick blacdok."

Wernicke's area is linked to Broca's area by a rope of nerve fibers called the *arcuate fasciculus*, which allows information to be transferred between these regions. This rope of fibers also makes contact with an area of the brain called the *angular gyrus*, which is located in the parietal lobe. These three areas form a highly interactional linguistic unit.

The Assimilation of Multiple Ideas and Associations

The angular gyrus is uniquely situated so that the areas of the brain that process visual, tactile, and auditory information are at its borders, where they all intercommunicate. That is, the angular gyrus sits at the juncture where touch is processed in the parietal lobe, where visual analyses are performed in the occipital lobe, and where sounds are analyzed in the temporal lobe. It is in the angular gyrus that auditory, visual, and tactile sensations are combined to form a multimodal representation of what is being experienced.

It is via information exchange in this border area that we are able to touch an object and are then able to visualize what it looks like and determine what it is called without even seeing it. Similarly, if we hear a sound we are also able to determine what the object might look like, how it might feel, taste, etc., as all these different associations are linked together in this area. In other words, through its interconnections with various brain areas, the angular gyrus is able to call forth ideas and relevant associations and then link them together to help form concepts and categories.

It is in this manner that when we read the word "Chair" we are able to visualize what it would look like, what the word sounds like, what it might feel like to sit in one, even how much one might cost. Moreover, we can think of a variety of chairs, all of which look and feel quite different. This is accomplished via the angular gyrus, which is able to call forth associations from other brain areas and then assimilates associations in a multimodal fashion. The angular gyrus is thus crucial in the organization of language and thought, and it is a structure unique to human beings.[6]

Together, these three brain regions—Broca's and Wernicke's area and the angular gyrus—along with other brain structures, allow for the expression and understanding of both spoken and written language. In this same way verbal thoughts are formed.

If we wish to describe something to a friend, these regions become activated: Wernicke's area begins to organize the grammatical order of what we intend to say, and the angular gyrus calls up associations from other brain regions so that what we say is properly labeled, so that we do not call a "spoon" a "stirrer" or a "fork" by mistake. All of this information is then transferred from the angular gyrus and Wernicke's area to Broca's area and our thoughts and ideas are expressed.[7]

Right-Brain Mental Functioning

There is enormous evidence that indicates that the right brain subserves a type of *awareness* that is considerably more ancient than, as well as qualitatively different from, that manifested by the left. However, this was not always the case. For many hundreds of thousands of years, the human brain, and original mind of our *Homo sapiens* ancestors, increasingly became more proficient in the hunting and stalking of prey, exploitation of the environment, and socializing as bands and tribes increased in size and complexity. With the exception of left-brain dominance for handedness, both halves of the brain performed quite similar functions for many millions of years.

With the development of language and writing, the more ancient specialties associated with the original mind probably became increasingly crowded out of the left half of the cerebrum and came to be represented only within the right brain. However, the evolution and elaboration of right-brain functions also continued unabated.

Among modern humans, the right half of the brain is responsible for discerning distance, depth, and movement; for recognizing environmental and animal sounds; and for controlling most aspects of emotion, social behavior, and body language, as well as the capacity to sing, dance, chase or throw something with accuracy, and run without falling or bumping into things. The visual, emotional, hallucinatory, and hypnagogic aspects of dreaming are also associated with right-brain mental activity.[8–14]

Music, Melody, and Visual Space

The right cerebral hemisphere is associated with nonverbal environmental awareness and the capacity to recognize emotional and environmental sounds, such as a chirping bird, a buzzing bee, a babbling brook, or a thunderstorm.[15] Related is the ability to sing and to recognize musical melodies.[16]

If select portions of the right brain were damaged, that person would be unable to recognize music, but instead might hear noise. If birds were singing or if rain drops were falling, he might be unable to recognize what he was hearing. He probably would never notice the sound. Moreover, this same right-brain-damaged patient might be unable to sing although he could still talk without much problem. He could say the words but have much difficulty singing them.

The right brain is also dominant in the perception of movement, speed, distance, and depth, as well as the geometric analysis of gestalts, angles, and visual relationships.[17] The right brain enables us to maneuver successfully about in space without getting lost, to detect and analyze the movement of others, and to determine how these actions and motions are interrelated. In this regard, the right brain was no doubt critically important in the survival of our ancient hunting ancestors, enabling them to detect, stalk, and dispatch various prey with alacrity and then to find their way home without the aid of street signs.

Because of these same right-brain visual-spatial capabilities, we can walk, run, or dance without tripping, can determine if something is near or far away, can catch or throw a ball with accuracy and can drive a car without bumping into things along the roadway, such as cars or pedestrians, while our "thoughts" are elsewhere. Likewise, if not for the proficiency of the right brain, most forms of sports, gymnastics, and athletic competition would be impossible.

Visual Closure

The left half of the brain is concerned with grammatical relationships, temporal-sequential organization (such as first to last), and the analysis of details and parts. The left brain cannot see things as a whole; it never sees the "big picture," only an assembly of parts. It cannot see the forest, only a sequence of trees.

The right brain can visualize and fully comprehend and perceive a complex visual array as an interrelated whole (or gestalt). Whereas the left brain may see "132 trees," the right brain sees a forest. Whereas the left brain sees a nose, a mouth, an ear, and lips, the right brain sees a face.

The right brain can thus "read between the lines" and infer what something may be or may mean, based on only a few features. It is able to determine how things and events may be interconnected and interrelated, so as to deduce their overall configuration and all-encompassing gestalt.[18]

As a consequence of the right brain's ability to "fill in the blanks" and deduce or determine what something may be or may mean, even when it is presented with, at best, scanty information, it is often thought of as the more intuitive half of the cerebrum.[19]

The Human Face

Partly because of its superior capacity to perform visual closure and to form gestalts, the right half of the cerebrum also mediates the capacity to recognize faces,[20] because the human face is not perceived as an assemblage of parts but as a whole. Indeed, if you were to focus on

Figure 12. An incomplete figure requiring visual closure for recognition (it is a rabbit). Reproduced with permission of author and publisher from: Gollin, E. S., "Developmental Studies of Visual Recognition of Incomplete Objects," 1960 11, 289–298. ©*Perceptual and Motor Skills* (1960).

Figure 13. Does this picture depict a scene that is happy, sad, neutral, or angry? An example of an item requiring visual closure and gap filling in order to be recognized, that is, the formation of a gestalt. The picture is of a man standing in front of an open coffin. Modified and reproduced with permission from A. E. B. Cancelliere and A. Kertesz, "Lesion Localization in Acquired Deficits of Emotional Expression and Comprehension," *Brain and Cognition* 13 (1991), 133–147.

the parts of a friend's face rather than on the face in its entirety, it would look unfamiliar.

When certain regions within the right brain are damaged, faces also begin to look unfamiliar. The patient may be unable to recognize the faces of friends or loved ones and may be unable to recognize his or her own face in a mirror, a condition referred to as *prosopagnosia*. As a consequence of this condition, some patients become paranoid and are convinced that their wife or husband has been replaced by an impostor.

The right brain also enables us to analyze facial expressions and facial emotion. Hence, we are able to tell if someone is upset, angry, sexually aroused, or disbelieving by the looks given us.

If damaged, the patient may lose this ability entirely or, with less extensive damage, may simply misread the emotions being expressed by facial gestures and eye movements.

Touch

The left half of the brain controls the movements of the right half of the body. The right brain controls the left side of the body. Similarly, the left brain perceives visual stimuli or tactile sensations that arise from the right side. However, in contrast, the right brain perceives touch and physical sensations that occur on both sides of the body. That is, whereas the left brain is limited to receiving information from only half of the body and its surrounding environment, the right brain is able to perceive sensations regardless of where they arise so as to map them to the body image. In this regard, the right of the brain is simply more sensitive to and aware of the body and its environment.[21]

The right brain is also more sensitive to touch, to pressure sensation, to painful stimuli, and even to how one is being touched, whether roughly or lovingly. In consequence, the right brain is more concerned with how others interact with us intimately and physically; it is the more cuddly, huggable, touchy-feely half of the brain.

The Body and Self-Image

If something is placed in your hand, this information is transmitted to a select region of the brain called the *parietal lobe*, where it is then analyzed for whether it is hard, cold, short, thin, pointed, rough, wet, and so on. In this way, you are able to recognize what you touch and what is touching you. The parietal lobe is specialized for analyzing sensory sensation regardless of where it occurs on the body. If an insect lands on your leg, nerve cells in the parietal lobe become activated, and you know that something has landed and is crawling, and you can localize the sensation as coming from your leg.

Because all sensations from the body are transmitted to the parietal lobe, a sensory representation or image of the body surface is maintained by the many specialized nerve cells in this brain region. Thus, all sensations from the arm go to the region of the parietal lobe that receives arm sensations, and all information from the leg go to the region where leg sensations are analyzed.

If we were to perform brain surgery and electrically stimulate the nerve cells in the parietal lobe where this sensory image of the body is maintained, the person would think that the corresponding part of the body was actually being touched and stimulated—when in fact it was not. That is, he may think someone was touching him on his arm when in fact it was his brain that was being stimulated. Since that part of the brain that represents a specific body part is activated, the brain assumes that the sensation must be coming from the body.

The greater the sensory importance of a particular body part, however, the more densely will it be represented in the brain. For example, there are a large number of nerve cells that receive information from areas of the face and hand. In consequence, there is a large area of the brain that corresponds to the fingers or lips as these areas of the body are very sensitive and important. By contrast, very little brain tissue is devoted to the representation of the forearm, as few events of sensory importance occur in this region and thus few nerve cells are needed. Thus, the image of the body that is mapped out and maintained within the brain is distorted.

If a person were involved in an accident and lost an arm or a leg, the sensory body image maintained in the brain would not be affected and the amputated body part would continue to be represented in the parietal lobe. This is why some amputees continue to *experience* and *feel* as if their absent limb were still present. Such a limb is referred to as a *phantom limb*, and a person may think she or he can feel a hand or arm even though it has been cut off long ago.

As we have two parietal lobes, one in the right brain and the other in the left brain, sensory body images are maintained in both. In general, because the left brain receives sensation from only the right side of the body, it maintains a sensory image of only that half of the body. Only half the body is represented in the left brain. In contrast, because the right brain receives sensation from both sides of the body, it maintains a sensory image of the entire body. In fact, the ability to imagine or visualize our bodies is made possible by the right brain. The left brain is thus comparatively limited insofar as body image and self-image are concerned.

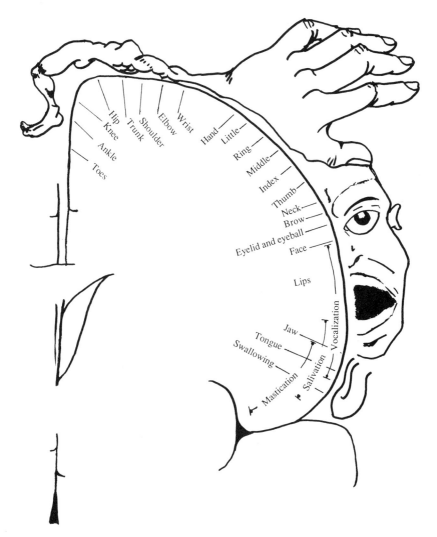

Figure 14. The map of the body as maintained in the right parietal lobe. Note that body parts are distorted so that those areas with the greatest sensory importance have the most extensive representation within the brain. From R. Joseph, *Neuropsychology, Neuropsychiatry, and Behavioral Neurology* (New York: Plenum Press, 1990).

Disturbances of the Body Image: Neglect and Denial

When the parietal lobe of the brain is damaged, physical sensory functioning and the ability to perceive touch may become grossly abnormal. Although sensory information continues to be transmitted toward the brain, if the brain area that is supposed to receive it has been destroyed, the signal will not register. In consequence, patients may experience peculiar disturbances in body image.

If damage is quite massive, awareness of corresponding body parts would be abolished. The body image would have been erased. However, because the right half of the body continues to be represented in the left brain, patients continue to be conscious of the right half of their body; unfortunately, the left side of the body is no longer acknowledged because its image has been destroyed.[21]

In consequence, patients with right-brain injuries may fail to perceive stimuli applied to the left side; may wash, dress, or groom only the right side of the body and ignore the left side; may confuse body-positional and spatial relationships; may misperceive left-sided stimulation as occurring on the right; may fail to realize it when their extremities or other body organs are in some manner compromised; and may literally deny that their left arm or leg is truly their own. They may, in fact, ignore all of the left half of space.

For example, I was asked one day to examine a woman who had suffered a massive right-brain stroke. When I walked into her hospital room, I noticed that she had brushed only the right side of her hair and had put makeup on only the right side of her face. Her robe had also been put on only partially, the right arm correctly through its sleeve, the left arm and left half of her body remaining undraped. Her left arm lay limply in her lap.

After I had introduced myself and shook her right hand, I asked her to raise her left hand. She smiled at me and raised her right hand:

Dr. J: No. That's your right hand. I want you to raise your left hand.
Patient (again raising her right hand): This is my hand.
Dr. J: Yes. That is your right hand. I want you to raise your left hand. Can you raise your left hand?
Patient (again raising her right hand): Yes. See?
Dr. J (picking up her left hand): No. I want you to raise your left hand. Can you raise this hand?
Patient (looking at her left hand): That's not my hand.

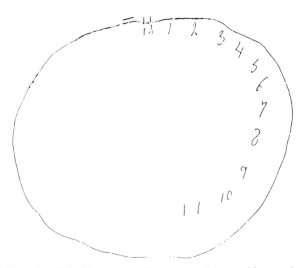

Figure 15. Example of left-sided neglect. The patient, a 54-year-old man who suffered an injury to the right half of the brain, was instructed to "draw the face of a clock and put all the numbers in it and make it say 10 after 11." From R. Joseph, *Neuropsychology, Neuropsychiatry, and Behavioral Neurology* (New York: Plenum Press, 1990).

Dr. J (still holding her left hand): Sure it is. This is your left hand. Can you raise it?
Patient (becoming upset): That's not my hand.
Dr. J (waving her hand in front of her face): It's not? Whose is it, then?
Patient: I don't know. Maybe it's yours?
Dr. J (showing her both his hands): No. It's not mine. I already have two hands.
Patient (becoming angry and irritable): Well, it's not mine. Maybe it belongs to my husband.

Although responses such as these are seemingly bizarre, these patients (that is, the talking, left half of their brains) are in fact telling the truth. The left hand belongs to the right brain, which in their case is damaged and no longer works properly. The body image of the left hand has been erased. The left (talking) half of the brain, being responsible only for the right hand and now having no information about the left hand (because this information has been deleted), thus answers correctly: "That is not my hand"; that is, it is not the left brain's hand.

Figure 16. Examples of left-sided neglect and distortion. Note the preservation of the right-sided details. The patient, a 33-year-old woman with damage to the right brain, was instructed to copy the "cross and the star." From R. Joseph, *Neuropsychology, Neuropsychiatry, and Behavioral Neurology* (New York: Plenum Press, 1990).

Why doesn't the patient remember that she has a left hand? Because those memories are stored only in the right half of the brain. The left brain has memories all its own. Hence, because the left brain also has no memory of a left hand, denying the ownership of a left hand does not seem preposterous to it. To claim a hand that you have no memory or image of would be odd.

This is yet another example of how the left brain likes to make up explanations (which it then believes) when information normally possessed by the right brain is not available to it. In severe cases, this is called *gap filling* and is also referred to as *confabulation*. That is, the left half of the brain makes up explanations to fill in the gaps in the information it receives, and then it believes its explanations.[22] Because the left half of the brain often has only incomplete access to certain forms of information, impulses, and desires, confabulation is an explanatory activity in which it frequently engages. Everyone confabulates at one time or another: Recall the supermarket experiment with the nylon stockings.

Vocal Melody, Intonation, Emotion, Meaning, and Inference

Have you ever had someone say to you, "It's not what you said. It's the *way* you said it"? Although you (or rather your left brain) may not have known what this person was talking about, your right brain certainly did. It is your right brain that controls the *way* you say things. It also controls your ability to hear things said in a certain tone or manner.

Although language is usually discussed in terms of grammar and vocabulary and is associated with the conscious mental processing of the left half of the brain, there is yet another major aspect of linguistic expression and comprehension by which a speaker may convey and a listener discern intent, attitude, feeling, context, and meaning. That is, language is both descriptive and emotional. A listener comprehends not only *what* is said, but *how* it is said—what the speaker *feels*. Feeling, intent, attitude, and related emotional states are perceived, processed, and expressed by the mental system of the right half of the cerebrum.

Feeling, be it sadness, anger, or empathy, is often communicated by a variation in the rate, amplitude, pitch, inflection, timbre, melody, and stress of the voice. Based on studies of normal and brain-damaged subjects, it is now well established that the right brain is superior to the left in distinguishing, interpreting, and processing these particular vocal nuances, including intensity cadence, emotional tone, amplitude, and intonation.

Through these abilities, the right hemisphere is fully capable of determining and deducing not only *what* another person *feels* about what he or she is saying, but *why* and in what *context* he or she is saying it—even in the absence of vocabulary and other denotative linguistic features. This process occurs through an analysis of the tone and melody of the voice.

Hence, if I say to you, "Do you want to go outside?" although both hemispheres are able to determine whether a question has been asked or a statement has been made, it is the right brain that analyzes the paralinguistic features of the voice to determine whether "going outside" will be fun or whether I am going to punch you in the nose; the left brain knows only that it is being asked to go outside. Indeed, it is through the perception of similar melodic and prosodic qualities that our dogs and cats are sometimes able to understand what we are saying and what we are feeling.

Determining Context and Drawing Inferences

Even without the aid of the actual words (if they were experimentally filtered out and deleted), and based merely on vocal tone, the right brain can determine the context and the feelings of the speaker. It can deduce if someone is being sarcastic or is feeling sad, or even if he is lost and asking directions, just by listening to the tone and melody of his voice. By an analysis of these same vocal features, it can tell if someone is being dishonest or is attempting to mislead us. For these and other reasons already noted, the right half of the brain is sometimes thought to be the more intuitive and inferential half of the cerebrum.

Correspondingly, when the right brain is damaged, the ability to process, recall, or even recognize these melodic and inferential nuances may be greatly attenuated.[23] For instance, right-brain-damaged patients have difficulty understanding context and emotional connotation, drawing inferences, relating what is heard to its proper context, and recognizing discrepancies. Hence, they are likely to miss the point, respond to inappropriate details, and fail to appreciate fully when they are being presented with information that is incongruent or even implausible. In other words, they fail to perceive the "big picture" and can no longer tell if "things don't add up," or if someone is being dishonest. This failure indicates not only that the right brain subserves these functions, but that the left brain does not have these capabilities.

These patients also tend to be overly concrete and literal. If told to read the sentence "He has a heavy heart" and then instructed to choose a cartoon that matches it, such as a picture of a man crying or a man weighed down by a large red heart on his shoulders, a patient with a right brain injury would pick out the picture with the heart.

If we didn't know that such patients had brain damage, or if the injury were quite mild (so that neither they, their family, nor their doctor were aware of it), we might think that they were simply quite dense. In this regard, the left brain is quite socially and emotionally dense. In fact, many of us have met seemingly normal people who are completely devoid of the ability to "read between the lines," who fail to infer what we mean, and who just don't "get the drift" of the conversation. Such people are too literal-minded to realize when others are kidding, being sarcastic, or being deceitful, or saying things that "just don't add up." They are just not using their right brain properly.

Discerning and Conveying Feelings

When devoid of melodic, intonational contours, language becomes a monotone, very bland, and a listener experiences difficulty discerning attitude, context, intent, and feeling. Conditions such as these arise

Functions of the Left and Right Brain

The left brain controls:	The right brain controls:
The right half of the body	The left half of the body
The right hand	The left hand
The right visual field	The left visual field
Talking, reading, writing, and spelling	Emotional and melodic speech
Speech comprehension	Comprehension of music and emotion
Temporal and sequential information processing	Insight and intuitive reasoning
Keeping score of a football game	Visual-spatial processing
Math	Throwing and catching a football
Marching	Riding a bicycle
Grammar	Dancing
Logical and analytical reasoning	Visual closure
Confabulation	Gestalt formation
Perception of details	Perception of environmental sounds
	Social-emotional nuances

after damage to select areas of the right hemisphere, or when the entire right half of the brain is anesthetized (e.g., during sodium amytal procedures) in preparation for neurosurgery.

If the right frontal part of the brain is damaged, the melody of one's voice is altered, and the ability to express one's feelings vocally is compromised. One's voice may become flat, like a monotone, or wildly distorted, so that friends and loved ones may have trouble understanding what is being conveyed. With right-frontal damage, the person may also lose the ability to sing and carry a tune.

James Brady, the former press secretary to President Reagan, had a problem similar to this after being shot in the right frontal portion of his brain. When speaking, he would sometimes lose control over his voice and would sound as if he were crying or wailing. Fortunately, he has since recovered significantly.

If damage occurs within the right temporal lobe, the person loses the ability to hear melodic nuances, including music. Thus, she may misperceive what is being conveyed.

Emotion

The right half of the brain is dominant for the perception and expression of most aspects of emotion. When most people smile or frown, the left half of their face (which is controlled by the right brain) is more emotionally expressive. If one were to cover up half of her or his face and look in a mirror, one would also notice distinct emotional differences in expression. When people speak or think emotionally, the eyes tend to dart to the left, and the left arm becomes more active.

Similarly, the right half of the brain becomes more electrophysiologically activated when a person is emotionally aroused. Even sexual orgasm and the experience of body pain are associated more with right-brain activation than with left-brain activation.

As noted before, the right brain is more adept than the left in reading emotional facial expressions, and in determining mood and emotional state by the analysis of a person's voice and body language. It is even more sensitive to the emotional nuances conveyed by touch.

The right brain is therefore highly involved in many aspects of emotional functioning. If it is injured, emotional functioning may become quite abnormal.

Though a person may become depressed with damage to the left or the right brain, right-brain damage may give rise to a wide variety of emotional disturbances, including mania, the loss of inhibitory restraint, impulsiveness, childishness, emotional lability or numbness, delusions of grandeur, and hysteria. Hence, the right brain is dominant over almost all aspects of emotion.[24]

Moreover, the right brain not only is dominant in the expression and perception of emotion but is also responsible for storing these experiences in memory. Thus, whereas the left brain stores language-related memories, the right brain memorizes and recalls emotional experiences.

Left-Brain Limitations

The right and left brains often act cooperatively. However, sometimes the left brain tries to dominate and tries to perform a task that is actually performed better by the right brain.[25] In some instances, the left brain interferes with, suppresses, and inhibits right-brain mental processing so that its capacities and abilities are not expressed. That is, they remain confined within the mental system of the right brain and their presence is not even hinted at. This is unfortunate, as two brains are better than one, and it is very self-limiting in that the left brain has difficulty processing and recognizing emotional signals, particularly those that are negative.

Sometimes, the left brain, acting in its dominant mode, so interferes with and attempts to inhibit and ignore the right brain that certain emotions and feelings conveyed by others are not perceived and recognized. The person thus becomes a social dunce or has problems in relationships. When the left brain interferes with the right and tries to dominate all interaction through logic and reason, it causes the person to completely miss the boat when it comes to successful social, emotional, and intimate interactions.

Sometimes, the left brain just does not want to know what is really going on emotionally as the information may be too painful or upsetting to consider consciously. Indeed, the left hemisphere may deny the significance of an intuitive conclusion drawn by the right half of the brain, even when someone is pointing it out.

Have you ever heard the saying "The spouse is always the last to know"? The spouse is the last because he or she (or rather the left brain) has ignored all the hints and clues normally picked up by the right half of the cerebrum.

This is compounded by the natural limitations in the ability to transfer or recognize information being processed by the opposite half of the brain. That is, some forms of information cannot be shared by the right and left brain, and some forms of information cannot be perceived or even recognized by the right or the left brain. For example, the left brain can't sing, and the right brain can't do math. This is the consequence of having two brains, each of which is specialized. Still, by having two brains (or, rather cerebral hemispheres), we can do twice as much.

Evolution: The Right and Left Brain and the Unconscious and Conscious Mind

It is important to emphasize that the right and left halves of the brain are not completely dichotomous and that they share many of the same abilities. Some functions are simply processed much more efficiently and effectively in one than in the other half of the brain.

This fact is not surprising because, for many millions of years, the right and left halves of the brain probably subserved many of the same functions and processed the same types of information in a similar fashion. Indeed, although the brain appears to have possessed some rudimentary temporal-sequential capabilities as long ago as two million years,[26] both the right and the left cerebral hemispheres were probably also equally adapted for processing visual-spatial information and environmental sounds. This adaptation, in turn, would have been reflected in the activities that ancient men and women probably engaged in fairly equally, that is, wandering over long distances, scavenging, hunting, and chasing small game, as well as gathering.

However, the capacity of the human brain to process visual-spatial, distance, depth, movement, and other environmental variables probably became enhanced even further with the onset of big-game hunting about 300,000 years ago.

Hunters had to be extremely aware of their environmental surroundings and able to read even subtle nuances that might inform them of the presence of an antelope herd, the nearness of a dangerous predator, or even the location of water. These ancient hunters would be able to distinguish via footprints and hoof marks, broken versus bent branches, the size, color, consistency, and texture of feces, the smell of urine, the manner in which a twig had been broken, or the color and size of a puff of hair hanging on a branch, and so on, what was left by a large versus small, male versus female, young versus old, sick, wounded, or feeble versus strong, powerful, and healthy animals and would then be able to plan the hunt accordingly.

In contrast, a modern man, wandering perhaps upon the same path, might not even know that a single animal, much less an entire herd of beasts, had passed the same way perhaps just hours before, even though all the same clues are right there before him, albeit in an invisible and unreadable form.

The ancient hunters traversing their own territory knew it as well as modern humans know their own backyard. Any alterations and every broken twig and vine stood out with significance and had a story to tell.

However, that story was in images, feelings, tastes, odors, expectations, memories, and emotions. Because our very remote ancestors did not read or write and did not speak in any way remotely resembling the grammatical utterances of modern humans, they had more brain space than modern humans to devote to performing what for them was central to their lives, functions that today are dominated by the right half of the brain and the limbic system. However, a major divergence in functional capability (i.e., handedness) possibly began to exert a profound effect on how the two brains processed and responded to information well over a half million years ago and particularly during the last fifty thousand years, with the invention and refinement of complex tool construction.

Over 80 percent of our modern-day population is right-handed. One reason that most of us become right-hand-dominant is that, during prenatal development, nerve cells and fibers in the left brain that control hand and arm movement begin to mature and grow more quickly than those in the right brain.[27] This development is genetically programmed. We become right-hand-dominant because the left brain gains a competitive advantage and obtains a head start in regard to motor control and development. Still, there is more to it than that.

The Knowing Hand

Handedness, Language, and Temporal Sequencing

It has been repeatedly argued by a number of investigators (e.g., D. Kimura, A. M. Liberman, M. Studdert-Kennedy, J. L. Bradshaw, N. Nettleton, and M. Kinsbourne) that language, and in particular its grammatical and syntactical components, is directly related to handedness and motor control. Among the majority of the population, the right hand is dominant for grasping, manipulating, exploring, writing, creating, destroying, and communicating. That is, although the left hand assists, the right usually is more frequently used for orienting, pointing, gesturing, expressing, and gathering information on the environment. The right hand appears to serve as a kind of motoric extension of language and thought in that it acts at the behest of linguistic impulses.

Moreover, while talking, most individuals display right hand/arm gestural activity which appears to accompany and even emphasize certain aspects of speech. When one is speaking the areas of the brain controlling right-hand use become activated, in part because of the spread of neural excitation from the speech area to the immediately adjacent cortical regions that control hand movement. Indeed, both occupy, to some degree, the closely aligned cortical space in the frontal lobe and rely on similar neural centers for programming. Hence, because they are so intimately linked, speaking triggers hand movement because of the spread of neural excitation.

Naming, Knowing, Counting, Finger Recognition, and Hand Control

A variety of theories have been proposed to explain the evolution of handedness and language. Nevertheless, from an evolutionary, phylogenetic, and ontogenetic perspective, handedness and temporal-sequential motor control probably preceded the development of language-specialized nerve cells.[28]

Indeed, it is first via the hand that one comes to *know* the world so that it may be named and identified. Hence, the infant first uses the hand to grasp various objects and to place them in the mouth for oral exploration. As the child develops, more reliance is placed solely on the hand (as well as the visual system) so that information is gathered through touch and manipulation.

As the child and its brain mature, instead of predominantly touching, grasping, and holding, it uses the fingers of the hand for pointing at an object and then naming it. These same fingers are later used for counting and the development of temporal-sequential reasoning; that is, the child learns to count on his or her fingers and then to count (or name) by pointing at objects in space.

In this regard, counting, naming, object identification, finger utilization, and hand control are ontogenetically linked. In fact, these capacities seem to rely for their expression on the same neural substrates, that is, the left inferior parietal lobe and the angular gyrus. Hence, when the inferior parietal region of the left brain is damaged, naming (*anomia*), object identification (*agnosia*—a term coined by Freud), arithmetical abilities (*acalculia*), finger recognition (*finger agnosia*), and temporal-sequential control over the hands and extremities (*apraxia*) are frequently compromised. Sometimes, in fact, all these symptoms may occur together and are referred to collectively as *Gerstmann's syndrome* (after the doctor who first recognized this association). The presumed link is not only the hand, however, but the processing of information in temporal sequences.

For example, a patient with damage in this area and suffering from apraxia and thus a loss of temporal and sequential coordination is unable to "pretend to open a tube of toothpaste, pretend to squeeze it out on a toothbrush, and then pretend to brush the teeth." The apraxic person confuses the order and may first pretend to brush the teeth and then squeeze out the paste and so on. Certainly, a person with apraxia would be unable to carry out the sequences involved in complex manual tool-making.

Relationships such as these lend considerable credence to the argument that, over the course of evolution, the predominant use of the right hand enabled the left brain to develop nerve cells specialized for counting, naming, temporal-sequential processing, and thus for the mediation of grammatical-syntactical speech and language.

However, it is equally important to emphasize that the social and emotional sounds of speech, those language capabilities more clearly associated with the right half of the brain and the limbic system, had no doubt been in complex use for at least a million years before the first tool was ever made. Social and emotional sounds were the first form of true vocal communication, and are still used by humans and animals today.

It is complex tool making and temporal-sequential, grammatical speech that are more unique to humans, functions that are still associated with the right hand and the left half of the brain.

Speculations on Tool Making and Language

There is some evidence that nonhuman primates and even some birds have functions that are more greatly concentrated in one than in the other half of their brains. However, among almost all other animals, about 50 percent tend to use their left limbs, and about 50 percent favor their right limbs. Animals are neither right- nor left-hand-dominant as are human beings. Our own human ancestors were probably also without preference until at least about two million years ago.

Once the right-hand preference became established, the left brain continued to evolve and adapt as it increased its proficiency in right-hand motor control. Tool making also evolved, as did the ability to make more efficient weapons and hunting implements. Indeed, around 2.4 to 2.6 million years ago, simple stone tools were being struck (by *Homo habilis*, the presumed ancestor of all species of *Homo*, including modern humans), and by 100,000 years ago, with the appearance of our remote *Homo sapiens* ancestor, the Neanderthal, humans were making a variety of very simple stone tools in abundance. However, it was not until about 40,000 years ago that tool making became literally an art and evolved beyond the use of rocks and bones.[29]

Tool making and tool factories, be they primitive or modern, require repetition and order as the implements are fashioned. First, you take your bone and scrape; then you heat; then you sharpen, and so on—so that a temporal sequence is established and the same type of tool can be made over and over again. Hence, by 100,000 years ago, the left brain not only was dominant over hand and motor control but was beginning to become increasingly proficient in temporal and sequential processing.[30] In part, this increasing specialization was also related to gathering activities, an intensive full-time affair that involved repeated temporal and sequential hand movements.

It is not unreasonable to assume that although quite primitive, the Neanderthal people had names for certain of their foods and tools, as well as for the implements they used to construct, gather, or prepare

Figure 17. Cro-Magnon art: The leaping horse from Bruniquel, France, carved from a reindeer antler. This artifact was possibly part of a spear thrower, or it may have served any number of unknown purposes—perhaps art for art's sake. Courtesy of Prehistoire de l'Art Occidental, Editions Citadelles & Mazenod, photos by Jean Vertut.

them.[31] Hence, if a chopper was needed, our primitive ancestor may well have been able to ask one of his compatriots to hand it to him by name.

Although admittedly such notions are speculative, it is seems likely that, by 100,000 years ago, the left brain not only was continuing to become adept in motor control and temporal sequencing but, at a minimum, was now using a limited vocabulary of words.[32] In fact, in addition to mimicry and its emotional-social language capabilities, the right brain was probably able to say the same words as well.

Nevertheless, although major differences had begun to occur, the right and the left halves of the brain were probably still similar in many ways, and language capabilities probably remained quite limited because of limitations in brain development and a poorly developed voice box.

The ability to engage in complex conversational speech probably remained severely limited until almost 40,000 years ago with the appearance of the Cro-Magnons.[33] Interestingly, the Cro-Magnons were probably as handsome, stood as tall, had fully developed vocal capabilities, and had a larger brain than present-day *Homo sapiens*.[35] These are the same people who 15,000–40,000 years ago left hundreds of paintings, drawings, etchings, and fine sculptures of bisons, deer, wild horses, and bears in the caves and rocky cliffs of eastern and southern Africa (where a few scholars have speculated that they may have appeared almost 100,000 years ago), Europe, Russia, the Middle East (where they began to appear in increasing numbers about 40,000 years ago), and possibly the Americas (where the remnants of hearths and base camps, dating anywhere from 20,000 to 38,000 years ago, have recently been discovered). The Cro-Magnons brought to the world complex spiritual beliefs, vivid and colorful imagery, a finely developed artistic expression, and pictorial language.[36]

A wild acceleration in the evolution of the mind and the brain may have begun to escalate about 40,000 years ago, at which time mental functioning became quite complex and profound.[37] Humans appear to have become self-conscious, and the two halves of the brain seem to have become increasingly adept in performing new functions. People maintained a home base, built large houses with stone foundations, and lived in villages, sometimes including as many as five hundred individuals. They sewed their clothes; wore makeup, jewelry, and personal decorations; believed in an afterlife; and had complex supernatural beliefs and rituals that they regularly performed. The Cro-Magnons buried their dead with flowers, clothing, ornaments, beads, headbands, necklaces, weapons, and offerings of food. They became extremely proficient in hunting, gathering and harvesting, tool making, and temporal-sequential processing; had names for each other and their tools; and were *working*—that is, engaged in food procurement—about two to four days per week on average. In contrast, those who had preceded them probably spent a good part of every day in the seeking of food. By 40,000 years ago, the Cro-Magnons were probably capable of engaging in complex conversations and perhaps because of the increased leisure time they now enjoyed, were creating paintings and sculptures in profusion.

With the appearance of language, profound artistic expression, self-consciousness, and right- and left-brain functional specialization, a

Figure 18. The skeletal remains of a Cro-Magnon man discovered in Russia, estimated to be over 25,000 years old. Note the beads, bracelet, and headband. The grave and body were sprinkled with ocher (a pigment). Courtesy of the Novosti Press Agency.

schism had formed in the psyche of humans. Whereas before there had been a more-or-less unified mind, now there was a new and additional form of mental processing, which also gave birth to a profound creative spirit as well as the capacity to reason and form complex thoughts.

By at least 100,000 years ago, with the arrival of the Neanderthals, a fragile and minimally developed linguistic consciousness probably emerged from what had been the original mind. As this linguistic consciousness evolved, so, too, did the left brain. As language, linguistic consciousness, and temporal-sequential processing increasingly came to dominate left-brain mental functioning, other capacities possibly became displaced. In consequence, functions that both the right and the left brain had expertly performed increasingly became the sole or dominant domain of the right half of the brain, whereas language and temporal-sequential processing became the hallmark of the left half of the cerebrum.

This original mind has not been discarded, however. Rather, as the left brain became increasingly associated with language and linguistic consciousness, this original mind appears also to have evolved and to have become more intimately associated with the right cerebral hemisphere. It is probably for these reasons that, among modern human beings, the right brain is associated with the presumably more primitive unconscious, whereas the left brain maintains what we have referred to here as the more recently evolved, language-dependent, conscious mind.

3

Right-Brain Unconscious Awareness
Socialization, Self-Image, Sex, and Emotion

Information Transfer and the Corpus Callosum

If we were to remove the top of someone's skull and look straight down, we would discover that the human brain is divided into halves by a large fissure called the *interhemispheric fissure*. These halves form the right and left brains or, rather, cerebral hemispheres. In other words, half a brain (or half a cerebrum) is a hemisphere.

If we were to squeeze our fingers down into this fissure, our progress would soon be interrupted by a large rope of nerve fibers that interconnects the two brain halves. This great bundle of fibers is called the *corpus callosum*.

The nerve fibers that make up the corpus callosum act as a passageway by which a portion of information from one half of the brain may travel to the other. In this way, the halves of the brains are able partly to communicate.

The other way through which a very limited amount of information can be shared is via a tiny tract of nerve fibers called the *anterior commissure*. There is no other "psychic corridor," and any other information exchange occurs, to a limited extent, by very indirect and circuitous routes through the very ancient limbic system and brain stem.

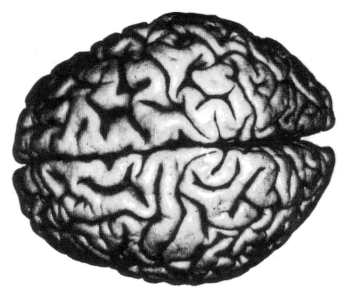

Figure 19. Overhead view of the brain. Note the interhemispheric fissure that divides and separates the two halves of the brain. From S. J. DeArmond, M. M. Fusco, and M. M. Dewy, *Structure of the Human Brain* (New York: Oxford University Press, 1976). Reprinted by permission.

Limitations in Right-Brain–Left-Brain Communication

Sometimes, information that is transferred from one half of the brain is miscommunicated and misinterpreted by the other half.[1,2] The transfer is not unlike the game of "telephone" sometimes played by children in school in which a single sentence is whispered from student to student becoming distorted along the way. By the time the message has been relayed to the last student it usually has undergone a complete transformation from the original.

The same thing sometimes happens when information has to be transferred from one half of the brain to the other. Something gets lost and distorted in the process. Moreover, since the two halves of the brain speak different *languages* (the right being emotional, visual-spatial, geometric, and tactual, and the left temporal-sequential, analytical, and

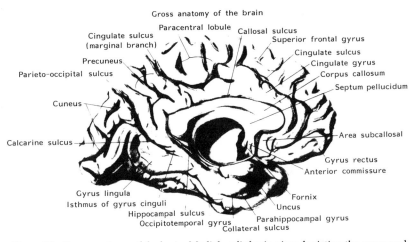

Figure 20. Gross anatomy of the brain: Medial, split-brain view depicting the corpus callosum. From F. A. Mettler, *Neuroanatomy*, 1948. Courtesy of the C. V. Mosby Co., St. Louis, MO.

linguistic), misinterpretation is made even more likely because of the need for interpretation.

Well-Kept Secrets: Some Information That Cannot Be Shared

Some information is never even transferred because each half of the brain is concerned with, perceives, analyzes, and memorizes different types of information. Just as an eye will never hear and the ear will never see, the two halves of the brain cannot perceive information they are not specialized to receive. No transfer of certain types of information takes place, as it cannot be perceived or recognized by each half's counterpart. Insofar as each half is concerned, certain types of information do not exist.

The ability to understand speech is localized in the left half of the brain in most people, whereas the capacity to recognize musical melodies is localized in the right half of the brain. The left brain has difficulty perceiving melody and is partially deaf to music; the right does not understand most aspects of speech and is partially deaf to words and sentences. Even if you yell your words or sing very loud, the "wrong" half of the brain will hear noise and garbled sounds and will misinterpret the information it receives.

Two Brains and Two Different Conclusions

Because memories, perceptions, and capacities are selectively represented in each half of the brain, the two brain halves may reach different and conflicting conclusions about what ostensibly seems to be the same piece of information. The reason is that each half of the brain is attending to different aspects of what is experienced.

When one is listening to music, the left brain attends to the rhythm (i.e., temporal sequences) and the words, and the right brain to the melody and the emotion being conveyed. Similarly, when one is talking to a friend, a loved one, or a salesperson, the right brain attends to the *way* things are being said, and the left brain attends only to *what* is being said. Two different messages may be perceived.

For instance, a husband may tell his wife that he is going to be working late. Her left brain hears only that he is going to be "working late" and accepts that on face value. Her right brain, however, hears the melody of his voice, notes the changes in his face and body language as he talks, and decides that he is up to something that does not involve work. How she reacts will depend on which half of her brain prevails, as well as on her past experience with her husband and his late-night sojourns. In either case, she is in conflict.

While listening to a politician, the left brain may hear, process, and become convinced by the words he uses, his promises, and the facts he uses to back up his statements. In contrast, the right brain might detect, via the analysis of tone of voice, body language, and facial expressions, that the politician is insincere and a liar. The listener who fails to use the impressions of the right brain and who instead relies only on the impressions made on the left half of the brain may be fooled and may vote foolishly.

Of course, the converse also occurs. Although the politician's insincerity may not be evident, his ridiculous promises may set off alarm bells in the voter's left brain. If the politician is truly effective in swaying the voter's right brain by speaking to the voter's "heart" and emotions, then, ridiculous promises or not, he may end up with the vote.

Well-Kept Secrets: Right-Brain Memories

The halves of the brain not only perceive things differently but have different memories triggered in response to those perceptions. Emotions, tactile sensations, the location of one's car keys, the melody of a certain song, and even the appearance of someone's face are stored in the memory banks of the right brain. Words, sentences, appointments, the name of one's dentist, the title of a book, stock quotes, and mathematical formulas are stored in the left brain.[3–11]

Based on the different way in which a single event may be perceived by the right and left brains as well as the different memories that may be triggered, each half of the brain may then act in an oppositional manner, one half attempting to do or believing one thing, the other half attempting to accomplish (or believing) something entirely different. In such circumstances, the person may feel in conflict, confused, mistrustful, or even paralyzed by indecisiveness, since certain memories cannot always be shared or transferred from one brain half to the other.

However, the halves of the brain may also work in harmony. For example, you receive a phone call and immediately recognize by the sound of the voice that it is your dentist (right brain). You remember his name (left brain), his face (right brain), what time your appointment is tomorrow (left brain), the terrible music he plays in the waiting area (right brain), how much it is going to cost (left brain), and how much you do not want to go (right brain).

In some cases, memories are triggered that are very emotionally traumatic. Still, the memory and all its attendant emotions may remain confined to the right hemisphere because of the differential organization of the brain. This is particularly likely if the emotional experience occurred during early childhood. (However, adults suffer traumas, too.)

When Carol was a little girl she was molested several times by her uncle. The first time, she was four years old and had been sitting next to her uncle on the couch when he began to stroke her hair and run his fingers through it. He continued this action while he cajoled and intimidated her into performing fellatio. He did this to her on ten or more occasions over a year's time, until he moved away. Somehow, she managed to forget all about this experience until many years later, when she was in college.

She was in bed with her boyfriend and they had just finished making love when he began to stroke her hair and run his fingers through it. All at once, she began to panic, became quite hysterical, and started crying and trying to strike her boyfriend. Then, grabbing up her clothes and quickly getting dressed, she ran from his apartment.

For the next several weeks, she refused to talk to him, hung up when he called her, and began to feel an overwhelming aversion toward men. She sought counseling, but to no avail.

It was only a year later, while she was watching a movie, that the entire memory of what had happened to her, so many years before, unraveled. In the movie, a man walked into a crying girl's bedroom and, while trying to soothe her, began to brush her hair and run his fingers through it. Immediately, Carol began to feel angry and upset, and then she remembered. The forgotten images began to fill her mind.

Essentially, the memory of what her uncle had done had been stored in the memory banks of her right brain and had been forgotten. Years later, when her boyfriend began to stroke her hair after they had had sex, the association between sex and having her hair fondled was reactivated. The right brain then began recalling the visual and emotional images of what had happened and became terribly upset.

Her left brain, however, having little linguistic memory of the tactile sexual act and its associated emotions, had "no idea" of why she was behaving this way. Nevertheless, her right brain prevailed, and she ran away in hysterics.

A year later, the left brain was clued in to what had triggered her hysterical reaction. When both the right and the left brains observed the man on TV brushing and fondling the little girl's hair, Carol's right brain became upset, and the images of what had happened to her spilled out. Looking at the movie and feeling the revulsion, the left brain was suddenly given a tremendous amount of information about what was bothering her and was then offered access to the horror that had so long been locked away, seemingly forgotten. Transfer was made possible because both halves of the brain were observing the same event and feeling the same feeling.

Indeed, I have frequently had patients tell me that long-forgotten memories of incest, molestations, and other traumatic events from childhood have been triggered by similar, seemingly innocuous occurrences. They may be watching children play, putting their own kids to bed, becoming angry when they have broken something, or merely overhearing a child saying something innocent in another room when suddenly a long-suppressed memory unfolds before their eyes. Essentially, the memory is being regurgitated by the right, and the left brain, cued by the emotional response coupled with the scene being observed, gains access to what has been a well-kept secret of the right half of the brain.

The Right- and Left-Brain Miscommunication and Psychic Conflicts

Overreliance on the Left Brain.   In many instances, particularly those involving social and emotional events, what occurs in the right half of the brain may never be transferred to or even recognized by the conscious system of the left half of the cerebrum. That is, the left brain may refuse to relinquish control over behavior and may ignore the perceptions, protests, and feelings of the right.[12–16]

A person's right brain might become aware of a potential problem in a relationship and may become concerned, upset, or alarmed. If the mental system of the left brain refuses or is unable to gain access to this information, the person may feel upset and may not know why consciously. This tendency to overrely sometimes on the left half of the cerebrum is unfortunate, as the left brain is very concrete and sometimes has difficulty fully appreciating the talents of the right. Some left brains, however, pride themselves on being logical and rational and in control. Unfortunately, this attitude creates many unnecessary difficulties. If the left brain maintains functional dominance at all times and relegates the right brain to second-class status, the person may be put at a tremendous disadvantage, as the right brain is so emotionally and socially astute whereas the left brain is not.

Right-Brain Social-Emotional Intelligence

The right brain is able to decipher a variety of interpersonal cues so as to determine sincerity, dishonesty, sarcasm, love, compassion, and purposeful attempts to mislead, as well as to deduce what is incongruous and implausible information. The right brain also has a better sense of humor.[17] The right brain, however, can also be fooled by a good actor. Nevertheless, for these and related reasons, it is often considered the more intuitive half of the cerebrum.

Unfortunately, some left brains don't believe in "intuition" and are quick to disavow "gut feelings," "warning signs," "red flags," and "alarm bells" when interacting with others. Some left brains prefer to rely on "logic," not "emotion." When this occurs, disastrous consequences sometimes follow.

Take for example a young couple, Lenny and Lucy, who have dated a few times and go to a party. Lenny is highly attracted to Lucy and is very hopeful that something serious may develop. However, having learned rejection as a child, Lenny strongly believes that feelings are dangerous and should be controlled. As a successful engineer, he has also long prided himself on his command of logic, rational thinking, and sober reflection. Hence, although he really likes Lucy, he has not admitted to himself how strongly he feels. He has kept his feelings in check (i.e., suppressed) because they make him very uncomfortable. That is, his desires, coupled with his fears, give rise to feelings of discomfort that his left brain prefers not to deal with.

These emotions remain for the most part confined to the right brain, which his left brain tries to ignore. Thus, he is not entirely conscious of these feelings and his desires and has not verbally expressed them. From a left-brain perspective, he is merely taking Lucy out to enjoy her company and because he likes talking to her. However, his right brain is hoping that a serious relationship will develop.

Once at the party, he begins to notice that Lucy repeatedly glances at a particular man (Jesse). There is something about the way she is looking at him that slightly bothers Lenny (his right brain), although his left brain is not sure why.

Later, he sees her talking with Jesse, and although Lenny can make out little more than the tone and melody of her voice, these and the way in which she is gesturing and smiling makes him (i.e., his right brain) feel peculiar, particularly in that she again has an odd look on her face.

Near the end of the evening he sees Lucy write something on a piece of paper as she and Jesse talk. She hands it to him and they exchange smiles.

On their way home, Lenny, feeling odd and uncomfortable about what he has observed between Jesse and Lucy, feels compelled (by his right brain) to ask her, "What was going on?" She laughs, calls him silly, and tells him that she and Jesse have a mutual friend, a girlfriend of hers whom he has not seen in a long time.

"Then what did you write on that paper?" Lenny asks, still feeling vaguely troubled.

"Her phone number," Lucy laughs.

Lenny nods his head. "That makes sense," he (his left brain) decides. And indeed, it is a logical and rational response. Yet something in her tone of voice bothers him (his right brain).

Arriving at her home, Lucy says she is terribly tired and does not invite him in as she had on their previous outing. She yawns and looks exhausted. Lenny is sent on his way.

However, Lenny, although satisfied with the *reasonableness* of her explanation, still feels uncomfortable. He argues with himself, *rationalizes* her actions, and again shrugs them off.

What is going on in Lenny's mind? Lenny's right brain, having attended to Lucy's facial expression, body posture, movements, tone of voice, and so on, is fairly convinced that he may have lost his girlfriend and that she has given Jesse her phone number. Insofar as his right brain is concerned, Lucy is lying and trying to deceive him—perhaps to spare his feelings. The right brain is upset and is justly aware of the fact that he should chalk this one up and not damage his ego by calling her again.

Lenny attempts to suppress these *irrational, nonlogical* thoughts, feelings, and impressions. They are countered by Lenny's left brain and its perception and rationalization of what occurred. All his left brain noticed was that she looked at and talked to Jesse. His left brain heard her explanations, and to his left brain they sounded perfectly *rational* and *reasonable*. The left brain, having great difficulty discerning social-emotional nuances, takes Lucy at her word rather than hearing her tone.

The left and right brain are in conflict. Lenny, however, manages to resist *consciously* acknowledging (linguistically) what he is unconsciously aware of and understands nonlinguistically. He represses what his right brain is fully aware of. He hides these implications from himself. He is *self*-deceived. His left brain has fooled itself and has failed to take right-brain feelings into account (this is his tendency anyway). He may also be influenced and sensitized by a past rejection that still affects him.

A few days later, although still feeling odd and a little insecure, he calls and asks Lucy out for the coming weekend. She tells him no, that she is planning on "washing her hair" that day, and thus makes a fool of both halves of Lenny's brain.

Hanging up, Lenny's left hemisphere, although still wanting to take her at her word, is now overwhelmed by the right. "Don't ever call her again!" it cries in its own unique language. But Lenny's left brain, not wanting to admit to what he is unconsciously aware of, is still tempted. A week later, he calls again, and Jesse answers her phone.

Although having two brains that perform so many different functions confers enormous advantages, it also unfortunately predisposes us to develop intrapsychic conflicts and sometimes to ignore warning signs and alarm bells when interacting with others. This may be particularly treacherous when dealing with members of the opposite sex. Often, the left brain fails to attend to or acknowledge what the right brain is fully aware of. However, sometimes the left brain is actively prevented from gaining access to this information.

Frontal Lobe Inhibition

It is sometimes very difficult for the two halves of the brain to cooperate and to exchange thoughts, feelings, or ideas. As a consequence of lateralized (right vs. left) specialization, certain types of information cannot be detected, much less recognized, by the opposite hemisphere. Even information that is transferred from one half of the brain to the other may be subject to interpretation and misinterpretation.

In addition, one brain half may be prevented from knowing what is occurring in the opposite half because of inhibiting and suppressive actions initiated by, for example, the frontal lobes. The frontal lobes are, in essence, the senior executives of the brain, ego, and personality.[18] They control behavior, attention, and information processing throughout the brain via inhibition, suppression, and censorship. A person whose frontal lobes have been severely damaged may say whatever pops into her or his head or may act on any impulse without thinking first and without concern about the consequences.

Because there are two frontal lobes (interlinked via the corpus callosum), one in the right brain and one in the left, they are able to assist greatly in determining what information will be processed within each half of the brain and what will be transferred between the hemispheres as well. They may act to prevent information from crossing over the "psychic corridor" (the corpus callosum) via inhibition. Thus, it can be said that they engage in censorship and stand guard on either side of the corpus callosum to determine what may pass.

Interestingly, whereas the left frontal lobe maintains inhibitory control over the left brain, the right frontal lobe exerts inhibitory influences on what is processed in either half of the brain. Indeed, the entire right brain appears to be dominant in regard to attention and arousal.[18-22]

Right-Brain–Left-Brain Cooperation

It is important to emphasize that a considerable degree of cooperative interaction occurs between the mental systems of the right and left halves of the brain. Many functions, in fact, require their dual and simultaneous interaction.

To read a book, I must keep my place on the page (in visual space) so that my eyes do not wander haphazardly from line to line (right-brain function). Simultaneously, I must read and decipher the linguistic symbols and letters that make up the words and sentences and must keep track of their temporal-sequential, grammatical order (left-brain function). Moreover, when I decide to go back to my book, I must remember where in my house I left it (right brain). If I ask someone where it might be, I must recall its title (left brain).

Making music also requires the dual interaction of the two brain halves. The left provides rhythm (temporal sequences), and the right provides the melody. Similarly, when we dream, the visual, emotional, hallucinatory qualities are provided by the right brain, and the left brain provides the accompanying verbal commentary or dialogue.

Indeed, optimal functioning as a human being requires full and harmonious use of both the right and the left halves of the cerebrum. Moreover, there is also considerable overlap in functional representation. Some abilities are maintained in both halves of the brain.

The Right Brain and Awareness

The mind and brain are intimately linked, for if one damages the brain, one damages the mind. That we possess two brains and that each brain has a mind of its own should not be at all surprising.

As language is highly interwoven with and related to consciousness, it seems quite reasonable to assume that, in most people, the conscious mind is clearly linked to the functional integrity of the left half of the brain.

In this regard, what so many authors have referred to as the *subconscious* and the *unconscious* appear to be more clearly associated with activity occurring in the right half of the brain. However, the right brain is also associated with much more.

The right brain predominates in the perception and identification of environmental and nonverbal sounds, the maintenance of the body image, the comprehension of melodic and emotional speech, and the perception of most aspects of musical stimuli. The right brain is also dominant regarding the analysis of geometric and visual space, including depth perception, position, distance, movement, and stereopsis— capacities essential for stalking game or playing football, tennis, or basketball.

Hence, it is the right brain that allows one to throw and catch a football with accuracy, whereas it is the left brain that keeps score. The right brain allows us to dance the night away while the left brain worries about the time. The right brain produces seemingly bizarre, sometimes highly emotional, dream imagery, whereas the left brain is unable to remember the dream in the morning. The right brain allows us to drive from Point A to Point B while our "thoughts" are elsewhere. The right brain likes to sing and listen to music, and the left brain keeps the beat and enjoys "talk" radio.

There are thus a myriad of activities at which the right brain excels and in which the left brain plays a minor supporting role (and vice versa).

Thus, the mental activity of the right brain is not something completely hidden or submerged and that defies scrutiny. Right-brain activity and unconscious mental activities are not synonymous. The right brain performs so much of its activity naturally and thus seemingly surreptitiously (e.g., enabling us to put our clothes on without thinking about it, or to drive a car, run, jump, or dance) that it does not draw much conscious attention. This situation could also quickly change if the right brain were upset.

Indeed, it is able to perform many of its functions without any assistance from the left brain (other than motor support), although the left brain is able to observe and even comment on much of what the right brain engages in. Of course, the left brain misses much as well.

The right brain provides much information that enables us to navigate through the environment and social relationships in a potentially successful way, and this same awareness allows us to reflect on our self-image and to appreciate the beauty, danger, excitement, and intrigue that make up our environment.

However, from a conscious (left-brain) perspective, much of what occurs within the domain of the right brain is not susceptible to conscious scrutiny or analysis, partly because it is nonlinguistic, cannot be labeled or categorized verbally, and is not temporal-sequential. In this regard, right-brain processing can be an unconscious phenomenon because it is nonverbal. It is unconscious because the left brain cannot verbally recognize it and may not know about these events until after they have occurred or until after it has lost control and the consequences have been experienced.

Sometimes, the left brain is completely overwhelmed by right-brain emotional outbursts, or by mental activity such as the visual imagery experienced in the form of a "daydream," which the left brain quite passively observes; similar to what occurs while dreaming at night. However, sometimes the left brain simply "blanks out" and pays no attention, or what the left brain observes does not register because it cannot be recognized.

Does this make these right-brain-mediated events and our associated feelings unconscious? Not necessarily. What seems to originate unconsciously is often amenable to conscious scrutiny and examination. We can talk and write about some of these experiences.

We are capable of becoming fully *aware* of what is occurring in the right brain, although it is nonlinguistic and nonconscious. We are aware of birds singing, we can hear a quiver of emotion in the voice of a friend or lover, we can judge distance when throwing or catching a ball, and we are aware of where our legs and arms are in space when we walk, skip, dance, or run; and we can do all these things without thinking about them. If we choose to, we can also become consciously aware of all them as they occur.

However, it is very difficult to put what the right brain experiences into words, and it is also not always necessary or useful. Instead of thinking, sometimes we need just to feel, experience, and observe without talking about it. If we only talk and think, all these beautiful, intriguing, and informative stimuli fade from conscious consideration, and we focus instead on single features of the environment or mistake words and labels for reality. If we stop talking for a few moments and stop our thoughts, the singing of the birds, the chirping of squirrels, and the distant laughter of children will suddenly come to the fore. If we consciously reflect on them, however, we will also know that we were aware of them the whole time.

There are thus a myriad of activities at which the right brain excels and in which the left brain plays a minor supporting role. However, because so much of what the right brain engages in does not require conscious assistance, it seems to be less important or even insignificant, at least insofar as the left brain is concerned. In fact, scholars as recently as the 1950s erroneously argued that the right brain was more like a well-trained automaton and that any mental functioning it assisted in was almost completely superfluous and irrelevant.

Moreover, because the left brain controls speech and the right hand, it has long been described by numerous scholars and scientists as the "dominant" half of the brain. Indeed, because the left brain talks and thinks in words and tends to label and categorize linguistically all that it experiences, this aspect of the mind *seems* to predominate in most psychic interactions as well; that is, it tells itself it is predominant.

Unfortunately, many of us practice a kind of left-brain jingoism and treat the right brain like some Third World country. Even our educational system stifles right-brain development by stressing and rewarding left-brain abilities, such as reading, writing, and arithmetic. Many right-brain capacities, such as drawing, painting, sculpting, music, dancing, sports, and gymnastics, are relegated to recreational status at best, as if studying algebra for two years while a student will serve most people better twenty years from now than having refined their various nonlinguistic skills.

Although right-brain capacities may be underused, ignored, and even suppressed, they nevertheless wield enormous influence over our behavior and our ability to lead successful, happy, productive lives. Because the two halves of the brain cannot fully communicate and process, store in memory, and recall different aspects of what appears to be the same event, considerable conflict may occur between the right and the left brain. When this happens, the left brain becomes alerted to the possibility of forces that are acting outside its control.

Thus, "Out of sight, out of mind" (that is, the mind of the left brain) does not prevent the fermentation of inner chaos and even revolution. The consequences of suppressing or underusing the right brain can be disastrous if its feelings and emotions unexpectedly erupt, thus greatly affecting the world that both halves of the brain share and occupy. When this occurs, the left brain is likely to cry out, "I don't know what came over me," or even "The devil made me do it!" as if sin and Satan were somehow associated with the right brain and the left side of the body.

Emotion

Probably one of the more important distinctions in the qualitative differences between the right and the left cerebral mental systems concerns emotion. As previously indicated, there is a considerable body of evidence indicating that the right brain is dominant in the expression and perception of visual, facial, and verbal affect, including the ability to make inferences or to determine another person's mood, attitude, and intentions through an analysis of posture, facial expression, body language, and vocal melodic and intonational qualities.

Psychiatric Disturbances

Because the right brain is dominant in controlling most aspects of emotion, when it is damaged a myriad of affective disturbances may result, including mania, depression, hysteria, gross social-emotional disinhibition, paranoia, delusions of grandeur, confabulation, denial, neglect, euphoria, childishness, puerility, emotional lability, or, conversely, complete indifference and apathy.[23-28]

Because of its dominance in perceiving faces, patients with damage, as we've seen, may fail to recognize the faces of friends, loved ones, or even pets. For example, a farmer who had a stroke in the right posterior part of his brain (where facial recognition is subserved) complained that he was having trouble recognizing his wife and could no longer tell his cows apart.

Some patients become emotionally upset when they have these and related perceptual difficulties, particularly because the right brain also controls emotion. If it is not apparent that they have had a stroke, doctors usually diagnose them as delusional, paranoid, and crazy.

For example, one patient went to the police and claimed his wife had been replaced by an impostor. Although he admitted they looked somewhat alike, he could tell there was something different about her face and voice. He also knew she was an impostor because, when she said she loved him, the tone of her voice wasn't right. He was sure she was plotting against him.

In many cases, patients have developed what appeared to be bizarre psychiatric disturbances when, in fact, they had suffered a stroke in the right brain. However, if the patient is not also paralyzed (which happens only when certain specific areas of the brain are compromised), most doctors fail to realize that the problem is due to brain damage. There are probably large numbers of such individuals under psychiatric care when their needs would better be served (at least in some cases) by a neurologist, a neuropsychologist, an occupational therapist, or a speech pathologist.

Hysteria and Body Image Distortions

Some patients with right-brain damage become delusional, produce bizarre confabulations, and experience a host of somatic disturbances, ranging from pain and hysterical body-perceptual distortions to seizure-induced sexual activity and orgasm.[29–33] The reason is that body image and self-image are maintained in the right brain.

Other patients complain of strange physical sensations along the left side or both sides of their body, such as electric-shock-like pain, swelling, abnormal tingling, and stiffness. However, when they are examined, no evidence of physical disturbance involving the body is apparent. The reason is that the altered sensation is an illusion caused by damage to the parts of the brain subserving those parts of the body image. Although initially many of these patients are thought to be hysterical or insufferable hypochondriacs, further study indicates right-brain tumors, strokes, or epileptic disturbances in the right parietal area where that part of the body image is represented. That is, the body image within the brain became abnormally activated or distorted due to these injuries thus causing these perceptual, seemingly hysterical symptoms.

There has also been some suggestion that individuals suffering from anorexia have damage or some unknown type of disturbance involving the right brain (i.e., the right parietal area where the body map is maintained). Although there is scant evidence to support this intriguing possibility, it may explain why these young people misperceive their bodies as fat when they are quite thin.

In any case, body image disturbance, be it neglect, distortion, or even denial of owning a particular body part, occurs almost exclusively with right-sided brain damage, because the parietal area of the right brain is dominant in maintaining the body image and the perception of physical sensation. The left brain can perceive only half a body.

Hence, if that area of the brain where the body image is maintained is damaged, patients (i.e., their left brains) may fail to recognize the left half of their bodies and may misperceive body sensations or fail to perceive them at all. When the brain's body map becomes distorted because of damage, perceptions also become distorted, when, in fact, there is nothing wrong with the body. In these instances, patients may be misdiagnosed as hysterical.

In fact, the right brain is so heavily involved in emotional and motivational functioning that individuals with massive right-sided lesions show less recovery and are more likely to die than those with left-sided destruction. Such damage affects the will to live.

Left-Brain Emotional Disturbances. In contrast, the range of emotional disturbances associated with left cerebral damage seems to be limited to apathy, depression, emotional blunting, and schizophrenia, although euphoria sometimes accompanies receptive aphasia and loss of comprehension.

Sex and Orgasm

One of the most fundamental associations with the will to live is reproduction. A species can survive only by reproducing itself, and this is accomplished through intimate physical contact, i.e., sex. Bodily sensations and the desire for sex, including the urge for sexual bonding, are mediated by the limbic system and the right half of the brain.

If the right brain is severely damaged, not only the will to live but sexual functioning may be severely affected. Indeed, patients with rare forms of right-brain damage have sometimes demonstrated or complained of epileptic-seizure-induced sexual activity and orgasm.[34]

One woman with a right-brain epileptic-seizure disorder would fall to the floor, make pelvic thrusts (as if she were having intercourse), and moan, "Do it. It feels so good," whenever she had an epileptic attack. Other patients with right-brain seizures have complained of experiencing painful orgasms when they have an epileptic attack.

Another patient had been arrested eight times for exposing himself in public. Even his parents complained that he would sometimes walk around the house exposing and grabbing his penis and would even on occasion urinate on the floor or on the wall.

Because he claimed he had no recollection of these events, he was referred to the VAMC/Yale Seizure unit where I was an intern at the time. Electrodes were placed on his head (to measure his brain activity via electroencephalographic recordings and to determine if he had epilepsy), and I began to examine him while he lay in bed. Suddenly, as we talked, his facial expression became blank, his left arm shot out to his side, he made a low moaning sound, and then his hands began playing with the buttons on his pajamas, working their way down until he had reached his penis. He then took it in his hand, began squeezing it terribly hard, and let loose with a stream of urine, which fortunately just missed me. He was unable to talk or respond to questions, and his stare was completely vacant throughout and for a few minutes after the episode. Later, when we looked at his EEG, it was apparent that he was having seizures coming out of his right brain.

Hence, he wasn't really exposing himself, he was having an epileptic seizure. His seizure activated those regions of his brain associated with taking hold of his penis. Instead of falling to the floor unconscious or suffering tonic-clonic spasms of the extremities, he would expose himself.

Numerous studies and observations indicate that the right brain is more involved in sexual activities than the left. Indeed, EEG electrode studies that have measured the brain's activity during sex have also shown that the right brain becomes more active during the sex act and particularly during orgasm.[35] Hence, the right brain may be considered the sexier half of the cerebrum.

4

Right-Brain Limbic Language and Long-Lost Childhood Memories

The Left Brain

The left cerebral hemisphere is associated with organizing and categorizing information into discrete temporal units. It also controls the sequencing of finger, hand, arm, and articulatory movements, and the perception and verbal labeling of material that can be coded linguistically or within a linear and sequential time frame.

The left brain is dominant in regard to most aspects of expressive and receptive linguistic functioning, including grammar, syntax, reading, writing, speaking, spelling, naming, verbal comprehension, and verbal memory. The left brain thinks in words and is very attentive to details and temporal-sequential organization, such as into first, second, and last.

Indeed, as already noted, within the left hemisphere one area largely controls the capacity to speak, and another region mediates the ability to understand speech and assists in imposing temporal-sequential order on everything that is heard and said. These regions are referred to as *Broca's speech area* (located along the lateral convexity of the left frontal lobe), the *angular gyrus* (in the left parietal lobe), and

Wernicke's receptive language area (within the left temporal lobe). Together these are referred to as the *language axis*.[1]

Nevertheless, the left brain has not always been dominant for language. Indeed, even written language was originally pictorial and may have originated in the right brain.

The Right Brain and Limbic Language

Language was originally a system of emotive and imitative sounds— sounds which expressed terror, fear, anger, love, etc., and sounds which imitated the noises of the elements: the rushing of water, the rolling of thunder, the roaring of the wind, the cries of the animal world and so on; and lastly those which represent a combination of the sound perceived and the emotional reaction to it.

C. G. JUNG

Limbic Language

Over the course of our evolutionary development, before the acquisition of complex speaking patterns, communication was no doubt accomplished through body language, gesture, facial expression, and, in particular, emotional sounds and mimicry. Language has not always been temporal-sequential or dominated by the left half of the brain.

In an infant, the right brain and the limbic system (a series of structures found in the depths of the brain) are initially dominant in regard to vocal communication.[2] An infant coos, gurgles, laughs, and expresses moods and needs by altering the amplitude and melodic qualities of its voice. Although the infant also babbles, this initial form of babbling is mostly reflex, the product of self-stimulation, and has no more communicative intent than an eye blink. This first form of emotional language has been referred to as *limbic language*, for it is mediated by the limbic system, and only later by the right brain as well.[3] This first form of language—that is, infant language—can communicate only diffuse feeling states.

By about three or four months of age, a purposeful form of babbling appears. This second form of babbling heralds the first real shift from emotional, melodic speech to temporal-sequential language. As the left half of the brain develops, it begins to separate and impose temporal sequences on the stress, pitch, and melodic intonational

245. LAUGERIE BASSE. CHAMOIS (?)

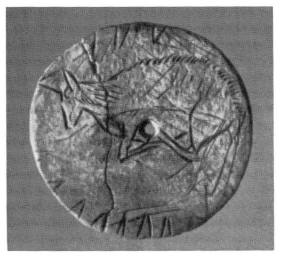

246. REVERSE SIDE OF FIG. 245. CHAMOIS

Figure 21. Cro-Magnon art: The possible first example of temporal-sequential pictorial images, depicting a story of an animal standing on one side and, on the other, showing the animal dead or sleeping on the ground. Carved on either side of a one-inch disk of bone. Courtesy of Prehistoire de l'Art Occidental, Editions Citadelles & Mazenod, photos by Jean Vertut.

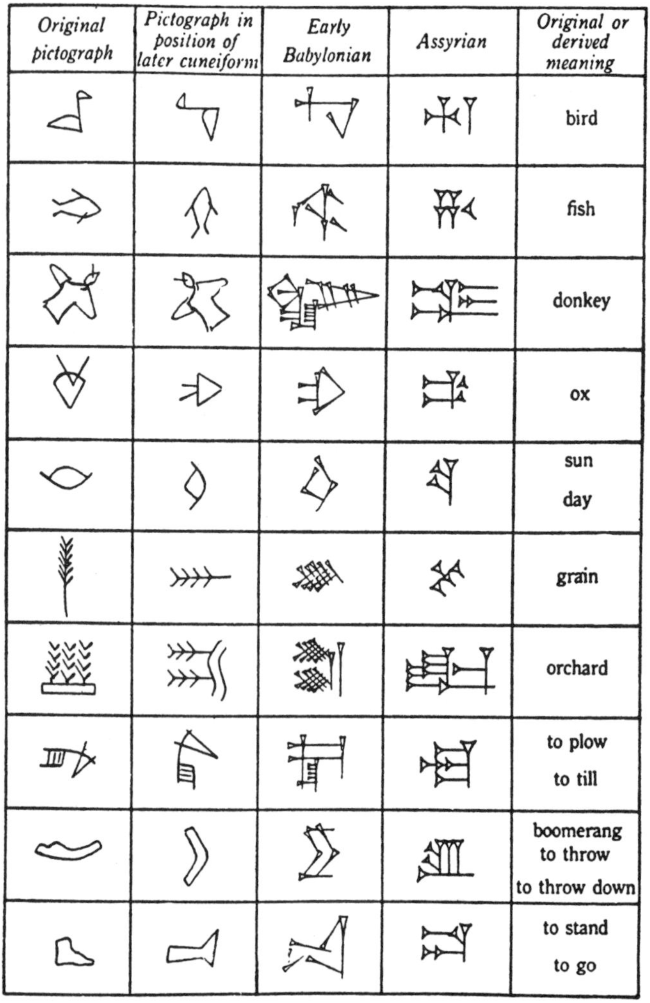

Original pictograph	Pictograph in position of later cuneiform	Early Babylonian	Assyrian	Original or derived meaning
				bird
				fish
				donkey
				ox
				sun day
				grain
				orchard
				to plow to till
				boomerang to throw to throw down
				to stand to go

Figure 22. The origin and evolution of writing from pictures used by the people of Sumer to develop cuneiform characters. The people of Sumer realized that pictures could represent not only images and ideas, but sounds, which gave their writing greater precision in meaning. Note that the pictures of animals and body parts originally faced to the right. From Edward Chiera, *They Wrote on Clay* (Chicago: University of Chicago Press, 1966). Reprinted by permission.

contours being produced by the right brain. That is, syllabication is imposed on the intonational contours of the child's speech by the left brain, so that the melodic features being produced by the right brain come to be punctuated, sequenced, and segmented, and vowel and consonantal elements begin to be produced. Left-hemisphere speech comes to be superimposed over the right-brain melodic language.[4]

The Right-Brain Melodic-Emotional Language Axis

Although left-brain speech eventually becomes preeminent in the expression of verbal thoughts and ideas, the right brain is dominant for melodic and emotional speech, perception, and expression. The right brain remains dominant for one's ability to discern and impart meaning, context, sincerity, and emotional intent.

In fact, as demonstrated by Eliot Ross, Kenneth Heilman, Don Tucker, B. E. Shapiro, M. Danly, and others, one area within the right brain mediates the ability to express melody and emotion vocally; it is located in the right frontal region. Another region in the right temporal and temporal-parietal area controls the capacity to hear these as well as environmental sounds.

Expression and Comprehension of Melody, Music, and Emotional Speech

A damaged right frontal region (opposite Broca's area) may produce speech release (or "motor mouth") rather than speech arrest or aphasia. Such patients talk at a very rapid pace, jump from idea to idea haphazardly, and become delusional.[5] Moreover, the tone and melody of their voice may become abnormal.[6] They certainly would have much difficulty holding a tune and would not be able to sing very well, if at all.

If the damage is in the right temporal area, the ability to comprehend or recognize emotional speech, music, and environmental sounds (e.g., birds singing, a knock at the door, or thunder) is compromised. These patients may lose their sense of humor and have great difficulty understanding or inferring what others were implying when engaged in a conversation. Unfortunately, when this occurs, the patients are likely to behave in an emotionally abnormal manner, as their emotional comprehension has become distorted.

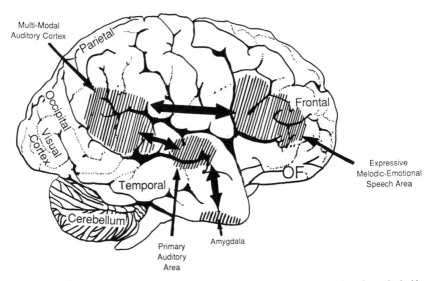

Figure 23. Schematic representation of the melodic-emotional areas within the right half of the brain. Auditory information is received in the primary auditory area, as well as within the amygdala of the limbic system (and other limbic areas). Emotional and related characteristics are discerned, comprehended, and/or assigned to the sounds perceived. When one is speaking emotionally or is singing or cursing, this information is transferred from the temporal-parietal and limbic areas to the right frontal area, which mediates the expression of the information. From R. Joseph, *Neuropsychology, Neuropsychiatry, and Behavioral Neurology* (New York: Plenum Press, 1990).

Together, the expressive emotional and melodic areas in the right frontal and right temporal areas make up a unit that provides emotional color to everything we say and hear, and that enables us to determine what others mean or imply when talking to us.[7] Similarly, if we are hearing music and want to sing along, the melody and tune are processed in the right temporal area. Through interaction with the right angular gyrus, these brain areas combine the sounds into a pleasing gestalt, which is then transmitted to the right frontal area, where the melody is vocally expressed. These areas are interconnected by a rope of nerve fibers, the arcuate fasciculus. If the left brain desires, it can provide the necessary temporal-sequential punctuation so as to provide the words and rhythm and thus sing along. However, it can also engage in foot or finger tapping if it cannot remember the words.

Lateralized Perceptions and Responses

Because each hemisphere is specialized in the type of material that it can receive, process, and respond to, some types of information cannot be transferred or even recognized by one or the other half of the brain. One half of the brain may perceive, process, and store in memory certain aspects of experience that the other half knows nothing abou'. This is particularly true regarding negative emotions and emotion. l traumas, which may be confined to the right half of the brain.

Because language, emotions, memories, and perceptions are lateralized, the halves of the brain may reach different and conflicting conclusions about what seems to be the same piece of information. Indeed, this is a major source of intrapsychic conflict, as the halves of the brain may not only perceive things differently but have different memories triggered and then, based on these different experiences, act in opposition. For example, Jane is in a store and asks a clerk a question. He *looks* at her as if she were an idiot, uses an insulting *tone* of voice, points with his finger like a parent lecturing a child, and says, "It is in aisle three."

Jane's left brain hears the directions and prepares to retrieve the desired item. Her right brain, having attended to his facial expression, tone of voice, and body language, begins to feel irritable and then suddenly remembers an incident in which someone important to her acted similarly and treated her quite miserably. Her right brain is irritated and upset. However, her feelings of anger are completely disproportionate to what is going on, as her right brain wants to strangle the clerk. Jane hesitates and then—heads down the aisle.

Lateralized and "Unconscious" Memories

Right- and Left-Brain Memory

The lateralization of specialized functions means that some abilities are found in a greater proportion on one side than on the other side of the brain. Functional specialization determines what type of material can be memorized, or even recognized, by each half of the cerebrum because the code or form in which a stimulus is represented in the brain and memory is determined by the manner in which it is processed and the ensuing transformations that take place. Because the right and left cerebral hemispheres process information differently, how it is represented in memory is also lateralized.[8]

It is well known that the left half of the cerebrum is responsible for the encoding and recall of verbal memories, whereas the right brain is dominant in visual-spatial, nonverbal, and emotional memory. When listening to someone who is angry, the left brain may encode the words he is saying. The right brain may perceive and store in memory the look on his face, his tone of voice, the angry gestures he uses, and the overall emotional gestalt of the situation, including one's own reactions to what is being said.

A badly damaged right brain (e.g., destruction of the right temporal lobe, which contains some of the memory centers), the patient (i.e., his left brain) would be unable to store in memory most aspects of emotional, visual-spatial, and related stimuli.

If the memory centers of the right brain were damaged, this particular person (i.e., his left brain) would not be able to remember where he laid his wallet or car keys, or how to get (drive through space) to the dentist's office, would fail to recognize the dentist if he ran into her in the parking lot, and would forget the argument he had had with her receptionist about an overdue bill. However, this person (his left brain) would be able to remember the dentist's name, his own phone number and address, the amount of the overdue bill, and the conversation he had with the receptionist about the bill.

If the memory centers of the left brain (which are located in the left temporal lobe) were damaged, this particular person would have exactly the opposite problems with memory. He would forget the dentist's name and the amount of the bill but would recall the argument, the dentist's face, and so on.

Responding to Hidden Memories

Lateralization of memory affects complex behaviors, for one half of the brain may experience and store certain information in memory and, at a later time, in response to a certain situation, may act on those memories, much to the surprise, perplexity, or chagrin of the other half of the brain. This is what happened to Carol, the young woman described earlier, who had been molested and whose hair had been fondled.

When Jeff was a child, his father frequently became angry and violent over nothing. As the father became enraged, he would begin pointing with his index finger and wave it up and down in Jeff's face, and then, his voice rising, he would sometimes slap Jeff across the face.

Twenty years later, Jeff was having a mild discussion with his new boss when this man suddenly raised his hand and began pointing with his index finger to emphasize a point. Jeff visibly flinched, was suddenly filled with anxiety and fear, and desperately wanted to leave the room. He didn't know (that is, his left brain didn't know) what was bothering him, but he was unable to pay full attention to what his boss was saying. From that point on, he felt an unreasonable fear of his new boss and soon began dreading going to work.

Unfortunately for Jeff, the link between his new boss and his father was not apparent to him—at least, not to his left hemisphere. It was only after a few counseling sessions that these early, seemingly forgotten memories came to the fore.

When one half of the brain learns, has certain experiences, and stores information in memory, this information may not always be available to the opposing cerebral hemisphere; one hemisphere cannot always gain access to memories stored in the other half of the brain. However, partial access can be obtained if observable clues are associated with internal feelings.

The visual image of his father's finger coupled with the anger that was followed by the slap was stored differentially in Jeff's right brain. When his new boss employed similar gestures (in a nonangry manner), this triggered in Jeff's right brain the recollection of his father and all the anger and fear associated with those early experiences. Hence, he became upset, but he (i.e., his left brain) did not know why. Insofar as his left brain was concerned, his boss's shaking his finger was insignificant, and in fact, he "hardly noticed it" as he was concentrating on what was being said. Therefore, his left brain had no idea what the real problem was. The associated emotional and visual memories were not available to it.

Some Memories Are Stored Only in the Right or Left Brain

Thus, in the normal brain, memory traces appear to be stored unilaterally (in one half of the brain) rather than laid down in both hemispheres (i.e., bilaterally) depending on whether they are visual-spatial, emotional, linguistic, and so on. For the opposing hemisphere to gain access to these memories, it has to activate the memory banks of the other brain half via the corpus callosum, the rope of nerve fibers that interconnects the left and right halves of the brain. As demonstrated by R. W. Doty and W. H. Overman, if the corpus callosum is severed, the memory remains confined to the half of the brain where it is stored, and the opposing hemisphere now knows nothing about it.[9]

In one study, G. L. Risse and M. S. Gazzaniga[10] injected sodium amytal (an anesthetic) into the left carotid arteries of various neuro-surgical patients in order to anesthetize the left half of the brain. After the left cerebrum was inactivated and essentially asleep, the awake right hemisphere, although unable to speak, was still able to follow and respond to commands, for example, holding and palpating an object with the left hand or looking at pictures. These two doctors then gave the right brain (via the left hand) a number of objects to hold and showed it several pictures.

Once the left brain recovered from the drug, as determined by the return of speech and right-handed motor functioning, so that both halves of the brain were now awake, none of the eight patients studied was able to recall *verbally* which pictures had been shown to the right brain or which objects had been held and palpated with the left hand, "even after considerable probing." Although encouraged to guess, most patients (i.e., their left brains) refused to try and insisted they did not remember anything. When they were shown several objects or pictures and asked to guess by pointing with the *right hand* (which is controlled by the left brain), patients continued to state they could not recall any of the pictures or objects, or they pointed at the wrong items. When they were asked to point with the *left hand*, most patients immediately raised the *left hand* and pointed to the correct object. The right brain, which controls the left hand, was fully able to recall what it had been shown, although this knowledge had not been translated into words and the right brain was unable to talk about it (talking being controlled by the left brain). The left brain was unable to gain access to these memories even though they were stored in the right brain. The right brain not only remembered but was able to act on its memories.

Surprisingly, once the person pointed to or grasped the correct object with the left hand, so that both halves of the brain were able to see what only the right brain had learned and remembered, the left brains of most of these patients immediately claimed to recall having been shown the item, just like the girl whose hair had been stroked while she performed fellatio as a child.

This indicates that, when exchange and transfer are not possible or are in some way inhibited, or if, for any reason, the two halves of the brain become functionally disconnected and are unable to share information, the possibility of information transfer at a later time becomes extremely difficult, and the information may be lost to the opposite half of the cerebrum. Nevertheless, although lost, these memories and their attached feelings may continue to influence whole-brain functioning in subtle as well as profound ways. However, once the right brain acts on its memories, the left brain may gain access to some of this information via observation and guess work.[11] If the left brain guesses correctly, the right brain may reward and reinforce it emotionally or may give it clues by generating various emotions until it guesses correctly.

In a very young child, however, sometimes these emotions and memories cannot be expressed or accessed by the left brain because of limited language ability (as well as the immaturity of the corpus callosum, as discussed below). Consider for example, a child who has been molested. Although she may never talk about it and may deny it if asked, she may nevertheless go to school and pull up her dress, pull down her panties, play with the genitals of her friends, or allow them to play with hers. In other words, she acts out the memories and experiences selectively stored within the right half of her brain. Because her right brain cannot talk about what has happened, it acts it out instead.

Long-Lost Childhood Memories

For most individuals, events that occurred before age four are very difficult, if not impossible, to recall. Several explanations may account for what appears to be an age-related amnesia.

One explanation is that information processed and experienced during infancy is stored in memory through cognitive transformations and retrieval strategies that are quite different from those used by adults. As the brain matures and new means of information processing are learned and developed, the way in which information is processed and stored is altered. Although early childhood memories are stored within the brain, the adult brain no longer has the means of retrieving them; that is, the key no longer fits the lock. Early experiences may be unrecallable because infants use a different system of codes to store memories, and adults use symbols and associations (such as language) not yet fully available to the child. Much of what is experienced and committed to memory during early childhood takes place before the development of linguistic labeling ability and is based on a pre- or nonlinguistic code. Hence, the adult, relying on more sophisticated coding systems, cannot find the right set of neural programs to open the door to childhood memories. The key does not fit the lock because the key *and* the lock have changed.[12]

Emotion and Split-Brain Functioning in Children

As we are now well aware, the developing infant is extremely vulnerable to early environmental and rearing influences, so that the infant's nervous system and behavior may be dramatically and permanently affected depending on how the infant is treated. Children and animals raised in a diverse environment with an abundance of physical, auditory, and visual stimulation are better able than those who haven't been to cope and are more resilient to the effects of stress and emotional extremes later in life. As has been demonstrated by M. R. Rosenzweig, F. L. Bennett, M. C. Diamond, W. T. Greenough, V. H. Dennenberg, and many others, those exposed to a diverse or enriched environment have brains that are heavier, with a neocortex that is thicker; their nerve cells are larger and more abundant; and the interconnections between their different brain tissues are richer and more extensive. Moreover, as R. E. Gallagher and I have shown, they are more intelligent, inquisitive, and capable of inhibiting irrelevant and even potentially dangerous behavior.[13] Interestingly, the right cerebral hemisphere seems to be more greatly affected by early experience than the left.[14] This is true in regard to right-brain development, its interconnections with other brain tissues, and its overall growth. In fact, the right brain is slightly bigger than the left half of the cerebrum.

The right brain is more greatly affected than the left by emotional extremes as well. Moreover, during these same early years, our traumas, fears, and other emotional experiences, like those of an adult, are mediated and stored in the memory banks of the right cerebrum.

Limitations in Right- and Left-Brain Communication

Much of what is experienced and learned by the right brain during the early years is not always shared with or available to the left hemisphere (and vice versa). That is, as in an adult, a child's two cerebral hemispheres are not only functionally lateralized but limited in their ability to share and transfer information.[15] More important, however, infants and young children have right and left brains that are not fully interconnected. Their right and left brains thus have considerable difficulty communicating, much more than in an adult. The reason is the immaturity of the corpus callosum.

The Immature Corpus Callosum

The great nerve fiber bundle, the corpus callosum, is the main psychic corridor through which information flows between the right and left cerebral hemispheres. The corpus callosum, however, takes over ten years to mature completely.[16] This immaturity before age ten greatly limits information transfer between the two brain halves, particularly in very young children.

There are good reasons for this slow maturation, as the developing brain does not need to be subjected to competing influences occurring on the other side of the cerebrum, nor does one hemisphere need to be flooded by activity or information other than that it is attempting to process and master. Otherwise, it would be overwhelmed, and its ability to become proficient in specialized tasks would be compromised. This also reduces competition for cerebral space, keeping certain functions confined to one or the other hemisphere. The immaturity of the corpus callosum ensures that information transfer will be greatly limited.

Nevertheless, as in adults who have undergone surgical splitting of the corpus callosum (called *split-brain surgery*), it has been shown by D. Galin, R. H. Kraft, D. S. O'Leary, A. Salamy, myself, and others[17]) that communication between the two halves of the brain is so poor that children as old as age four have difficulty transferring tactile, cognitive, or complex visual information between the hemispheres, for example, describing complex pictures shown to the right brain, or indicating with the right hand complex objects and shapes that have been felt by the left hand (and vice versa).

Indeed, as my colleagues and I have demonstrated, when a child is questioned about a picture shown selectively to the right brain, the left (talking half of the) brain may respond with information gaps that are erroneously filled with confabulatory explanations. That is, the left brain makes up explanations as to what was seen by the right because the left brain does not know.[18]

Thus, the left brain of a very young child has incomplete knowledge of the contents of and the activity occurring in the right, and the stage is set for differential memory storage and a later inability to transfer information between the cerebral hemispheres. That is, just as in the experiments of Doty, Overman, Risse, and Gazzaniga, when the two hemispheres are unable to communicate and one brain half learns and stores certain experiences in memory, the other half of the brain cannot gain access to those memories even when communication between the two brains is restored or established.

In young children, the two brains cannot fully communicate and cannot share information because of corpus callosum immaturity. Later in life, although the corpus callosum has matured and now allows more efficient communication, these early memories still cannot be shared.

As discussed earlier, if one hemisphere learns, and at that time memory and learning transfer is not possible, transfer later in life becomes very difficult. The information and memory become stored in only one half of the brain. They remain well-kept secrets.

When Well-Kept Memories Become Activated Secrets

Because of lateralization and limited exchange, the effects of early "socializing" experience has potentially profound effects. As a good deal of this early experience is likely to have unpleasant if not traumatic moments, it is fascinating to consider the later ramifications of early emotional learning occurring in the right hemisphere unbeknownst to the left—learning and associated emotional responding that may later be completely inaccessible to the language axis of the left cerebral hemisphere.

Although (as in the adult brain) limited transfer in children confers advantages, it also lays the groundwork for the eventual development of a number of very significant psychic conflicts—many of which do not become apparent until much later in life. The reason is that the two brains not only have different memories that cannot be shared, but the hemispheres may independently recall certain experiences and all associated feelings and may then act on them. Therefore, a person may respond nervously, anxiously, angrily, or fearfully and may not know what is really bothering her or him or "what came over" her or him. Moreover, because of the immaturity of the corpus callosum, children can frequently encounter situations in which the right and left brains not only perceive differently what is going on but are unable to link these experiences so as to fully understand what is occurring and to correct misperceptions.

This is what sometimes happens with young children who are molested. Not only are they confused about what is happening to them (or, as explained by some, they pretend it is happening to someone else), but sometimes, they seemingly forget what had happened, or even that it had happened at all, although they may be plagued by the associated trauma for years. Indeed, the molestation may stay a secret of the right brain until the memory is accidentally triggered by some action or event witnessed later in life by both halves of the brain.

Nevertheless, similar confusions, miscommunications, misinterpretations, failures to transfer information, and differential storage occur even in less severe cases of abuse. These experiences or traumas simply remain secrets of the right half of the brain.

Take for example a young divorced mother who has some ambivalent feelings about her young son. She knows she should love him, and of course, she does. She wants to be a good mother and makes herself go through the motions. However, she also resents her son because he restricts her freedom, he is a financial burden, and his presence may hinder her from finding a desirable mate. She is confronted by two opposing attitudes, half of which are unacceptable to the image she has of a good mother. Like many of us, she must prevent these feelings from reaching consciousness or from being acted on. This does not, however, prevent them from being expressed nonlinguistically or tactilely through the right brain. That is, both sides of her feelings are expressed.

Her son, who, of course, also has a right brain, perceives her tension and ambivalence. His right brain notes her stiffness, sometimes, when she holds or touches him and is aware of how she sometimes looks at him. Worse, when she says, "I love you," sensing the tension and tone of her voice, his *right* brain correctly perceives that what she means is, "I don't want you" or "I hate you." His *left* brain hears, however, "I love you" and notes only that she is attentive. He is in a "double-bind" conflict. Indeed, they both are.

This little boy's right brain feels something painful when the words "I love you" are spoken. When his mother touches him, he becomes stiff and withdrawn, for his right brain, through its analysis of facial expression, emotional tone, tactile sensation, and so on, is fully aware that, on some level, she does not want him.

Later as an adult, this same young man has one failed relationship after another. He feels he can't trust women and often feels rejected, and when a woman says, "I love you," it makes him want to cringe, run away, or strike out. As an adult, his left brain hears, "Love," and his right brain feels, "Pain." Because the two halves of his cerebrum were not in communication during early childhood, his ability to gain insight into the source of his problems is greatly restricted. His left brain cannot access these memories. It has "no idea" of the cause of his conflicts. However, if his right brain could talk, it might fully regurgitate all that was really bothering him: His mother sometimes pretended to love him when she really didn't even like him. If his own mother didn't want or love him, how can anyone else?

Thus, this curious asymmetrical arrangement of brain function and maturation may well predispose the developing individual later to come upon situations in which he finds himself responding emotionally, nervously, anxiously, or neurotically, without linguistic knowledge, or without even the possibility of linguistic comprehension of the cause, purpose, eliciting stimulus, or origin of his behavior.

As children or adults, we may find that we are faced with behavior in ourselves that is mysterious or embarrassing, and the real cause and origins of any misery and unhappiness we may feel may remain a well-kept secret throughout our lives.

5

The Split Brain
Two Brains, Two Minds, and the Origin of Thought

The Right Brain and the Unconscious Mind

There is now indisputable evidence that the right brain is associated with a highly developed form of mental processing that can be loosely referred to as an *unconscious awareness*. From the perspective of the left brain and conscious mind, this dimension appears to represent the unconscious *per se* because it is nonverbal. As compared with the conscious mind, which is heavily influenced by language and verbal thoughts, the right-brain mental system does not rely on linguistic forms of analysis; thus, it is quite difficult for the language-dependent left brain to be conscious of what is occurring in the other half of the cerebrum. That the right brain, through the right frontal lobe, exerts inhibitory influences on information reception within the left brain only adds to its influences outside conscious control.

Moreover, because of its specialized abilities, the left brain cannot always understand or even recognize actions, memories, and ideas produced by the right (and vice versa) so that, again, they appear to be unconsciously mediated. The reason is, partly, that the left brain prefers to consider things linguistically and in order, one at a time, rather than simultaneously and all at once as the right hemisphere does, and is thus unable to grasp or comprehend anything that cannot be sequentially categorized or coded verbally. Indeed, even consciousness of the body image and the surrounding nonlinguistic environment is also fragmented in terms of parts and details, whereas the right brain not only maintains a bilateral image but mediates the expression of emotion through body language. By contrast, the left brain is socially and emotionally quite dense. Thus, from a language and left-brain point of view, much of what occurs within the domain of the right brain occurs outside of verbal consciousness and is thus potentially unconscious because it is nonverbal and nonlinguistic.

Nevertheless, we are certainly capable of being *aware* of much of what takes place within the right brain and the functions it mediates: singing, dancing, drawing, the appreciation of music, the recognition of thunder or water running, the expression of emotion, the perception of social nuances, and so on. These functions sometimes just seem to occur unconsciously only because we can engage in these as well as a variety of other behaviors without the aid of language or self-conscious participation of the left half of the brain. However, the conscious mind and the left brain keep on thinking about what is occurring so as to provide explanations.

The Origin of Thought

Thinking is clearly a form of communication, and one form of thought—that is, linguistic thought (thinking in words)—is a form of inner language, an organized hierarchy or train of associations that appear in one's "mind's eye."

The "train of thought" often explains things and usually appears as a progression or an associative advance with an initial or leading idea, which is then followed by a series of related words and ideas. However, it can also be a means of questioning, deducing, planning, and forming goals, as well as explaining, be it an idea, action, thing-in-the-world, or upcoming event so that it may be understood, communicated, and possibly acted upon or prevented. It is through linguistic thought that the conscious mind is able to better understand and manipulate the world.[1]

Verbal thinking often serves the function of explaining things. Yet the need to communicate and explain things to oneself seems paradoxical. It might be asked, "Who is explaining what to whom?" Apparently the "I" that I am explains and communicates these things to the "I" that I am. That is, I serve as both the explainer and the explainee. A functional duality is thus implied in the production and reception of some aspects of verbal thought.

This also implies that some forms of verbal thought are often based on information and knowledge that are already in existence in the mind but in a nonlinguistic, imaginal, emotional, sensory, or nonorganized form. That is, they exist in a form that cannot be recognized or understood by the conscious mind. For the left brain to gain an understanding of this information requires that it be transformed into a linear, temporal sequence of language-related ideas and images.

That is, the only way this information can be understood by the left brain is if it is transformed. The only way it can be transformed and comprehended linguistically is by thinking about it in a verbal manner. Thus, linguistic thought sometimes serves to explain and communicate something that we are already aware of nonlinguistically and unconsciously. It communicates it to the left half of the brain, and it is comprehended and understood via the aid of Wernicke's area and the angular gyrus.

Thinking in words is associated with the conscious mind and the language axis (Broca's and Wernicke's areas and the angular gyrus) of the left cerebral hemisphere. In this regard, thought sometimes serves as a means of organizing, interpreting, and explaining impulses that arise in the nonlinguistic portions of the nervous system, so that the language-dependent regions may achieve understanding.

Verbal thinking thus seems to act as a conscious inner language that organizes and assigns verbal labels and grammatical order to our "not-thought-out" ideas, which are not fully conscious, into an organization which may then be understood consciously.

Egocentric Speech and the Internalization of Thought

Thinking can take the form of images, physical sensations, melodies, and so on. However, insofar as thinking is verbal, it is clearly associated with the functional integrity of the left half of the brain.

Thinking often serves as a means of covert self-expression. That is, our thoughts occur within the privacy of our own heads and are meant for no one other than ourselves. Although we may listen to our thoughts, we usually listen alone, unless, of course, we are thinking "out loud." However, even when we think out loud, or talk to ourselves, our thoughts are usually meant for our ears alone and no one else's.

Thinking and "talking to oneself" are ontogenetically linked. Before children can think in words, they must be able to *hear* words, so as to know what they sound like. Because of this, in part, people first learn to think out loud as children. It is only over the course of the first seven years of life that verbalized thoughts become progressively internalized as the private dialogue that we all experience. This internalization, in turn, corresponds to the increasing maturation and development of the corpus callosum, which carries out information transfer between the right and left brain.

Children first begin to think out loud, talk to themselves, and verbalize their thoughts at ages two to three. This form of communication has been referred to by Jean Piaget[2] and L. S. Vygotsky[3] as "egocentric speech." Up to this age, children think in visual images, songs, feelings, desires, emotions, rhymes, and an occasional word and may frequently engage in daydreaming, but they do not yet think in connected words and sentences. That is, left-brain verbal thinking has not yet developed.[4]

Egocentric speech is a form of thinking out loud and talking to oneself, and it usually takes the form of commenting on and explaining one's actions. Children use egocentric speech both when they play alone and when they are in groups. At its peak, almost 40 percent of all language production in children is egocentric.

When children engage in egocentric speech, however, they do not appear to be concerned about the listening needs of those who are nearby, as their words are meant for their ears alone. Moreover, children engaged in an egocentric monologue seem oblivious if someone responds to them by asking a question or making a comment. In fact, several young children may be observed playing together, all of them engrossed in their own monologues and none of them listening to the others.[5]

When egocentric speech first makes its appearance (around age 2–3) and a child first begins to comment on his actions, the monologue is initiated only after a behavior has been completed. For example, a child draws a picture and, once it is finished, tells himself what he has done and explains the drawing. As the child ages, the egocentric monologue begins to occur at an earlier point in the action. He may explain his creation when he is still drawing. Finally, as the child matures (around age six or seven), he announces what he will draw and then draws it.[6]

Essentially, as the child ages, he appears to receive more advanced warning of his intentions until, finally, the information is available before rather than after he acts. First the child comments on an action after it occurs, then at an older age while the action is still occurring, and then finally the child has reached an age where he describes what he is going to do before he performs it. It is soon after the child reaches this stage, at about age seven, that egocentric speech becomes almost completely internalized as verbal thought.[7]

The Left Brain, Egocentric Speech, and the Corpus Callosum

That egocentric speech initially appears only after the action has occurred suggests that the left-hemisphere mental system is responding to impulses and actions initiated outside its immediate realm of experience and understanding. It seems that the left brain, in the initial production of egocentric speech, is attempting to organize the results of its experience (the observation of its behavior) into a meaningful verbalized description or explanation, which it then linguistically communicates to itself, out loud, but only after the action has occurred.[8] However, that it can describe what has happened only after it occurs indicates that the impulse to act originated elsewhere.

Egocentric speech is predominantly a function of the left brain's attempt to label verbally and make sense of behavior initiated by the right half of the brain.[9] However, because of the limited communication between the two brains, which is more pronounced in children, the left brain uses language to explain to itself the behavior that it observes itself engaging in. Later in life, in response to certain actions and extremes, it may simply proclaim, "I don't know what came over me."

When a child performs a "bad" behavior and is then confronted by an adult, she may blame the actions on someone else, such as an imaginary friend, or she might conjure up some fantastic imaginary explanation. When the left (speaking) half of a child's brain denies responsibility for something she has obviously done, it may in fact be telling the truth, and it may be as upset as the mother or father. That is, the child's left brain didn't do it. It was her right brain all along.

However, blaming others or making up explanations is not egocentric speech; it is confabulation.

As previously indicated, the corpus callosum is so poorly developed in young children that only a limited amount of information can be shared by the halves of the brain. Thus, when the right brain initiates an action, the left brain of a very young child has little advance warning and receives little or no information about what the right brain is up to—hence, the egocentric speech in which the child comments on her or his behavior after it has occurred. When the corpus callosum begins to mature and thus allows more information to be shared, the left brain begins to comment egocentrically on an action while it is going on, as it gets information from the right brain via the corpus callosum. When the efficiency of the corpus callosum transmission begins to approximate that of an adult (which is itself limited), the child can explain or comment on her or his actions before they occur. (The explanation may not be correct, however.) The left brain is now able to gain access to this information internally rather than via external observation. In consequence, it begins to create its explanations internally as well; it begins to think thoughts as well as speak them.

As the child ages, the production of egocentric speech appears to diminish. However, as argued by Vygotsky and Piaget, it does not disappear; it just goes underground and eventually becomes completely internalized as verbal thought.

Right-Brain Thinking

When the right brain produces and communicates ideas, the left brain is sometimes completely perplexed about where this information originated and what it means. The reason is that functions mediated by the right brain are not always easily amenable to linguistic description or analysis. Nevertheless, the right brain is responsible for the production of diverse and highly developed forms of mental activity, such as a flash of insight, intuition, creative leaps of the imagination, daydreams, and a variety of images, feelings, and thoughts that seem to envelop the mind, only to disappear as quickly as they arrive.[10] Often, the left brain has no idea as to where these images and feelings came from, since they originated in the right half of the cerebrum.

Frequently, the left brain, and the conscious aspects of the mind that it controls, is a passive spectator of these right-hemispheric events.

Indeed, in many instances, we engage in very complex activities without the aid of the left brain or the conscious mind. In these instances, the behavior is being controlled by the right brain. Unfortunately, we also engage in other complex behaviors that are sometimes emotionally upsetting, self-destructive, insulting, or inappropriate and that also originate within the unconscious part of the mind and the right half of the brain. We may respond emotionally to some event or in response to what someone has said, or we may become upset by the selective recollection of a traumatic emotional memory for reasons that the left brain cannot discern. Thus, we may become anxious, nervous, or upset and "not know why," say the wrong thing, use the wrong tone of voice, forget to perform some important action, be overcome with rage, or be blinded by "love," while the left brain is simultaneously without a clue as to what happened.

Lateralized Attitudes and Information Processing

Because language, emotions, memories, and perceptions are lateralized, the two halves of the brain, as we've seen, may reach different and conflicting conclusions about what seems to be the same piece of information. The two halves not only may perceive things differently but may have different attitudes regarding what they witness, hear, and observe. Indeed, this can be a major source of intrapsychic conflict, as not only different attitudes but different memories may be triggered. In response to these different perceptions, feelings, and memories, the right and the left brain may act in opposition, one half responding emotionally and attempting to do one thing, the other half attempting to accomplish something entirely different.

Indeed, because each half of the brain maintains an independent mental system, it is not at all unusual for each to have its own likes and dislikes, hopes for the future, goals and aspirations, social values, and political affiliations, as well as its own unique attitudes regarding their personal life. These two mental systems can sometimes act independently of one another. Each may purposefully initiate behavior, guide response choices, and recall and act on certain desires, impulses, or situations—without the aid, knowledge, or active self-conscious participation of the other half of the brain.

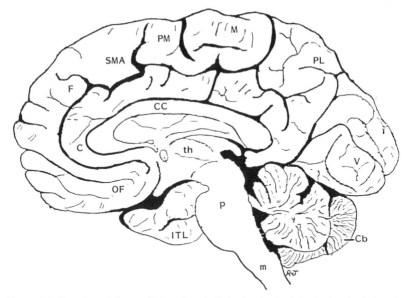

Figure 24. Drawing of the medial surface (split-brain view) of the brain, depicting the corpus callosum (CC) and other structures.

The Split Brain

The corpus callosum acts as the passageway through which information from one half of the brain travels to the other half. If the corpus callosum is traumatically injured or surgically severed, the ability of the two halves of the brain to communicate is almost completely attenuated.

In the 1940s, a unique neurosurgical procedure was developed and introduced for the purposes of controlling the spread of epileptic seizure activity from one brain half to the other via the corpus callosum. This procedure was called *corpus callosotomy* (also referred to as a *split-brain operation*), and it involved the surgical severing of the connections between the two brain halves, that is, the corpus callosum. When the callosum was surgically severed, seizures that could not be controlled by medication were confined to one brain half. This surgery lessened and in many cases stopped epileptic seizures in these patients. However, the ability of the right and left halves of the brain to share information was also abolished, as was the unity of the mind. The right and left brains were now freed to act more independently, and with some unusual consequences.

Following the surgical splitting of the brain by this procedure, some patients began to complain that the left half of the body sometimes did unusual and cruel things that they (or rather, the left, speaking half of the brain) did not understand and could not control. One patient reported that, when he was reaching inside a drawer to retrieve a sock with his right hand, his left hand had slammed the drawer shut on it. Another "split-brain" patient complained that his left hand (right brain) would not allow him to put on a particular shirt and kept unbuttoning the buttons and trying to pull the shirt off. Apparently, the right brain didn't care for that particular item of clothing.

One split-brain patient complained that, while he was shopping, after he had placed a few items in the shopping cart with his right hand, his left hand would reach in and take those items out of the cart, place them back on the shelf, and retrieve something entirely different. Presumably, his right and left brains did not like the same kinds of food.

Another split-brain patient complained that his left and right hands would struggle over which clothes he would put on in the morning. Again, it appeared that this man's right and left brains had different ideas in regard to what was fashionable.

Some twenty years later, these split-brain surgical procedures were greatly refined, and the patients subject to these operations were extensively studied by J. E. Bogen, M. S. Gazzaniga, J. Levy, R. W. Sperry, and their colleagues.[11-18] Dr. Sperry, in fact, received the Nobel Prize for his work with these patients. These and other investigators found, however, not only that the right brain acted independently of the left brain but that after surgery these patients, like those operated on twenty years before, behaved as if they had two separate psyches. According to Sperry:

> Everything we have seen so far indicates that the surgery has left these people with two separate minds, that is, two separate spheres of consciousness. What is experienced in the right hemisphere seems to lie entirely outside the realm of consciousness of the left hemisphere. This mental division has been demonstrated in regard to perception, cognition, volition, learning and memory. . . . In the right hemi-

sphere . . . we deal with a second psychic entity . . . that runs along in parallel with the more dominant stream of consciousness in the left hemisphere. Each brain half, in other words, seems to have its own largely separate cognitive domain with its own private perceptual, learning, and memory experiences, all of which are seemingly oblivious to corresponding events in the other hemisphere.[19]

Behavior Following Split-Brain Surgery

Following split-brain surgery, there are initially no grossly observable changes in the patient's personality, temperament, or general intelligence. However, when split-brain patients are stimulated tactilely on the left side of the body, their left brains do not register and fail to report feeling any sensation. The patients are unable to name objects placed in the left hand (if they are hidden from view) and cannot see or describe pictures or objects placed in the left half of their visual field. In other words, the left brain appears to have no knowledge of anything that is occurring on the left side of the body (which is monitored by the right brain). Nor can the left half of the cerebrum gain information from the right brain, including right-brain memories.

However, by raising the left hand, the disconnected right brain is able to indicate when the patient is tactually or visually stimulated on the left side. If an object is placed in the left hand (so that neither the hand nor the object can be seen; e.g., they are hidden behind a screen), although the patient is unable to name it, he or she can point to the correct object with the left (but not the right) hand if given multiple choices. When simple written words are presented to the left of the visual midline (so that the pictures goes to the right brain), although patients are unable to name the words, the right brain, when offered multiple visual choices, is able to lift the left hand and point correctly to the word viewed.

In one experiment, the word *toothbrush* was presented so that the word *tooth* fell in the left visual field (and was thus transmitted to the right brain) and the word *brush* fell in the right field (and went to the left hemisphere). When offered the opportunity to point to several words (e.g., *hair, tooth, coat,* and *brush*), the left hand usually pointed to the word viewed by the right brain (i.e., *tooth*), and the right hand

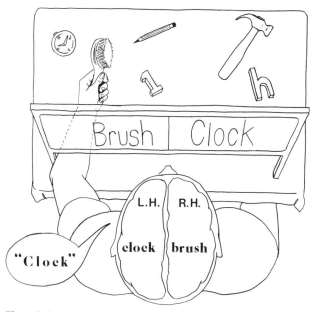

Figure 25. The split-brain patient is facing a screen on which is flashed (via a tachisto-scope) the words *brush* and *clock*. He is instructed to reach inside with his left hand to retrieve the object that corresponds to the word he has seen. Because the word *brush* is viewed by the right brain, he retrieves the brush with his left hand. However, because the word *clock* is viewed by the left (speaking) brain, when asked what he saw, he states, "Clock."

pointed to the word viewed by the left hemisphere (*brush*). If offered a verbal choice, the speaking (usually the left) hemisphere responded, "brush," and denied seeing the word *tooth*.

In many such experiments, however, when the left brain was asked to guess what had been shown to the right brain and it guessed wrong, the right brain tried to give it a clue. If a spoon was placed in the left hand and the left brain of the person who was asked to guess said, "fork," the right brain might shake the head no. The person's left brain would then usually say, "Sorry, I meant, uh. . ." and would guess again.

Lateralized Attitudes

In other experiments, split-brain patients were asked to give a "thumbs-up" or "thumbs-down" sign if they liked or disliked something that was being asked of them or shown to them. One young man who had undergone split-brain surgery was asked if he liked President Nixon. He gave a thumbs-up with his right hand and a thumbs-down with his left hand. Another split-brain patient was asked if he missed his girlfriend (he had just broken up with her). His right hand gave a thumbs-down and the left hand a thumbs-up. When one patient was asked if he liked to smoke, he gave a thumbs-up with his right hand and a thumbs-down with his left.

Indeed, the right and left brain may have not only their own likes and dislikes, which may be contrary to the values maintained by the other half of the brain, but oppositional attitudes, goals, and interests. For example, Reuben Gur described to me one split-brain patient whose left hand would not allow the patient to smoke and would pluck lit cigarettes from his mouth or his right hand and put them out. Apparently, although his left brain wanted to smoke, his right hemisphere was trying to quit.

Another split-brain patient examined by me experienced conflicts when attempting to eat or watching TV, his right and left brains apparently enjoying different TV programs and types of food.[20] Nevertheless, these difficulties are not limited to split-brain patients, for conflicts of a similar nature often plague "normal" people as well.

Right-Brain–Left-Brain Cooperation

Because the brain of the "normal" as well as the split-brain person supports two psychic realms, it is not at all surprising that a considerable degree of conflict may occur between the halves of the brain during the course of everyday activity. This conflict results because the two minds do not always share the same goals, interests, and memories.[21] Frequently, however, as in the case of the "split-brain" patient, LB, described below, the right half of the brain, although isolated, is fully willing to assist the left in a myriad of activities, and in fact, many split-brain (as well as normal) individuals are only seldom plagued by these problems. The surgically disconnected right hemisphere may attempt to aid the left or provide it with clues when the left (speaking) brain is called on to describe or guess what information or knowledge the right brain has or what type of stimulus has been shown to the right brain. For example, a picture of a face may be selectively flashed to the right brain (via the left visual field), and the left brain may be asked to name or describe the picture. When the left brain guesses wrong (which it always does), the right brain will try to give it a hint. This may involve nodding the head, clearing the throat (so as to indicate to the left brain that it has guessed incorrectly), or attempting to trace or write an answer with the left hand on the back of the right hand.

For example, after one split-brain patient's right hemisphere was shown a picture of Hitler, the patient signaled "thumbs-down" with his left hand:

Dr.: That's another "thumbs-down"?
LB: Guess I'm antisocial.
Dr.: Who is it?
LB: Soldier, G.I. came to mind, I mean . . . (The patient at this point was seen to be tracing letters with the first finger of the left hand on the back of his right hand.)
Dr.: You're writing with your left hand; let's keep the cues out.
LB: Sorry about that.[22]

Right-Brain Perversity

Right-Brain–Left-Brain Conflicts

Nevertheless, the behavior of the right hemisphere is not always cooperative, and sometimes it engages in behavior that the left brain finds objectionable, embarrassing, puzzling, or quite frustrating. This is true in the normal as well as in the split-brain individual.

In 1945, A. J. Akelaitis described two patients with complete corpus callosotomies who had experienced extreme difficulties in making the two halves of their bodies cooperate:

> In tasks requiring bimanual activity the left hand would frequently perform oppositely to what she desired to do with the right hand. For example, she would be putting on clothes with her right and pulling them off with her left, opening a door or drawer with her right hand and simul-

taneously pushing it shut with the left. These uncontrollable acts made her increasingly irritated and depressed.[23]

Another patient's left leg would not allow him to go for a walk. A recently divorced male patient noted that, on several occasions while walking about town, he found himself *forced* by the left half of his body to go some distance in another direction. Later, Dr. Akelaitis discovered that this diverted course, if continued, would have led the patient to his former wife's new home (although the patient's left hemisphere was not *conscious* of this at the time). Apparently, his right brain wanted to go for a visit.

According to Norman Geschwind, one split-brain patient complained that his left hand on several occasions suddenly struck his wife—much to the embarrassment of his left (speaking) hemisphere. In another case, a patient's left hand attempted to choke his own throat and had to be wrestled away. The patient expressed complete shock and surprise regarding these incidents and claimed that his left hand had acted of its own accord.

Another split-brain patient, described by S. J. Dimond, remarked that, on several occasions when she had overslept, she was suddenly awakened by her left hand, which had slapped her until she woke up.

J. E. Bogen reported that almost all of his complete commissurotomy patients manifested some degree of intermanual conflict in the early postoperative period. One patient, Rocky, for years complained of difficulty getting his left leg to go in the direction he (or rather his left hemisphere) desired. Another patient often referred to the left half of her body as "my little sister" when complaining of its peculiar and independent actions.[24]

Indeed, the French neurosurgeons Brion and Jedynak reported that this type of independent left-sided (right-hemisphere) activity was common in their split-brain patients and termed it the "alien hand."

Similar difficulties plagued one of the split-brain patients I examined.[25] Following callosotomy, Carl was frequently confronted with situations in which his left extremities not only acted independently but engaged in purposeful and complex behaviors—some of which he (or rather, his left hemisphere) found objectionable and annoying.

Carl complained of instances in which his left hand would perform socially inappropriate actions (e.g., attempting to strike a relative) and would act completely opposite to what he expressly intended; it would turn off the TV or change channels even though he (or rather, his left hemisphere) was enjoying the program. On one occasion, he had turned the TV on, found the channel that he wanted to watch, sat down on the couch, and suddenly found himself pulled off the couch by the left half of his body, which dragged him back to the TV. The left hand then changed the channel to something else, which Carl claimed he (or rather, his left brain) did not want to watch.

Once, after retrieving something from the refrigerator with his right hand, his left took it, put it back on the shelf, and retrieved a completely different item—"even though that's not what I wanted to eat!" he said. On at least one occasion, his left leg refused to continue "going for a walk" and would allow him only to return home.

In the examination and testing room, he often became quite angry with his left hand, striking it with his right hand, yelling at it, calling it names, and expressing hate for it. On one task, both hands, while out of view, were simultaneously stimulated with either the same or two different-textured materials (e.g., sandpaper to the right, velvet to the left), and Carl was required to point (with the left and right hands simultaneously) to an array of fabrics hanging in view on the left and right of the testing apparatus.

Almost always, the right and left hands pointed to the fabrics that they had experienced. However, because his left hand pointed at something different than that experienced by the right hand (the left brain), he would become angry and cry out, "That's wrong!" Repeatedly he reached over with his right hand and tried to force his left extremity to point to the fabric experienced by the right (although the left hand had responded correctly! His left brain didn't know this, however). His left hand refused to be moved and physically resisted being forced to point at anything different. One time, a physical struggle ensued, the right hand fighting and wrestling with the left.

In these instances, there could be little doubt that his right brain was behaving with purposeful intent and understanding, whereas his left brain had absolutely no comprehension of why his left hand (right hemisphere) was behaving as it did. It is also unlikely that the right brain was simply mirroring the left, as the oppositional attitudes described above occurred only infrequently and seemed limited in scope, as in the selection of food, television programs, and the like, or in purposeful actions such as correctly perceiving a textured object or picture and then pointing at the correct items repeatedly, without error.

Goals, Attitudes, Memories, and
Uncooperative Mental Activity

It is again important to note that split-brain patients are generally only infrequently plagued by these very profound and obvious behavioral difficulties, as the two halves of the brain are generally cooperative. However, even in the "normal" intact brain, situations often arise in which one brain half has little or no knowledge of what is occurring in the other and behaves uncooperatively.

The confusions and conflicts that are a part of everyday experience are, in part, a function of each half of the brain's being specialized to receive and process different types of information. Some stimuli cannot even be recognized by one or the other half of the brain, nor can all this information be shared. Thus, even when analyzing the same stimulus, each half of the brain may interpret and process it differently and then reach its own conclusions, unbeknownst to the other side of the brain. This split, in turn, may give rise to the formation of different memories in each half of the brain.

Perceptions and memories that are stored differently in the right and the left brain may also generate distinct and opposite goals, plans, hopes for the future, and attitudes that can oppositionally affect behavior.[26] Even when their goals are the same, the halves of the brain may produce and attempt to act on different strategies.[27] Differential memory storage also influences how we interact with others and even how we treat ourselves. A person does not need to undergo a split-brain operation in order to choke on self-hate.

It is not at all surprising that when we are involved in highly complex social interactions or even when we are trying to perform everyday activities, we sometimes feel in conflict, confused, or even paralyzed by indecision. When opening the refrigerator and trying to figure out what we want to eat, or when staring at the closet and trying to decide what to wear, we may have difficulty making up our mind or find ourselves making one decision and then quickly changing our mind. Sometimes these mental balancing acts are literally due to a "changing of the mind" as the right and then the left brain strives for control over our thoughts, feelings, and behavior. However, unlike the split-brain patient, the battle goes on in our head.

6

The Limbic System and the Most Primitive Regions of the Unconscious

In 1966, Charles Whitman, a former Boy Scout and Eagle Scout, a model husband and overall "great guy," climbed a tower at the University of Texas and began to indiscriminately kill people with a high-powered rifle. Earlier in the day, he had brutally murdered his wife, whom he claimed, even after her death, to have "loved dearly."

For several weeks, he had been troubled by violent thoughts and headaches. He found himself thinking about killing and hurting people and had been increasingly overwhelmed by violent urges. As these were completely contrary to his nature, he became extremely worried and made an appointment with a psychiatrist. Unfortunately, after a two-hour session, the therapist was unable to provide him with any insight or relief.

Feeling increasingly frightened by what was occurring in his mind, and fearing that he might act on these violent impulses, he tried to turn himself in to the police. Because they couldn't charge him with any-thing, they sent him home. The impulses began to grow in intensity, and he finally sat himself down and wrote a letter:

> I don't really understand myself these days. Lately I have been a victim of many unusual and irrational thoughts.

107

These thoughts constantly recur, and it requires a tremendous mental effort to concentrate. I talked to a doctor once for about two hours and tried to convey to him my fears that I felt overcome by overwhelming violent impulses. After one session I never saw the Doctor again, and since then I have been fighting my mental turmoil alone. After my death I wish that an autopsy would be performed to see if there is any visible physical disorder. I have had tremendous headaches in the past.

Later he wrote:

It was after much thought that I decided to kill my wife, Kathy, tonight after I pick her up from work. . . . I love her dearly, and she has been a fine wife to me as any man could ever hope to have. I cannot rationally pinpoint any specific reason for doing this.

That evening he killed his wife and mother and wrote:

I imagine it appears that I brutally killed both of my loved ones. I was only trying to do a good thorough job.[1]

The following morning, he climbed the university tower carrying a high-powered hunting rifle, and for the next ninety minutes, he shot at everything that moved, killing fourteen people and wounding thirty-eight, until he was finally brought down by a hail of gunfire.

Nothing in Whitman's history could explain his violent behavior. He was well liked, responsible, and easygoing and was a genuinely caring person just trying to get ahead in life. Nevertheless, at his autopsy, doctors opened up his skull and found a giant tumor the size of a walnut compressing part of the amygdala of the limbic system. Was it damage to this old cortical nuclear mass, the amygdala, that had triggered the murderous and violent feelings that soon overcame him as the tumor grew in volume?

The amygdala is part of the limbic system and is highly involved in the ability to experience moods and emotional feelings, including the capacity to discern the emotional and motivational significance of all sensory events.[2] For example, it is the amygdala that enables you to detect if someone is behaving in an angry or a loving manner, to determine if something may be particularly desirable to eat, to feel fear, love, hate, anger, desire, and attraction, and to form social and emotional bonds.[3] It provides the emotional color to all that we perceive and feel and enables us to feel close to others. However, it also subserves the capacity to feel, and act on murderous rages.

Unfortunately for Charles Whitman, it appears that his abnormal amygdala, infiltrated by the tumor, began to color his world in tones of violence and rage. Although he tried to rationally resist his irrational impulses, as often happens to most normal individuals, he became overwhelmed and then acted on them. However, due to the abnormality within his brain, he behaved in an abnormally violent manner.

The Limbic System

The Origins of Emotion

Many members of our Western civilization experience emotion as a potentially overwhelming force that warrants and yet resists control—as something so irrational that it can overwhelm or even happen to an individual ("you *make* me so angry!") that rational control is lost. Indeed, we can be overcome with rage or be blinded by love while remaining consciously aware that we should not be acting this way and helplessly wishing to control ourselves.

This schism between the rational and the emotional is due in part to the raw energy of emotion, which has its source in the cortical nuclei of the ancient limbic lobe, the limbic system. The limbic system controls the most archaic and reflexive aspects of emotional and motivational functioning, as well as the ability to experience feelings such as love. It is concerned with the monitoring of internal homeostasis, the fulfillment of basic body and tissue needs such as hunger and thirst, the control over predominantly biological, emotional, and motivational states such as sexual behavior, and the ability to experience and express feelings and emotions such as pleasure, happiness, satisfaction, aversion, fear, anger, anxiety, and rage.

The limbic system constitutes part of the old brain and is thus composed of old cortex—a series of nuclei and cell assemblies in the centermost portion of the cerebrum which first make their phylogenetic and evolutionary appearance long before human beings walked on this earth.[4] In fact, within the brains of apes, monkeys, dogs, rodents, reptiles, and even certain fish, and performing many of the same functions, can be found some of the same limbic system structures that are possessed by humans. The limbic system is so ancient and, in some respects, primitive that some have referred to it as the *reptilian brain*. However, this designation is correct only for the most primitive aspects of the limbic system such as the hypothalamus, as primates, including humans, have limbic nuclei not found in reptiles.

Although over the course of evolution a new brain (neocortex) has developed and has become superimposed over the old cortex, we remain creatures of emotion; we have not completely emerged from the phylogenetic swamps of our original psychic existence. The old limbic brain has not been replaced, for it remains buried within the depths of the cerebral hemispheres, where it continues to exert its influence much as it has for the last 100 million years.

The Nuclei of the Limbic System

The structures and cell assemblies that make up the limbic system include the hypothalamus, the amygdala, the hippocampus, and the septal nuclei as well as other structures that will not be reviewed here. Unlike the neocortical tissues (also called *gray matter*) that make up the outermost portions of the right and left halves of the cerebrum and that are fully developed only in humans, some of the same nuclei that make up the limbic system (such as the hypothalamus) can be found in reptiles and even in sharks. The emotions expressed by this region of the brain are very primitive indeed.

Unlike sharks, however, certain highly intelligent, social animals such as dogs, monkeys, apes, and humans have limbic systems that, although still primitive, also contain more evolutionarily advanced nuclei, such as the amygdala. It is because these and so many other animals have a limbic system similar to ours that they are also so similar to humans in their need for love, affection, and physical closeness.

The *hypothalamus*, located at the very center of the brain, controls and monitors hunger, thirst, and the ability to feel extreme pleasure, displeasure, and even rage. It is the most primitive part of the limbic system and is the source from which all emotions originate as raw, undirected, but powerful feelings. The hypothalamus represents the emotional core of our being.

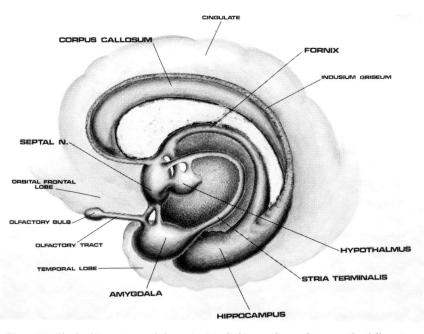

Figure 26. The limbic system and the major interlinking pathways between the different limbic nuclei. From R. Joseph, *Neuropsychology, Neuropsychiatry, and Behavioral Neurology* (New York: Plenum Press, 1990).

The hypothalamus is also intimately involved in all aspects of sexual behavior, including the assumption of sexual postures, ejaculation, and hormonal secretions such as those involved in pregnancy and the menstrual cycle.[5,6] Because the hypothalamus in females serves different sex-related functions, it is, not surprisingly, somewhat different in structure and complexity from that of men.[7–9] Males have a hypothalamus that has a distinctly male pattern of interconnections between nerve cells, a consequence of the influence of testosterone on the brain during neonatal development. Females have a hypothalamus with a more rich and complex pattern and with more intricate interconnections between nerve cells. It is more complex because of the complexities involved in the reproduction and the nurturance of infants. It has also been recently reported that some male homosexuals (those who

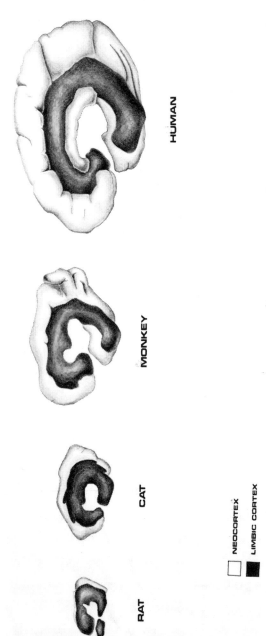

Figure 27. Schematic diagram of the limbic system in four species. Note the relative similarities in location and the increased representation of neocortex (new brain) in humans. From R. Joseph, *Neuropsychology, Neuropsychiatry, and Behavioral Neurology* (New York: Plenum Press, 1990).

died of AIDS and were brought to autopsy) may have hypothalamic nuclei that more closely resemble the female pattern.

Thus, the hypothalamus is sexually differentiated so that there are male and female limbic systems. At the very core of our emotional being lies one of the origins of sex differences in behavior and feeling.

As the source from which feelings of pleasure arise, and through the generation of feelings of reward and punishment, the hypothalamus is able to exert tremendous influence over the rest of the brain so that its desires are met. However, although extremely powerful, the hypothalamus (at least in humans) is controlled, in part, by the frontal lobes of the brain and the later developing limbic nuclei such as the amygdala.

The *septal nucleus* is involved in the capacity to form emotional and social bonds with others. It also appears to exert dampening effects on mood. It is able to tap into the emotional reservoir of the hypothalamus, with which it maintains rich interconnections, so as to exert emotional influences on the rest of the brain.[10] It is also intimately associated with the hippocampus (and thus memory functioning) and in some ways acts to counteract the amygdala.[11]

The *amygdala*, as already mentioned, discerns the emotional significance of all aspects of experience.[12–14] It adds emotional color to one's thoughts and enables one to feel complex emotions such as love and anxiety. The amygdala (also referred to as the *amygdaloid nucleus*) is very sensitive and responsive to touch, regardless of where it occurs on the body, and is involved in the ability to feel emotions when one is held. It is also involved in memory and is intimately interconnected with the hypothalamus, the septal nucleus, and the hippocampus.

The amygdala, located in the temporal lobes, also receives visual and auditory perceptual information and therefore can emotionally influence how we view and even think about things (e.g., threatening or inviting). Indeed, this structure, when damaged, can make a person not only misperceive or fail to perceive the emotional qualities of the environment, but in very rare cases of temporal lobe epilepsy, it can cause them to view things with mystical awe and religious reverence. An individual on LSD demonstrates maximal activation and arousal of the amygdala and the surrounding temporal neocortex, particularly that of the right half of the brain.

The *hippocampus* is highly involved in all aspects of memory and learning and becomes activated during dreaming. The hippocampus in the right brain is concerned with visual, emotional, tactile, and nonverbal memories; in the left brain, it stores verbal and mathematical memories. Through its interconnections with the amygdala and the rest of the limbic system, the hippocampus stores in memory events that are of emotional and motivational significance, or that are arousing and thus interesting and important.[15–18] The amygdala alerts the hippocampus to store emotionally significant experiences in memory, particularly those that are visual.

If the hippocampus (which, like the amygdala, is located in each temporal lobe) is damaged, such as by stroke or hypoxia, memory functioning is severely affected. Verbal memory is affected if the left hippocampus is destroyed, and visual, emotional, and related memories (where something is left) are impaired with right-sided destruction. If both hippocampi are destroyed, the person is unable to remember or learn anything following the injury. However, everything already stored in memory (that is, prior to the injury) remains recallable. If you introduced yourself, left the room, and then returned a few minutes later, a patient with bilateral destruction of the hippocampal nuclei would have no memory of having met or spoken to you.

One well-known individual, Henry, who underwent neurosurgery and had both hippocampi removed, has worked with Brenda Milner for over twenty years, and yet she is an utter stranger to him. He is so amnestic for everything that has occurred since his surgery that, every time he rediscovers that his favorite uncle died (which happened after his surgery), he suffers the same grief as if he had just been informed for the first time.

Nevertheless, despite his lack of memory for new information, Henry has adequate intelligence, is able to recall what occurred before his surgery, is painfully aware of his deficit, and constantly apologizes for his problem. "Right now, I'm wondering," he once said, "have I done or said anything amiss? You see, at this moment everything looks clear to me, but what happened just before? That's what worries me. It's like waking from a dream. I just don't remember. . . . Every day is alone in itself, whatever enjoyment I've had, and whatever sorrow I've had. . . . I just don't remember."[19]

The Frontal Lobes and Limbic System Control

Whereas the right and left brain mental systems take years to develop, many regions within the limbic system, such as the hypothalamus, are almost fully functional at birth and are responsible for almost all aspects of psychic functioning and emotional responding that occur during infancy. A baby is, in fact, a bouncing bundle of limbic, hypothalamic needs and feeling states. Consciousness, like language, develops much later.

Fortunately, as the brain of the child develops, the frontal lobes (the senior executive of the brain, ego, and personality) also mature and grow extensive interconnections with the limbic system, a maturational process that may take over fifteen years. Thus, as we age, the frontal lobes not only control or direct information processing within the brain but increasingly exert inhibitory influences on our emotional states as well. Unlike the limbic system, however, the frontal lobes are minimally developed in all other species except humans.

Two Levels of Unconscious Mental Functioning

Conscious and Unconscious Awareness

The human mind consists of two broad regions that we have referred to as the *conscious* and *unconscious mind*. The left half of the brain subserves and mediates linguistic or verbal aspects of consciousness.

Consciousness and the left brain are interlinked insofar as both rely on language to classify, think about, and describe the world. The left half of the brain also controls the right hand, which is dominant in the majority of the population for fine and repetitive motor functioning.

Through language and use of the right hand, the left brain is able to explain, explore, label, describe, write about, count, classify, manipulate, and exert control over the environment. This ability makes the left brain the conscious and seemingly dominant half of the cerebrum. However, due to the linguistic and language-dependent nature of the conscious mind, at least as defined in this book, the right brain is not associated with consciousness. Instead, the right brain is associated with a highly developed form of mental awareness that can be referred to loosely as an *unconscious awareness*. Functions mediated by the right brain are not easily amenable to linguistic description and seldom require language, or even active conscious participation. When we dance, run and catch a football, skate, or ride a bicycle, the left brain,

Functions of the Left and Right Brain

The left brain controls:	The right brain controls:
Linguistic consciousness	Unconscious awareness
The right half of the body	The left half of the body
The right hand	The left hand
Perceives right half of body	Perceives faces and both halves of body and maintains the body image
Talking, reading, writing, spelling	
Speech comprehension	
Linguistic and verbal thinking	Emotional and melodic speech
Verbal intelligence	Singing and swearing
Verbal memories	Comprehension of music, emotion, body language, and environmental sounds (chirping birds, buzzing bee, babbling brook, thunderstorm)
Dreaming in words, thought dreams	
Rhythm, temporal and sequential	
Information processing	
Keeping score of football game	
Marching	Visual, emotional, musical, creative, and geometric thinking
Math	Emotional and childhood memories
Typing	
Grammar	Insight and intuitive reasoning
Logical and analytical reasoning	Seeing the forest
Confabulation	Reading between the lines
Perception of details	Perceiving the overall "big picture"
Seeing the trees	Visual-spatial processing
Broca's expressive aphasia: Loss of speaking ability	Throwing and catching a football
	Riding a bicycle
Wernicke's receptive aphasia: Loss of language comprehension	Dancing
	Visual closure
Agraphia: Loss of writing ability	Gestalt formation
Alexia: Loss of reading ability	Initial self-concept
Acalculia: Loss of math ability	Dreaming in visual-emotional images
Apraxia: Loss of ability to perform skilled temporal-sequential movements	
	Amusia: Loss of musical abilities
	Prosopagnosia: Loss of facial recognition
	Anognosognosia: Left-sided neglect and denial, disturbances of the body image

and the conscious aspects of the mind which it controls, is often a passive participant and spectator.

This does not mean to imply that all behaviors or abilities controlled by the right brain occur unconsciously, for we are often completely *aware* of the activities we are engaged in and of what is going on around us. They are unconscious only insofar as the left brain does not know about them, cannot label them, or is not actively paying conscious attention. Nevertheless, it is only in regard to certain emotions, intuitions, inferences, desires, impulses, and other feeling states that right-brain functions are truly unconscious. This is particularly true of emotional and traumatic memories, which the left brain and the language-dependent conscious mind are completely unable to gain access to or to understand. For the left brain, these particular memories and impulses do not have a conscious origin and are thus unconsciously mediated.

Nevertheless, some of what occurs within the right brain is well within the realm of at least partial conscious scrutiny. Moreover, both the left and the right brain are often similarly and simultaneously activated, and when this occurs, we may say that we are conscious and aware.

Whereas the left brain is associated with language, math, verbal thinking, and consciousness, the right brain is associated with a mental state best described as *awareness*. Right-brain *awareness*, as defined here, is emotional, intuitive, inferential, visual-spatial, tactile, melodic, musical, artistic, pictorial, imaginal, geometric, gymnastic, environmental, and nonlinguistic. When we are staring out at the ocean and listening to the waves and the birds crying overhead and are not engaged in conversation or producing verbal thoughts, we are in a right-brain mode of psychic functioning and can be said to be completely aware of what is occurring around us.

If while at the seashore we are reading a book, talking to a friend, or thinking about our job, we are in a left-brain mode and are using the language-dependent conscious mind. When in this mode, we may not be conscious of the sounds of the waves or the birds, although we may be aware of them.

Thus, awareness, as mediated by the right brain, is a distinct mental dimension, as in "Although I never gave it any thought, I was aware of it the whole time." Nevertheless, insofar as something is not conscious, it can be said to be occurring nonlinguistically and thus unconsciously (or subconsciously). In this way the functions of the right brain are associated with an unconscious awareness.

The Limbic Unconscious

We have identified two mental realms: a conscious mind and an unconscious awareness. Both are highly developed, unique to human beings, relatively accessible, and associated with the left and right cerebral hemispheres, respectively.

The unconscious mind consists of two levels. The more accessible aspect of the unconscious consists of the highly refined social-emotional, melodic, environmental, visual-spatial, pictorial, and nonlinguistic awareness of the right brain. There is, however, a relatively inaccessible aspect of the unconscious that is associated with the very primitive, very ancient region of the brain, the limbic system. As noted, the limbic system is a "system" of nervous tissues and cell assemblies located and buried within the depths of the centralmost portion of the brain. The most ancient structure within the limbic system, the hypothalamus, is almost fully functional and completely dominant over the brain at birth.[20]

From a phylogenetic and evolutionary perspective, the appearance and development of the limbic system predates the emergence and differentiation of all other brain structures that aid in consciousness, awareness, and thought. It constitutes the most primitive, archaic, reflexive, and purely biological aspect of the mind. However, with the development of the amygdala, the limbic system also became capable of generating very complex and refined emotional states as well.

This aspect of the brain and mind is almost completely unconscious, in the more popular sense of the word. It is concerned with rudimentary and reflexive mental activities, which are difficult to analyze, control, or even recognize, as well as with complex emotional states, including the desire to form emotional bonds. This is the mental region most deserving of the term *unconscious,* as its functions originate and often occur for the most part independently and outside of conscious awareness.

The limbic level of the mind could be referred to as the *primary unconscious,* as its appearance precedes all other aspects of mental functioning. Given the widespread existence of limbic cortical nuclei in primitive animals and their long evolutionary history (compared to the more recently developed neocortex), it could also be argued that this region of the mind is in some way associated with the "collective unconscious," particularly in that parts of the limbic system are not only ancient and commonly found in mammals and even reptiles, but extremely important in the formation of memories and dreams. However, we will not argue that position here.

Because the right brain mental domain is slower to develop and matures long after the emergence of limbic system psychic functioning, it can be referred to as mediating the "secondary unconscious." The limbic unconscious is the bedrock on which the foundations of the mind will be laid and is thus the primary unconscious, because it is there first.

Impulses, desires, and related feelings and motivational states originating in the limbic system often occur completely outside conscious awareness. It is very difficult for the conscious mind to scrutinize or understand this aspect of the psyche. In fact a great deal of what transpires in this region of the mind is biological or reflexive and never enters conscious awareness at all, except after the fact or in the form of primary or urgent impulses, needs, hungers, desires, dreams, and memories.

Impulses that originate in the limbic unconscious need not be transmitted first through the secondary unconscious maintained by the right brain in order to achieve consciousness, however. Moreover, these impulses are able to discharge directly, independently, and simultaneously into the stream of conscious awareness associated with the left and right halves of the brain. Thus, the left and right halves of the brain and the mind can, in fact, be influenced and overwhelmed by these forces simultaneously, independently, or in isolation. These impulses need not be transmitted first through the unconscious mental system maintained by the right brain to be acted on because the limbic system maintains rich interconnections with both halves of the brain and the mind. Half the limbic system is located within the left, the other half within the depths of the right cerebral hemisphere. We have two amygdalas, two hippocampuses, and so on. Nevertheless, whereas these impulses are almost completely foreign to the language-dependent conscious mind, and the left half of the brain, the right brain, being more involved in emotional functioning, is often (but not always) able to discern and recognize these limbically induced feeling states and desires for what they are.

The Hypothalamus and Limbic Interactions
with Conscious Awareness

The hypothalamic region of the limbic system is the most primitive and important part of the brain. The hypothalamus serves the body tissues by attempting to maintain internal homeostasis and by providing for the immediate discharge of tensions, almost reflexively. It appears to act in an almost on/off fashion so as to seek or maintain the experience of pleasure and escape or to avoid unpleasant, noxious conditions. Hence, feelings elicited by this part of the brain are very short-lived. This part of the brain cannot bear a grudge, and the feelings generated may disappear completely after just a few seconds, although they may also last much longer.

This portion of the limbic system's direct contact with the real world is quite limited and almost entirely indirect, as it is largely concerned with the internal environment of the organism. If there is a need for nourishment, the hypothalamus, sensing this, becomes activated and demands to be fed. If there is a physical sensation of arousal or discomfort, it senses this and demands relief.

Emotions elicited by the hypothalamus are often triggered reflexively and without concern about or an understanding of the consequences. It seeks pleasure and satisfaction, and whether the stimulus is thirst or sexual hunger, the basic message of the hypothalamus is "I want it now!" It is unable to consider the longer-term consequences of its acts, and it has no sense of morals, danger, values, logic, or "right and wrong."

Although quite powerful, hypothalamic emotions are largely undifferentiated, consisting of feelings such as pleasure, displeasure, aversion, anger, rage, happiness, hunger, thirst, and so on. The more recently evolved regions of the limbic system, such as the amygdala and the septal nucleus, are more emotionally advanced, specific, and refined, have specific likes and dislikes, and even enable us to feel moods or affection, as well as the rudiments of love. Indeed, through the limbic system, we are able to like or love someone. The limbic system responds with considerable feelings of pleasure when we associate with individuals we like and find attractive, and it makes us laugh with pleasure or, correspondingly, weep from sadness.

Limbic System Emotional Abnormalities

If the limbic system is damaged, feelings of like, love, hate, and affection, and even sexual responsiveness and desire, may be abolished or may become severely abnormal. Presumably, this is what happened to Charles Whitman. His limbic system (i.e., the amygdala) had become infiltrated by a giant tumor that caused him to feel extreme rage and violent impulses. These impulses, in turn, completely over-whelmed the mental systems of both the right and the left brains and caused the perceptions of both to be colored by violent emotions.

Violence and Rage. Robert, a man of the clergy, had recently begun developing raging, violent impulses. He even checked himself into a local psychiatric unit, where he rapidly decompensated and became extremely vicious and wild. After his hospitalization and until he died (about two weeks after the onset of his symptoms), he acted almost like a wild animal. By his final days, he was growling and baring his teeth in response to any sound or the approach of anyone and was howling with anger. He was thought to have suffered a "nervous breakdown," although some whispered of demonic possession. However, when he suddenly died and was brought to autopsy, a massive and apparently fast-growing tumor was found to have destroyed wide areas throughout the limbic system.

One need not suffer damage to this area to feel these types of responses. Some people become ferociously angry at their friends and loved ones, to the point where they kill them brutally. Electrical stimulation of the limbic system can also cause feelings of violence, including facial contortions, baring of the teeth, dilation of the pupils, widening or narrowing of the eyelids, flaring of the nostrils, growling, irritation, anger, and, finally, rage, which seems to build up gradually until the person will attack anyone within range.

Uncontrolled Sexual Desire. Electrical stimulation of the hypothalamus and the amygdala may produce involuntary sexual behavior. In females, this includes ovulation, uterine contractions, lactogenetic responses, and posturing their bodies for sexual penetration. Similarly, electrical stimulation of the limbic system in males may cause penile erections and pelvic thrusting followed by an explosive discharge of semen even in the absence of a sexual partner. Indeed, a marked increase in sexual desire and copulations occurs.

If certain regions in the limbic system (i.e., the amygdala) are damaged, there can result heightened and indiscriminate sexual activity, including excessive and almost constant masturbation and genital manipulation and repeated, indiscriminate attempts to have sex with anyone or anything. Animals that have suffered surgical destruction in this area will even attempt to have sex with animals that might kill them. For example, one small male cat repeatedly made highly aggressive attempts to copulate with a male dog, even when the dog made it quite clear that this behavior was not at all appreciated. Hence, when the amygdala is damaged, the ability to control selective sexual behavior, to discern what constitutes acceptable sexual activity, or to perceive sexual nuances so that appropriate partners can be obtained is completely lost.

Thus, the amygdala makes it possible for us to feel sexual arousal and to act selectively in the choosing of partners. However, due to its interconnections with the hippocampus, its choices are also greatly influenced by memory, and emotional memories in particular.

Uncontrolled Laughter. The limbic system mediates and controls the ability to feel pleasure and displeasure.[20] Although the limbic system probably does not have a sense of humor, there is some evidence that it generates not only the pleasure we feel in the company of others, or in general, but even the pleasure that can give rise to laughter.

Peter was a serious young man who was quite close to his mother. Thus it came as a terrible surprise to him and family members that, while attending his mother's funeral, he was seized at the graveside with an attack of uncontrollable laughter, which embarrassed and distressed him considerably. He laughed so much that he had trouble catching his breath. He finally walked away feeling humiliated.

About an hour later, he again had an attack of laughter, which lasted longer than the first. At short intervals, he had another and then another attack of laughter and finally dropped to the ground and died laughing. At autopsy, it was discovered that he had suffered a hemorrhage that had severely damaged part of the hypothalamus and had compressed other portions of the limbic system. As the blood continued

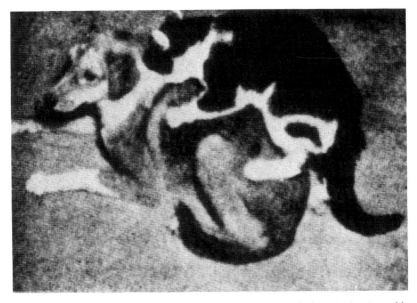

Figure 28. When both sides of the amygdala have been destroyed, the organism is unable to discern the emotional or social significance of what it experiences. This cat has suffered bilateral limbic destruction and is attempting to mount and have sex with a dog. From L. H. Schriener and A. Kling, "Behavioral Changes Following Rhinencephalic Injury in the Cat." Courtesy of *Journal of Neurophysiology* 16 (1953), 643–659. Reprinted by permission.

to escape within his brain, the excess fluid finally compressed vital centers, and this compression killed him.

Another unfortunate person was suddenly overcome by nonstop laughter. She was in fact shaking with laughter and could not stop. The laughter was so continuous that, quite quickly she began having trouble breathing as she had only enough time between laughs to take in a shallow breath. She fell to the floor laughing, trying to breathe, became cyanotic, continued laughing in little gasps, and finally lost consciousness and fell into a coma. Within twenty-four hours, she was dead. Like Peter, described above, she had died of a massive hemorrhage in the limbic system; and the massive buildup of blood had triggered, via compression of these vital centers, the uncontrolled laughter.

Eating. A major function of the hypothalamic portion of the limbic system is the monitoring of internal homeostasis, including the need for nourishment and the intake of food and liquid. Not surprisingly, electrical stimulation of the hypothalamus can make a person eat "like a pig" or conversely lose her or his appetite. Animals and humans with destruction of certain areas of the limbic system may suffer extreme eating disorders and, in some cases, eat until they have doubled their weight and are ready to burst.

The Pleasure Principle

The limbic system mediates a wide range of simple emotions.[21] Because it actually controls the ability to feel pleasure and displeasure, it is able to generate and use these emotions to meet a variety of its needs, be they sexual, nutritional, or emotional. That is, it can reward or punish the entire brain for meeting or not meeting its needs by generating feelings of pleasure or aversion that the whole brain then experiences.[22] If, for example, the hypothalamus experiences pleasure, it will generate rewarding feelings so that the person continues engaging in the activity desired by the limbic system. If it begins to feel displeasure, it will stop. However, if limbic needs continue to go unmet, the person will experience depression, anger, or rage.

The limbic system, via the hypothalamus, is similarly able to reward or punish others if they meet its needs or not. That is, it may strike out or reinforce others by expressing feelings of rage or happiness. This is how the limbic system initially communicates with "mother" and is able to motivate her to take care of its needs and desires, that is, via rage or smiles of pleasure and happiness.

As the hypothalamus of the limbic system maintains internal homeostasis through its ability to reward or punish the organism with feelings of pleasure or aversion,[23] it tends to serve what Freud called the "pleasure principle." The pleasure principle is a concept that represents the tendency of all organisms to strive for pleasure and to avoid noxious situations. As described by Freud, the pleasure principle not only serves to maximize pleasant experiences but acts to keep the psyche as a whole free from high levels of tension and excitation (whether pleasurable or unpleasant). It acts to reduce the tensions of unmet needs (e.g., hunger, thirst, or sexual desire) by promoting their fulfillment. For example, the person who feels sexual desire and abstains may feel tension. Paradoxically, the only way to reduce this tension is by increasing it until orgasm and thus tension release is obtained. Fortunately, increasing sexual tension results in feelings of pleasure. Thus, there is no paradox or contradiction, at least insofar as Freud was concerned.

Like the hypothalamus, the pleasure principle, or the drive to fulfill needs and obtain pleasurable satisfaction, is present from birth. Indeed, for some time after birth, the search for pleasure is unrestricted and intense, as there are no forces other than "mother" to counter it or help it achieve its strivings. Again, its basic message is "I want it now!"

If a baby is hungry, thirsty, or uncomfortable, the hypothalamus becomes activated, and the baby begins to cry and then goes into a rage—even if it is the middle of the night and mommy is totally exhausted. When the baby is taken care of, the hypothalamus rewards the baby and its mother with feelings of pleasure; the baby stops crying, smiles, and goes back to sleep.

Because of the infant's helplessness, the hypothalamus has no way of reducing tension or acting on the baby's needs. If the baby's needs (e.g., hunger or thirst) go unfulfilled and feelings of unpleasantness are generated, the only response available to the limbic system is to cry and make raging vocalizations, the first rudiments of limbic language. When satiated, the limbic system responds with a state suggesting pleasure or at least quiescence. However, sometimes the limbic system, via the amygdala and the hippocampus, creates a hallucinatory memory image or what is desired in the form of a dream. This has been referred to as the *primary process*, the original means of "wish fulfillment."[24]

The baby can't talk, and the neocortex of the right and left cerebral hemispheres are so immature that the limbic system cannot tap into its resources so as to make the rest of the brain act on its needs. Thus, "higher" and more refined emotions and feelings are not yet available or subject to the influences of the limbic system. Emotionally, the infant is capable only of screaming, crying, or demonstrating very rudimentary features of pleasure, that is, an attitude of acceptance or quiescence.

It is only with the further differentiation and maturation of the amygdala and the hippocampus and of the right and left cerebral hemispheres that the infant begins to achieve some awareness of external reality and begins to form memories as well as to differentiate and associate externally occurring events and individuals so as to influence them more directly.

With the development of the right and left cerebral hemispheres, the limbic system is able to motivate the rest of the brain to act at its behest. When the rest of the brain begins to mature more fully and begins to take control over behavior, the limbic system is able to generate feelings of pleasure and aversion and thus to motivate the brain to engage in this behavior or that, whatever it takes to satisfy its desires. This does not mean that the limbic system controls or directs the rest of the brain. For the most part, the limbic system is subject to the massive inhibitory influences of the rest of the cerebrum, the frontal lobes in particular. Limbic moods, desires, and impulses sometimes overcome the right and left brains, and for the most part, they strongly influence, although they do not control, conscious awareness.

The Misinterpretation of Unconscious Urges

Some desires and impulses from the limbic unconscious can reach conscious awareness only in the form of undefined and sometimes mysterious feelings, urges, cravings, and tensions. Sometimes the right and left cerebral mental systems may be unable to identify or decipher these vague feeling states and emotional sensations, so that some limbic impulses are guessed at correctly and then acted on, or they are misinterpreted and go unfulfilled. Just as the mother originally had to figure out what the baby's limbic system needed, now one's own mature brain must do the same.

Because the limbic system mediates hunger, thirst, anger, sexual desire, pleasure, and a variety of needs and feelings states, one desire may be misinterpreted or confused for another by a different region of the psyche, for example, the conscious mind. Indeed, sensations of tension (or need) are commonly mislabeled as hunger or even anger when, in fact, some other need may be going unfulfilled. In fact, because a variety of needs are mediated by this one small area, the limbic brain, one need or desire may trigger a wholly different desire or need simply because both are limbically linked; that is, different feelings are mediated by brain areas that are adjacent, and they may trigger each other because excitation spreads to adjoining nerve cells.

If sexual tensions often go unmet, a person may eventually begin to feel and think he is angry when in fact he is "horny." However, he can also feel angry because his desire for increased arousal has mobilized the brain to act aggressively so as to procure a temporary mate. In most instances, the frontal lobes of the brain exert inhibitory control and keep a lid on such feelings.

When limbic needs are misinterpreted, it is usually because the three mental systems (consciousness, unconscious awareness, and the limbic unconscious) speak and understand different *languages*. Thus, the limbic impulses must undergo various transformations and interpretations as they are transmitted to the rest of the brain in order to be understood or acted on. From the perspective of conscious awareness, we may feel "something" but not know what it is, or why we are feeling it. We just know that we feel aroused. When we are beset by impulses from the limbic regions of the mind, not knowing what they are, we may try to interpret them or, failing that, misinterpret them and decide that we feel stressed, tense, nervous, hungry, depressed, or "antsy," and then act accordingly or try to figure out "why" we feel the way we do.

Some limbic impulses are not only misinterpreted but purposefully suppressed and inhibited. Although the limbic system may proclaim, "I want it now!" the rest of the brain may, for whatever reasons, have different ideas and strongly disagree with this demand. The frontal lobes then act to inhibit the impulse. Of course, some limbic impulses do not require the cooperation of the rest of the brain to be fulfilled; the entire cerebrum may simply be overwhelmed, and the person finds himself or herself compelled to eat, search out a sexual partner, masturbate, or explode with rage.

Love and Attachment

Human beings can feel love not only for one another but can also love and form deep emotional attachments to animals and even inanimate objects. Likewise, some animals return even meager scraps of affection with devotion and loyalty, if not love. Certain highly social animals, such as chimpanzees, even form friendships and have best friends with whom they chum about on a daily basis. However, humans can feel love and feel so much hate for the ones they love that sometimes they end up killing them.

However, love, devotion, intimate friendship, and a desire for intense social contact with others are not prerequisites for humanness but characterize the interactions of many primates and higher mammals, all of which share the same or very similar limbic systems. In fact, the ability to experience these feelings, including emotions so basic as maternal affection, is mediated by the amygdaloid nucleus of the limbic system.

If the amygdala is destroyed, the ability to experience these emotions or to perceive social nuances being expressed by others is abolished. For instance, one young man whose amygdala was surgically destroyed seemed to become completely devoid of emotional feeling and was unable to recognize the feelings being expressed by others. He preferred to sit alone and isolated (often masturbating), as he no longer had a need for human contact. He even shunned contact with his mother, to whom he had formerly been quite close.

However, he was still able to speak rationally and to answer questions, as well as to feed himself and take care of his personal needs. Unfortunately, he was no longer able to recognize the emotional, motivational significance of his behaviors, or even what he ate. Indeed, whatever he chanced to see and touch, he would put in his mouth, much as an infant does. Infants often respond similarly because their amygdala is quite immature and is functionally limited for many months.

Free-roaming humans, monkeys, and apes often form intense social bonds as well as dominance hierarchies in which certain members not only have a higher status but can exert social control over their social inferiors. If the amygdala is destroyed, highly dominant animals will abandon their hard-won social status and complacently accept a position at the bottom of the hierarchy, even passively accepting threats and abuse from the most lowly of their former subordinates.

In fact, highly social monkeys that have had both the right and the left amygdala destroyed cease to respond socially and seem perplexed by the entreaties of former friends that desire their company; they act as if they are completely devoid of the ability to read social and emotional nuances. If approached, they withdraw, and if followed, they flee. The majority of the adult primates that underwent this procedure left the social group all together and lived out their days in isolation.[25,26]

As might be expected, even maternal behavior is severely affected. As reported by A. Kling, primate mothers that have had both amygdalas removed behave as if their "infant were a strange object to be mouthed, bitten, and tossed around as though it were a rubber ball."[27]

Interestingly, otherwise normal primate mothers that have been deprived of mothering and close social contact during their own infancy often treat their babies similarly. As described by Harry Harlow, neurologically intact, non-brain-damaged chimpanzees that were raised with only a terry-cloth-covered wire frame with attached milk bottle (i.e., a "terry-cloth mother") completely lost the ability to respond in even the remotest motherly fashion.

> After the birth of her baby, the first of these unmothered mothers ignored the infant and sat relatively motionless at one side of the cage, staring fixedly into space hour after hour. As the infant matured desperate attempts to effect maternal contact were consistently repulsed. Other mothers brutalized them, biting off their fingers or toes, pounding them, and nearly killing them until caretakers intervened. One of the most interesting findings was that despite the consistent punishment, the babies persisted in their attempts to make maternal contact.[28]

The Limbic Need for Physical and Intimate Association with Others

Physical, social, and emotional interaction and contact during infancy are critically important to the child's well-being as well as to her or his neurological, sensory, cognitive, intellectual, social, and emotional development. Indeed, babies need all the love and attendant physical and emotional interaction they can get.

The more an infant is held, stroked, and spoken to, and the greater the visual divergence of its surroundings, the greater will be its resilience and capability in adapting to negative emotional and physical onslaughts and in withstanding stressful extremes later in life. In fact, the very cells of the nervous system will prosper by growing larger and more complex.

Attachment

So great is the need for stimulation that, until six to seven months of age, most infants eagerly and indiscriminately seek social and physical

Figure 29. A baby monkey that was raised with a terry-cloth-covered surrogate mother. Animals and humans who do not receive adequate mothering become extremely abnormal later in life and have much difficulty interacting appropriately with others. Courtesy of Harry Harlow Primate Laboratory.

contact from anyone, including complete strangers. Indiscriminate social interaction is not merely a manifestation of friendliness; it serves a specific purpose, maximizing opportunities for social and physical contact and interaction. Like hunger and the desire for food (which are mediated by the hypothalamus), there is a physical drive and hunger for social, emotional, and physical stimulation (which are mediated by the amygdala).

At about seven months of age, the infant becomes more discriminate in its interactions, and a very real and specific attachment (e.g., to the mother) becomes progressively more intense and stable. This does not mean that, before seven months, the mother is not highly important to the infant; however, maximal social interaction takes precedence during the first critical months of life. In other words, the baby needs more contact than one person (e.g., the mother) can provide.

After a specific attachment, such as to mother, has been formed at about seven months, most children show increasing anxiety, fear, and even flight reactions at the approach of a stranger. By age one year, 90 percent of children respond aversively to strangers. This response also serves a purpose, for it maximizes the bond with the mother and ensures that a child who can crawl and maneuver through space will not indiscriminately attach itself to and wander off with a stranger.

Thus, the infant's initial seeking of indiscriminate social contact is followed by the seeking of progressively narrowed contact. These stages of emotional development coincide with the maturation of the amygdala and the septal nuclei.[29]

The amygdala and the septal nuclei sometimes act in opposition. As noted earlier, if both halves of the amygdala are destroyed, the ability and desire to respond socially or emotionally are almost completely abolished. In contrast, if the septal nuclei are destroyed, the desire for close physical contact with others becomes overwhelming because this function is mediated by the now-unopposed amygdala.

Indeed, humans with disturbances involving the septal nuclei develop a condition sometimes referred to as *stickiness*; that is, they inappropriately seek close contact even with those who have clearly indicated they do not want it. Other mammals that have suffered bilateral destruction of the septal nuclei respond similarly but more extremely and even seek contact with animals that may try to eat them. In some respects, the behavior of these animals with septal lesions is much like that of Harlow's infant chimpanzees, which continued to seek love from their neglectful mothers even when the mothers responded with abuse.

If several animals with septal lesions are placed together, extreme huddling results. If no others animals are available, these unfortunate creatures seek out pieces of wood, old rags, or even bare wire frames, so intense is their need for intimate physical contact. Again, this reaction is due to the amygdala.

The above discussion demonstrates the opposition of the amygdala and the septal nuclei in the involvement in the seeking of social and intimate contact seeking. That is, the *normal* amygdala promotes social contact seeking, and the *normal* septal nuclei act to inhibit and restrict these tendencies so that they are directed and focused, rather than generalized and indiscriminate. These two regions of the brain are thus highly interactive and crucially important in the formation of our first and earliest attachments, as well as of those later in life. When the influence of one is removed (because of damage or immaturity), the other is released from oppositional pressure.

The amygdala, which is quicker to mature (as compared to the septal nuclei), drives the infant to seek emotional and physical contact indiscriminately. The septal nuclei, which develop and mature later, cause the infant to increasingly narrow its responsiveness until only select attachments are maintained and the bonding with the mother is maximized. Hence, the amygdala and septal nucleus interact so as to maintain selective social bonds and intimate contact, and they act to counteract one another.

Later in life, these same limbic nuclei are involved in the ability to feel love for (as well as hate for and anger at), and attachment to, a loved one. That is, the limbic system controls the basic aspects of emotion, such as love, hate, anger, rage, fear, pleasure, and the desire to bond, as well as biological drives, including hunger, thirst, and even the capacity to experience orgasm during sex. All these impulses and needs, at one time or another, become associated with the mother or the primary caretaker and later in life (to a considerable degree), with the spouse.

Because of limbic attachment, the actions of a mate elicit limbic reactions, including infantile feelings of rage and abandonment, and even the desire to kill one's spouse just as the presence or absence of the mother elicits these responses, e.g., rage, when the infant is not being held or fed. Indeed, loss of love, such as occurs when a relationship ends, second only to jealousy and money, is a prime elicitor of such murderous feelings and is due to the high involvement of the limbic system in all affairs of the heart. That is, if a person who has met another individual's primary needs for love, affection, and physical intimacy wants to end the affair, the limbic system may respond in characteristically infantile fashion, with frustration, anger, and rage. The amygdala, striving to maintain the bonds of love, responds with rage when the bond is severed, and the hypothalamus responds similarly. If the hurt and anger are sufficient, the frontal lobes and the rest of the brain may be overwhelmed, and the person may act on these primitive needs, in either an extremely dependent ("Don't leave me or I'll kill myself") or a violent manner ("Don't leave or I'll kill you").

Abnormalities in Love and Socialization Skills

If contact and interaction with others is restricted during the early phases of infant development, the ability to interact successfully with others at a later stage in life is retarded.[30] That is, the infant and child must experience love and nurturance, or the limbic nuclei will not develop normally,[31] and gross abnormalities may result. Children will lose the ability to form emotional attachments with others, sometimes for the rest of their lives.

This is even true of the so-called lower animals. Kittens that are not handled or stroked by humans soon become "wild" and unapproachable, even when they have otherwise been exposed to people on a daily basis. Similarly, infants and young children who are separated from their parents and who fail to maintain a mother-child bond or to receive the necessary stimulation are also affected adversely. They have difficulty forming emotional attachments, and their brains may not even develop properly. If not adequately physically and emotionally stimulated, a child may even die.

In other words, if a child is not firmly attached to a mother figure and has been neglected early in life, his or her ability to form attachments increasingly narrows and then disappears, possibly forever. The child becomes attached to no one, and his or her ability to form loving attachments later in life will be abnormal if drastic countermeasures are not taken.

One reason is due to abnormal learning. Another is that cells in the brain and amygdala, not receiving sufficient and appropriate stimulation, begin to die and atrophy from disuse; just as a muscle does if unused: "Use it or lose it." Once these limbic neurons die, or if certain interconnections of different regions are not maintained, they are no longer able to respond appropriately to physical, emotional, and social interaction. Interestingly, the right amygdala appears to be more greatly affected by early rearing experiences.[32]

Misinterpretation of the Need for Association

When social and physical interaction with others is inadequate, the limbic system begins to generate feelings of tension, which is often labeled (by the left brain) as *loneliness* once the child grows older.

Like many other needs, the tension of loneliness is sometimes misinterpreted.

Heather never knew her father, because he had died before her birth. Consequently, her mother, who had to work, could take only a few weeks off after Heather was born. Heather's mother loved her, however, and always made it a point to hold her baby in her arms during every feeding. She would even sometimes come home during her lunch hour and retrieve Heather from the baby-sitter long enough to hold and feed her.

This pattern continued for several years. As Heather grew a little older, however, her mother no longer held her; she would make snacks so Heather would have something to eat whenever she got hungry. She also prepared elaborate meals for Heather to eat for breakfast, lunch, and immediately on coming home from school.

Heather hated being left with various baby-sitters, none of whom she ever felt close to. Her only solace for the pain of loneliness she felt for her mother was to eat whatever the mother had prepared for her during the course of the day. This eating afforded Heather a sense of security as well as a feeling of being loved. However, looking at the food also made her sad because it reminded her that she was alone. Besides, she missed her mother terribly and could not understand why she was always going away. It made her feel unwanted. Being left with different neighbors or, in some cases, relatives did not help matters, as what Heather wanted most was her mother.

Regardless of her mother's good intentions, from a limbic and psychological perspective Heather was being emotionally neglected by her mother (who, because of her financial situation, could not attend to Heather's needs for love and affection). Consequently, Heather felt hurt, rejected, unlovable, and unworthy of the concern of others. These feelings, in turn, affected her ability to interact successfully with others.

From a limbic perspective, Heather was experiencing the tension of loneliness; that is, the need for social and physical stimulation. At a conscious level, Heather also felt tension. However, she did not always consciously recognize the tension for what it was; she sometimes interpreted it as hunger and would head straight for the refrigerator when the tension mounted.

Unfortunately, Heather began to confuse one need for the other, and years later, whenever she felt the tension of loneliness, she interpreted it as a need for food. She began to interpret her loneliness and hunger for friendship as a craving to eat and would stuff herself. By the time she was a teenager, she was obese.

From a limbic perspective, eating and loneliness had been associated in Heather's mind since early childhood. Just as Pavlov's dogs learned to salivate each time they heard a bell (which always rang just before they were fed), Heather had long ago associated food with her feelings of loneliness and of being unwanted. Indeed, her mother had inadvertently substituted food for love, and the limbic system, not being very intelligent and having no ability to think about what occurs in reality, linked and associated these two very different needs—an association that unfortunately drove the conscious aspect of Heather's mind to act as well.

The Limbic Linkage of Rudimentary Needs

Whenever a need or desire is fulfilled, the limbic system responds with feelings of pleasure. In fact, if two very different needs are being simultaneously experienced and only one is satisfied, the pleasure generated is sometimes enough to fool the entire limbic system into thinking that all its needs have been met.

Because the limbic system mediates hunger as well as the need for contact, it may generate a feeling of happiness in a lonely person who eats. The person feels a temporary pleasure, and the unpleasant tension of loneliness is temporarily dampened. Soon, the lonely person eats even when not hungry. Thus, some people eat incessantly because they are trying to fill up an empty space deep inside caused by loneliness, and eating makes them feel good. Others eat (or cease to eat) when they are depressed, angry, or upset.

Similarly, a person who is feeling sexually aroused may experience considerable tension if this need is not met. If the arousal level becomes exceedingly high, other limbic feelings or needs may be inadvertently triggered by the excessive level of tension and excitation. The person may then drink unremittingly or even misinterpret his or her aroused feelings of tension as feelings of anger and go looking for a fight— whatever it takes to reduce the level of tension and arousal.

Nevertheless, if the original need continues to go unmet, the limbic system will again motivate the person to take some action, any action. If a wrong action satisfied the need just hours or days before by reducing the level of tension and inducing feelings of pleasure, this same erroneous behavior will be engaged in again and again. The person may eat to the bursting point or may choose to go out and fight with anybody who looks at her or him cross-eyed.

Emotional Trauma during Infancy and the Limbic Unconscious

Many early experiences and traumas, such as those occurring during infancy, take place long before the development of linguistic consciousness or the highly refined form of emotional awareness mediated by the right brain. They are stored in the memory banks of the limbic system by means of a code that is entirely nonlinguistic. As a result, many of these very early impressions and feelings remain unidentifiable. And yet, these same early experiences may exert profound influences on the adult cerebrum.

Take for example an infant who was repeatedly punished by her mother for touching her genitalia and masturbating; behavior that is common among many infants (as well as adults) and that is mediated by the limbic system. Soon, through the repeated association of physical punishment with urges to touch herself and masturbate, the urge becomes linked to punishment, as Pavlov's dogs linked a bell with food. However, instead of salivating in response to the sound of a bell, this little girl learned to feel pain in response to the urge to engage in self-manipulation. Because the tension that gave rise to genital stimulation and exploration was often followed by punishment (the mother's slapping the little girl's hand and yelling, "No"), soon the urge to touch became, in itself, punishing, and the body expected to be punished. Hence, when the urge arose, it led to an expectation of punishment. The infant ceased to touch her genitalia, and the urge was extinguished soon after it emerged. Before it could even be recognized as "sexual" in origin or intent, it gave rise instead to feelings of anxiety in anticipation of physical pain.

Unfortunately, the child's mother was successful in extinguishing this rather normal behavior. Hence, when this punished impulse was triggered, it was immediately followed by anxiety and physical displeasure in anticipation of punishment, even long after the child reached adulthood. Her sexual impulses were not recognized as such (by the left or the right brain); instead, they were interpreted as nervousness, anxiety, fear, and so on.

Because the behavior, which was mediated by the limbic system, was extinguished and punished before the advent of a proper verbal or emotional label, the impulse cannot now be recognized for what it is. The hippocampus and the amygdala act together so that the emotional memory is of a vague impulse originating in the vaginal area, followed by pain, and the result is a sexually disturbed individual who not only fails to recognize her sexual needs but responds in a rigid, distanced, and nonsexual manner whenever she feels even slightly aroused. The amygdala and the hippocampus act to call forth associated memories and feelings, and then the amygdala generates a desire to avoid physical and, in particular, vaginal contact.

For this woman, sex is something unpleasant and too disturbing to think about, much less to do. Consequently, she avoids men who "turn her on" sexually because they make her feel uncomfortable, though she does not know why. If she ever gets married, she may find sex unpleasant, upsetting, and perhaps painful, and her husband, no doubt, will probably describe her as cold, selfish, and frigid.

The Limbic Unconscious

The limbic system represents a relatively inaccessible aspect of the mind that is concerned with primitive, rudimentary, biological, and reflexive mental activities that are difficult to analyze, control, or even recognize. Because it controls the very foundations of emotional experience as well as memory, all aspects of social and emotional functioning and the ability to relate successfully to others may become severely abnormal if an individual is repeatedly traumatized or severely neglected during infancy.

The limbic system controls the ability to experience pleasure, sexual desire, anger, and rage, as well as hunger and thirst. Sometimes, these different needs are confused by the rest of the brain. When this occurs, a person may decide he or she is angry or hungry when he or she is, in fact, sexually aroused or lonely.

Unfortunately, because the limbic system represents the most inaccessible regions of the unconscious, its secrets not only are well kept but may become permanently hidden from the rest of the mind. Nevertheless, this does not keep the rest of the brain from trying to guess at and satisfy these needs, which often demand to be met "now."

7

Speculations on the Evolution of Mind, Woman, Man, and Brain

Sex Differences in Language and Communication

It has frequently been noted in the popular press and in sociological literature that many men tend to look to one another in terms of status, physical and intellectual dominance, and control. In contrast, many books suggest that women speak and interact alike in regard to the family, interdependence, and social intimacy.[1] Although such notions do not apply to all men or all women, there does appear to be a gender-linked difference in how boys, men, women, and girls tend to interact and speak.[2]

Many boys tend to play in groups in which there is a recognized leader and hierarchical order of followers. The leader often wins this position based on physical strength and capability, risk taking, and his ability to control or bully others. Similarly, the hierarchy of followers is arranged along these same competitive lines. When playing, boys often engage in very physical, aggressive games; wrestling, tripping, and pushing each other in fun are part of the activity.[3]

Many girls tend to play in much smaller groups or in pairs of best friends. Although hierarchies may also form, they tend to be based on personality, articulatory skills, and physical attractiveness (e.g., who is the "nicest," prettiest, or the most fashionably dressed) rather than on physical competitiveness (although it is often a factor).[4] Of course, many boys and professional men seem increasingly concerned with these issues, too.

In contrast to boys and men, who are more likely to become involved in physically aggressive team sports in which there are clear winners and losers, young girls are more likely to engage in cooperative activities that focus on friendship, intimacy, sharing, talking, imaginativeness, and being liked. Challenges and competition between girls are more likely to be subtle and indirect, whereas cooperativeness, at least overtly, is the glue that binds them together.[5]

Although both girls and boys are often concerned about being the best and tend to use force in order to get their way, boys are more likely to use threats or actual physical violence.

It is important to emphasize, however, that there are numerous exceptions, as many girls are quite physically competitive and aggressive, and many boys seek close social and emotional intimacy with their best friends. Probably most members of both sexes fluctuate between these modes of interacting.

These general patterns of interaction tend to color the way in which men and women interact as adults as well, including how they view the actions and even the speech of others. That is, in very general terms, men and women are often concerned about different aspects of the same experience because they sometimes have different priorities.

A recent spate of books contend that these sex-biased viewpoints affect the very way some men and women speak, so that when they are having a conversation, they may focus on different aspects of the same information and emphasize quite different features.[6]

Many males use language not only to impart information, but to establish status and superiority. Among men in white-collar and professional positions, this language use may involve considerable posturing about intellectual and financial superiority. Jockeying for status may also entail derogating other men through teasing, sexual jokes, or direct insults. Indeed, sexual remarks and teasing are frequent among adolescents and some, but by no means most, men in blue-collar jobs and construction (fields in which I worked while a student).

Men's comments about one another are often quite graphic and go well beyond innuendo, including remarks about sexual inadequacy or potency and even challenges to other males to serve as willing orifices.

Although sexual comments and insults may seem to be made in fun and may seem to be accepted as such, they are usually meant to achieve dominance over other men.[7] When a man teases, comments on, or interacts with a woman on a similar level, even if the comments are considerably toned down, she is likely to be offended, much to the bewilderment of the man, who sees his behavior as normal.

Females are much less likely to insult one another in fun, to deride each other's supposed sexual shortcomings (at least while face to face), or to demand in a jocular tone that other women serve them as sexual objects. Such conversation is not seen by women as promoting rapport or dominance. Moreover, when women seek status or seek to put one another down (again, at least while face to face), they tend to be less confrontational, more subtle, and less aggressive; they appear to be more concerned with social harmony and with establishing a mutual, friendly understanding and rapport.[8] They are more likely than men to try to smooth things over and to be more artful and subtle rather than overtly or physically aggressive, even when they are jockeying for positions of dominance. Of course, like men, many women have no difficulty in cutting up a competitor.

Thus, men and women tend use language differently and to convey different messages. Although both use language to convey information, many women are more interested in discussing what they or others feel, what has happened to them or their friends, who said what, and how various relationships are going. For many women, talk serves as a means of maintaining intimacy and friendly interaction as well as of conveying details on business, finance, politics, and a variety of other subjects.[9] These personal details are of little importance to many men:

Ginger: That business lunch went just great. You should have seen Ruth. She was so happy after getting that big bonus, and she had on the most gorgeous outfit. Really quite stylish and . . .

Andy (*yawning and picking up a newspaper*): So what happened?

Ginger: Put down that paper so I can tell you. Anyway, she has lost so much weight. She must really be feeling more confident about herself, with all the money she has been bringing in. Even Sally was impressed, and you know how Sally feels about that.

Andy: About what?

Ginger: Haven't you been listening to me?

Andy: Yeah, sure. You were telling me about some business deal that went down at lunch.

Ginger: Anyway, and oh, did I tell you? Sally's little boy has been getting so big. She had these pictures . . .

Andy: I thought this was a business lunch.

Ginger: It was. You should have seen the look on Ruth's face when she saw those pictures. She wants to have a baby so bad . . .

Andy: What has this got to do with your job? What did your boss have to say?

Ginger: Sally is my boss!

Andy: And she brings pictures of her kid to show you?

Ginger: What's wrong with that?

Andy: I don't see what that has to do with your job.

Ginger: It has everything to do with my job. If you work with people, you should be interested in what is going on in their life. People don't live and work in a vacuum. What goes on at home affects their work. Haven't you learned *that*?

Andy (yawning and reaching for the newspaper): Learned what?

For many men, when language and talk is personal it is viewed either as irrelevant or as power-oriented. Indeed, their talk often becomes personal when they feel their status or authority may be threatened.

Jake has come home and begins to complain to his wife, Ruth, about a problem at work. He is very irritated and upset:

Jake: I don't know what's wrong with my boss. He hired this new girl, and she is not working out!

Ruth: What does she look like?

Jake: Very ugly. Anyway, she was supposed to type this stuff, and she never even got started.

Ruth: How old is she?

Jake: What's that got to do with anything? She's not doing her job. She was supposed to type these papers for me, and she didn't do it.

Ruth: Well, she's new. Is she married?

Jake: Why are you asking me all this irrelevant stuff? Who cares what she looks like or how many times she's been married. She is incompetent.

Ruth: Maybe if you were nicer to her . . .

Jake: Hey! Whose side are you on? She's the one who is screwing up. Not me.

Ruth: I didn't say you were.

Jake: Bullshit! I don't know why I even bother talking to you. All you do is put me down.

Ruth: Well, excuse me for being curious.

Men tend to use language to describe what they are going to do and how they are going to do it. It serves more as a form of achieving mastery over their environment and establishing their status and position in the world than as a means of revealing their feelings or intimate personal details about their lives.[10] Many women use language similarly, but less aggressively and overtly. Women are more likely than men to use language as a means of maintaining social intimacy and are much more likely to interweave professional talk with friendship. Women and girls use language as a means of maintaining a relationship, of feeling close and involved.

Men are more likely to focus on accomplishing something together, or talking about the accomplishments of others, such as those who won a certain football game and those who excel or need improvement in their performance. Indeed, men are more likely to use speech as a means of maintaining or establishing status and to leave out details concerning their emotions or personal problems.[11]

Women tend to become more interested in the details of a person's life and in discussing the details of their own lives, as well as their thoughts and feelings. When women are together, this tendency forms a bridge that maintains their rapport. They expect to be informed. However, when women are with men, they may feel that males act disinterested, are not communicative, or somehow lack the ability to express their feelings. Indeed, it is a common female complaint that men do not express their feelings and are not communicative, at least not in the company of women.

In fact, the man often is uninterested and wonders why the woman is talking about this or that and asking him all these "irrelevant" questions. Many men do not share their feelings and other personal details because to do so may put them at risk with other men, who, being competitive, may use such information against them or make fun of them.[12] In fact, men who have the "gift of gab" are often viewed by other men as a "bullshitter," or even as lacking status or as posing a possible threat because they "talk too much" or have "big mouths." Moreover, knowing that men are not very interested in the social and personal aspects of another man's life, even when discussing family and children men usually keep these topics quite brief as they seem irrelevant to the task at hand; to conquer the world, make more money, get that promotion, or to discuss those who are now conquering the world.[13] These are viewed as topics worth discussing at length.

Even when it comes to a personal difficulty, men are more likely to keep it to themselves and answer questions about it in as few words as possible:

Andy: So how you been getting along since that divorce?
Jake: You know how it is.
Andy: Yeah.

By contrast, women are more likely not only to discuss these issues at length, but to discuss them again and again with all their friends. Many men see no competitive advantage in discussing personal difficulties, and discussing them is sometimes seen as an admission of personal weakness or as a source of information that can be used to create an imbalance in power. Indeed, in my practice, I have often heard men bitterly complain about how their spouse or girlfriend goes around telling people, including relatives, very personal details about their relationship. Many men view this as betraying information that is best kept just between the partners.

For many women, keeping these details to themselves is a breach of friendship and intimacy. Their friends and mothers may insist on knowing the details of a relationship and derive considerable enjoyment from hearing and discussing them. Hence, women achieve closeness by revealing intimate details about themselves and others, whereas men lose power by doing so.

However, when it comes to plans, actions, future goals, or even past conquests, the discussion between men is likely to be quite extensive, particularly if it enhances their status:

Andy: So, getting any lately?
Jake: I'll say. You remember that blond that used to work at the . . .

Nevertheless, even in these contexts, men are less likely to discuss what was said, how they felt about it, how romantic the evening was, how their partner felt, and so on, because these details are viewed as irrelevant. They want to talk about their actions and accomplishments, whereas women tend to focus on the relationship. This difference, of course, can create considerable difficulty when men and women talk together.

Girls and women thus tend to focus on the social, supportive, familial, and communal aspects of interaction. Men tend to focus on the individual and the struggle for dominance over the community, the family, and each other. In this regard, when women interact they tend to be less confrontational, as they are more sensitive to social and emotional nuances, and as one of their main interests is to maintain the cohesion of the group or the friendship.

Again, however, it is important to emphasize that there is much variability in these seemingly differential attitudes toward the use of language as a means of social interchange. Some men are more interpersonally inclined, and many gossip among themselves about sports, politics, and each other, although they do not call it gossip. Similarly, many women are much more interested in the nonpersonal, political, business, and informational aspects of communication and would rightly take extreme exception to the notion that they spend their time talking predominantly about family or relationships.

Sex Differences in Brain Functioning

Can all differences in language and communication be due to cultural forces or the different ways in which boys and girls are raised? Environment certainly plays a tremendous role, as does biology,[14] but not in the way that most people might think. Some of these differences are due to the interactions of the environment and our biological makeup and the subtle differences in the structure and organization of the male and female nervous systems. However, these differences developed over several million years and reflect not only sex differences in endowment, but the division of labor that generally characterized the two sexes, particularly over the course of the last several hundred thousand years (i.e., big-game hunting vs. gathering).

Although men and women vary widely in physical attributes and ability, males are on the average bigger, stronger, faster, and more athletically skilled than females in almost every single sports activity, even at the Olympic level. This is not a sex-biased comment; it just happens to be the truth. However, it is important to emphasize that many women far exceed the average man in sports ability and achievement.

Women have their own superiorities. They are constitutionally superior to males in that they can withstand stressful extremes far better than men, such as extremes of fatigue or illness, or even starvation.[15] Probably, this superiority is a direct consequence not only of their role in carrying and giving birth to babies, but of their role as the main providers for their children and mates for the last half million years. Hence, they have had to be hardy. Not only do females live longer than males, but males are so much more delicate that they are more likely than females to be aborted during times of stress, to die *in utero*, and to die at birth, and from birth to old age, they die at a higher rate than females.

Even at a genetic level, the female has an advantage. For example, gender is determined by one's chromosomes. Females have two X chromosomes, whereas maleness is determined by an X and a Y chromosome. It is the double X that gives the female a greater genetic importance in the maintenance of the species and in the transmission of traits, whereas the Y chromosome (which some have referred to as a broken X) is more restricted to simply determining maleness. The tiny Y chromosome is also quite feeble. In fact, no matter how many Y chromosomes are possessed by a cell, if there is no X chromosome, the cell and the organism will die. X chromosomes don't need a Y in order to prosper. The female chromosome is thus superior.

Sex Differences in Cognition

Based on a variety of independent studies conducted by both men and women researchers, it is well established that, in general, males have far more elaborately developed and thus superior visual-spatial and spatial perceptual skills than females.[16] Only about 25 percent of females exceed the average performance of males on tests of such abilities.

Some of these differences are present during childhood, including a male superiority in recalling and detecting geometric shapes, and of figures that are embedded in an array of other stimuli, in constructing three-dimensional figures from two-dimensional patterns by visually rotating or recognizing the number of objects in a three-dimensional array, in playing and winning at chess (which requires superior spatial abilities), in directional sense and geographic knowledge, in solving tactile and visual mazes, in aiming and tracking such as in coordinating one's movements in relationship to a moving target, in coordination in aiming and throwing, and in comprehending geometrical concepts.[17] These are all skills associated with the functional integrity of the right half of the brain.[18] Moreover, many of these abilities are directly related to skills that would have enabled an ancient hunter to track game efficiently, to throw a spear, and to dispatch various prey without getting lost and while maintaining a keen awareness of all else occurring within the environment.

However, women do not lack these abilities; many women far exceed the average male in these capacities, and in other right-brain capabilities, such as discerning and expressing social and emotional nuances, females are often considered much more sensitive and adept. Females also tend to display many superiorities in left-brain skills.[18] In contrast to males, females vocalize more as infants and speak their first words at an earlier age, they develop larger vocabularies at an earlier age, and their articulation skills improve at a faster rate. Among children, the speech of females is easier to understand, and women excel over males on word fluency tests, naming more words containing a certain letter, or words belonging to a certain category. Females also tend to excel over males in fine-motor skills, such as those involving rapid temporal sequencing.[19]

In clinical practice, it is well known that following strokes that cause aphasia, women tend to recover language more quickly and more fully. However, some have argued that the reason is that men and women suffer different types of strokes (e.g., thromboses vs. embolisms). Nevertheless, men are also far more likely to suffer language-related disorders and to lose language abilities at a faster rate than women as both age.

These and the other findings mentioned above suggest possible neuroanatomical differences in the brains of males and females. Indeed, not only the hypothalamus and the limbic system, but possibly the language axis of the left hemisphere as well, are sexually differentiated. In the anatomy of the brain, it is apparent that those areas that subserve language (e.g., the left superior temporal lobe) are larger in the left vs. the right brain. However, there are some studies which indicate that when these differences are statistically analyzed, these differences are maximal in females. Women appear to have more brain space devoted to language functioning. However, some well-respected neuroscientists have also claimed that there are no gender differences in the size of the left superior temporal lobe. Nevertheless, as both sexes age, males tend to lose more cortical tissue in the brain areas subserving language, a loss indicating that this modality is more fragile in males.

Complicating the issues even further, there is some suggestion that functionally Broca's area is better developed in females than in males, whereas the left temporal parietal language areas are functionally more developed in males. Although I believe that this suggestion is purely speculative, it is interesting to note that Broca's area is associated with temporal-sequential motor control and that, when it becomes activated, so does the right hand.

As suggested earlier, many of these differences in verbal versus visual-spatial capabilities are related to environmental pressures and the presence or absence of the male hormone testosterone, during the period in which the brain becomes sexually differentiated.[20] However, these differences are also rooted in our evolutionary past and in the different activities that men and women have engaged in for well over half a million years.

Speculations on Hunting and Gathering

Food, Sex, and the Big Brain

Foraging, scavenging, and chasing and hunting small game have probably been dominant activities of human beings and their ancestors for several million years.[21] Naturally, our brains have been tremendously influenced by these activities and have evolved accordingly.

Both the male and the female of the species probably engaged almost equally in scavenging, gathering, and hunting small game until just a few hundred thousand years ago, with the onset of big-game hunting. It was possibly with this event, coupled with the rapid and progressive development of humans' big brain several thousand years earlier, that a divergence in the mind of man and woman began to occur. The hunting of large game animals appears to have become the dominant domain of males,[22] although they probably continued to assist in gathering on occasion. Presumably, females continued to gather and, to a lesser extent, to fish and hunt small game, and only occasionally assisted males during the hunting and stalking of large animals.[23]

Nevertheless, with this general division of labor, one might suppose that, over the course of several hundred thousand years, tremendous differential influences on mental and brain functioning resulted due to selective evolutionary pressures on survival. That is, those who were best able to adapt were most likely to pass on their genetic traits to the next generation. In consequence, the brains and minds of men and women were probably adapted to and molded by the activities that each were best at.[24]

As gathering and the harvesting of wild foods involves perseverative temporal-sequential hand movements, those who were most successful at these activities probably had brains that were adapted to these tasks. If the right hand is being used predominantly for picking and gathering, and the left hand is used only to hold a receptacle, one might suspect that the left half of the brain would become more proficient in temporal-sequential processing.

In contrast, as the searching for and the stalking and killing of prey requires good visual-spatial skills and sensitivity to environmental and nonverbal nuances, those who were most successful in these activities would continue to develop and would pass these same capabilities on to the next generation.

The question might arise, however: What caused this division of labor? Why, over the course of evolution, did females remain engaged in gathering, whereas males pursued large game? In large part, for the development of the skills necessary to be successful at big-game hunting, the human brain had to become more complex. Over the course of human evolution, the human brain and the human head became increasingly larger.[25] The progressive and significant increase in brain size, in turn, required an adaption in the pelvic opening of the female. A bigger brain comes in a bigger head, and eventually, the hips of the human female became wider as well so as to accommodate a larger-brained baby. With a wider pelvic opening and with wider hips, a big-brained baby could be delivered without becoming damaged or lodged inside the womb unable to emerge. If that were to happen, the baby, the mother, or both would die.

As a consequence of these changes in the hips and pelvis, the gait and balance of human females became altered over time so that they would wiggle when they walk and were no longer able to run as fast as males, and their mobility became slightly restricted as well. This slow evolutionary change began to exert significant limiting influences on the female perhaps around 500,000 years ago or longer, at which time the brain of our remote ancestor *Homo erectus* appears to have reached its maximum size (approximating that of a small-brained modern-day human).[26]

The human brain, of course, is not fully developed at infancy; it is only about a third the size of an adult brain. If it were close to the adult size, it would be impossible for an infant to be born. As a consequence, the brain has to do a tremendous amount of growing after birth, a process that continues for the first ten years of life and then slows over the course of each ensuing decade. Because brain development is prolonged, the period of helplessness became lengthened as well, which in turn necessitated prolonged child care. Hence, in contrast to most mammals and all other forms of life, which remain helpless and unable to fend for themselves for only a few months at most, humans remain helpless for four to six years, four to six times longer than monkeys, and twice as long as all other apes, which, like their female counterparts, invest a considerable amount of time in child care.[27] Thus, the human infant, including the infant Cro-Magnon, Neanderthal, and *Homo erectus*, required almost the full-time presence of a mother for many years. This increased dependency reduced female mobility as well. Hence, at a minimum, human females have been engaged in prolonged child care for at least half a million years.

In consequence, having a baby or several children to watch over, and being unable to run as fast as males and being more clumsy in the attempt, adult human females were not as good and reliable hunters as men. One can't carry an infant while stalking prey because one's hands and arms must be free to carry weapons, and one must be quiet. The sudden cooing or crying of an infant is likely to result in an empty stomach. Hands and arms must also be free to carry weapons. Hence, a number of forces conspired to limit females' ability to excel as big-game hunters. Indeed, their smaller stature also made them less formidable foes and possibly even tempting prey.

Thus, females probably wandered less than males, stayed closer to the home base, and engaged in activities compatible with nurturing infants for years at a time. The ultimate consequence of these particular limitations was that the human female probably remained more involved in gathering and the occasional hunting of small game closer to the home base, whereas males were free to engage in more wide-ranging activities, such as big-game hunting.

Again, one might naturally assume that to be successful as a hunter required certain adaptive changes in the human brain, alterations which would be maximal among males. Conversely, gathering, an activity involving prolonged and rapid temporal and sequential hand movements, would in turn exert its maximal selective pressures on the brain of the female. That is, those who were best at these activities were more likely to survive and pass on their genetic contributions, and thus their adaptive skills, to their progeny.

Sexual Bonding

A woman, or a band of women, could probably live and flourish without the aid of the bigger and more muscular male and could certainly eat quite well without ever savoring a steak from a big game animal, but the smaller size of the female and of her children certainly put her at a disadvantage in dealing with males. Indeed, as our ancestors may have been slaughtering each other since the time of the Australopithecines,[28] one might speculate that it would certainly have been to the female's advantage to have at least one male around to aid in her protection as well as the rearing and education of the children.[29] She may often have needed a male to protect her from other males who might rape her or kill her and her children. There were probably other advantages to having a man about the house (or cave) as well, such as being able to partake in the high-protein diet that meat afforded.

This bonding of a man to a woman was aided, in part, by the establishment of a base camp and the maintenance of a family social group, events that appears to have first occurred during the time of *Homo erectus* (who lived from 1.9 million to about 300,000 years ago). However, from an evolutionary perspective, it was necessary to provide the male with some incentive to return home and thus maintain the unity of the family. Such an incentive was provided by an appeal to the limbic system: food and sex. To ensure that the male would return to the base camp and share his spoils with the female and her infant required that males become bonded to females or at least attracted enough to return again and again. One source of this attraction is human females' being sexually receptive and sexually available twenty-four hours a day, 365 days a year. Indeed, the human female is the only female, regardless of species, that is able to have sexual intercourse during times when she is not biologically prepared to become pregnant. In all other species, sex serves only as a means of reproduction, whereas among humans, sex is also a source of pleasure and serves to bond a man and a woman.

As is well known, for most other mammals there are long periods in which females are not sexually receptive and this is characterized by long periods of nonsexual activity in which males show little or no interest in females. When the female becomes sexually receptive there is a short period of frenzied sexual activity, and the males will literally go sexually crazy for days until the female ends her estrus, or "heat." Of course, when a female mammal goes into heat, all other activities cease or greatly suffer.

It may have been during the time of *Homo erectus* that this great change occurred and the female estrus disappeared altogether. It was certainly during the time of *Homo erectus* (or the immediate ancestor of *Homo erectus*) that the first base camps came into existence. However, even if we assume that this sexual revolution did not come until the end of the reign of *Homo erectus*, around the time when the first primitive *Homo sapiens* arrived on the scene, it would still appear that full-time sexual availability had its onset between 300,000 and 500,000 years ago. Nevertheless, in consequence of that momentous event, the human female has been the sexiest of all other females ever since.

The woman's continuous sexual availability freed men and women of the purely biological and hormonal influences that drive and control sexual behavior in other animals. They could now postpone sexual activity and could now decide for themselves when and if they were going to have intercourse, and they could decide where it would take place. Moreover, unlike the apes, which only occasionally and briefly enforce a selective preference, humans could decide with whom they would have sex and could enforce their preference through denial and bonding.

All in all, the availability of a full-time sexual partner, coupled with the ability to make personal choices and to act on preferences, made human sexual relations more enduring, along with social relationships between men and women, all of which probably promoted the development of language and the art of conversation. Men, still functioning at the behest of the limbic system, could now have one almost insatiable need repeatedly satisfied. However, as noted, a second insatiable need, food, was also provided by the female because gatherers provide the bulk of the food supply.[30] There were thus two very good reasons for males not only to come home but to stay home for prolonged time periods as females were not only providing for their sexual needs, but gathering and probably preparing their supper as well. Man was becoming domesticated.

Food and the Original "Breadwinners"

With the exception of those who lived in the very coldest climates, where the gathering of edible vegetable matter would have been very difficult, for at least the last 100,000 years until perhaps about 10,000 years ago, females, and not males, appear to have been the main providers of food.[31] Big- and even small-game hunting has always been (except in the much colder far northern climates) only a supplementary means of acquiring an adequate food supply. Even in the great majority of the very few hunting-and-gathering societies still in existence, spoils from hunting account for only about 35 percent of the diet.[32] Gathering, which we assume has been the predominant domain of the female for the last 300,000 years or so, accounts for the remainder. Even among the Cro-Magnon, for whom hunting was the center of religious and artistic life, 60 to 80 percent of the diet consisted of fruits, nuts, grains, honey, roots, and vegetables,[33] which were probably gathered mostly by the females. Hence, it was probably the female, and not the male, who wielded economic dominance for a significant portion of our history.

Given women's dominant role as gatherers and their general responsibility for both reproductive and most subsistence activities, males could almost be considered coproducing dependents.[34] That is, women were probably the original "breadwinners." Perhaps it is this ancestral tendency to nurture and provide that enables some modern women not only to tolerate but to support men who sponge off them.

Not surprisingly, males recognized that females were a tremendous economic asset, and in many societies, dominant males tried to accumulate as many wives or lovers as possible.[35] Although sexual availability may have been the major incentive for this pattern during ancient times, the fact that females were dominant producing partners meant that, with many women, the man would be freed from spending all his time in the pursuit of food, in its preparation, or even in the maintenance of the dwelling. The more females he had, the more leisure time he had to engage in recreational and artistic tasks.

Indeed, such a relationship, at least for dominant males, may have been rather ideal. For the first years of their lives, they were provided for by their mothers, and then later in life, they were cared for by their wives. On the other hand, one could argue that not only were women the main providers, but that they were being exploited for the last 100,000 years as well.

With increased leisure time and the evolving complexity of social relations, tremendous selective pressures were put on the brain of human beings, and only those who could adapt to these changes passed on their genes. Not only did the right and left brains evolve new capabilities associated with sex differences in socializing, gathering, and big-game hunting, but about fifty thousand years ago, the frontal lobes of the brain expanded tremendously in size. With the development of the frontal lobes came the capacity to engage in long-term planning, in the formation of goals, and in the inhibition of one's immediate desires and impulses to serve future goals.[36] This great change came about with the appearance of the Cro-Magnons, and social relations expanded beyond small bands to large tribes.[37]

Hunters: Bringing Home the Bacon

As food gathering was such an important part of economic existence for so many hundreds of thousands of years (until very recently), it might be asked: Why has so much emphasis been placed on the importance of man the hunter, and on the false notion that he was the main provider?

To better understand the role of providing meat in the assumed dominance of men as the most important providers, let us consider our nearest living evolutionary cousins, the apes. When a chimpanzee captures another living creature for the purpose of consumption, others will rapidly and excitedly gather around and beg and beg for just a morsel of the meat.[38] Nothing like this happens when they forage for vegetables or insects. Indeed, the hunting and capturing of meat has an immediate, highly excitatory effect on the whole band, all of which gather around in hopes of being given a tiny morsel. Moreover, chimps and baboons chase after and kill small animals, but they do not eat dead ones. That is, they do not scavenge.[39] For them, the pleasure of the meat is tied to the hunt, the capture, and the killing. However, it is the male that predominantly engages in these acts. Thus, among the chimpanzees, for some reason, the procurement and eating of meat promotes considerable social excitement and food sharing. In contrast, they do not share and show no interest in sharing or receiving any of the vegetable matter that another chimp may have found and is eating.[40]

Similarly, among hunter–gatherer societies, the procurement of meat promotes food sharing, and the eating of meat is thus a very social activity.[41] Indeed, when the hunter returns, he is likely to be met by most members of his band, all of whom eagerly seek a share of the spoils. The hunter naturally gives the largest pieces to his own relatives and the smaller shares to friends and other band members. These individuals, in turn, give a share of what they have received to their own special friends and relatives. Hence, a certain degree of group cohesion and bonding occurs when meat is caught and shared, and the sharing reinforces the social bond. Meat thus becomes an important medium of exchange.

In contrast, when females bring home the vegetable matter that they have gathered, there is no excitement, and there are no begging hands eagerly seeking a share. What the female procures is shared only with her own immediate family.

Indeed, this same emphasis on the importance of meat can be found repeatedly in Genesis and the books of Moses, as God demanded that his sacrifices be of living flesh. Hence, when Abel offered God "the firstlings of his flock, the Lord had respect unto Abel and to his offering." But when Cain, a tiller of the soil, made an offering of his vegetable produce, the Lord God "unto Cain and to his offering he had not respect."

Because of the social and even religious and spiritual importance of meat, a successful hunter is looked on as a very special and powerful individual who has great prestige among his band. Moreover, the successful hunter who gives out shares of meat in accordance with his own prerogatives gains power over the group as well as respect. In consequence, the successful hunter is sometimes rewarded with more than one wife (or has numerous extramarital affairs), presumably because he can provide for more than one, and because he is actually sought out by females as well as prospective fathers-in-law.[42]

The early men who were the best hunters were no doubt the most intelligent of their times, at least insofar as survival was concerned. Supposing that the most intelligent members of the race were more likely to breed with each passing generation, there was a manifold increase in the numbers of intelligent people. That is, the best hunters would have the most wives and thus the most children, whereas those who were not proficient in hunting either died out or bred so sparsely that their descendants and genetic contributions simply ceased.[43] Moreover, females tended to be attracted to the males who held a more dominant social position than they themselves, and males tended to establish relations with females who were socially subordinate. Hence, our ancient spear throwers, like modern athletes today, often had their pick of willing females, which further increased their particular genetic contribution to the race.[44]

As noted earlier, however, the ancient female was also a prized possession[45] because of her sexual availability and gathering skills, so males had to compete with each other for her favors and attention, particularly for those of the females who were the most desirable, that is, similarly dominant and intelligent.

Males also had to fight and sometimes even kill to maintain their dominant position, to escape a position of inferiority, and sometimes for sexual access to females, which was also a source of status. (Females were not subject to these pressures.) The same behavior is evident among nonhuman male mammals.[46]

Hence, not only successful hunting, but general intelligence and the ability to compete successfully with other males for status, dominance, and access to women were tremendous concerns of the human male for several thousand years. One supposes that his brain adapted accordingly. Does this explain, at least in part, the tendency of modern men and adolescent males sometimes to denigrate, even in a teasing manner, the intellectual, physical, and financial abilities or even the sexual prowess of one another? Are these jibes the remnants of an age-old means of achieving dominance?

Language and Tool Making

Tools, Language, Gathering, and Shopping

The basic skills necessary in the gathering of vegetables, fruits, seeds, and berries and in the digging of roots include the ability to engage in fine and rapid temporal-sequential physical maneuvers with the arms, the hands, and particularly the fingers. As gathering was a dominant activity for such a long time in our prehistory, it is not surprising that the brain has possessed rudimentary temporal-sequential capabilities for several million years.[47]

To aid them on their daily foraging trips, women carried large pouches which were made either of leather or of the stomachs or bladders of various animals and in which they could deposit what they gathered.[48] Such a pouch probably hung over the woman's shoulder, almost like a long-strapped purse. Women have thus been gathering (shopping?) and carrying purses for at least 100,000 years, and perhaps for more than 500,000 years.

The pouch was also highly beneficial in the development of social relations. If they had not collected food, women would probably have simply eaten what they found on the spot and would have had nothing to bring home.[49] In fact, humans are the only primates that engage in gathering. All other primates eat their vegetables as they find them, although a few highly social animals, such as wolves and wild dogs, take food back to the den in their stomachs and then regurgitate it for the young, or even for the old, injured, or feeble. Birds also take food to their young.

In addition to gathering, women made much use of tools and may have been the first tool makers. Tool making, like gathering, may have been a major aspect of their lives, and one they may have spent considerable time engaged in.[50] In fact, the first tools were made not for hunting but for gathering and rooting plants, that is, if we suppose that gathering predated hunting and scavenging. For example, in grubbing for roots and bulbs, gatherers needed a digging stick, which they probably had to sharpen periodically by using stone flakes. They would also have carried a hammer stone for cracking nuts and for grinding the various kinds of produce collected during the day.

In addition, as the female also sometimes hunted small game (probably with the help of a big, friendly dog), she may have carried an ax and spear for the purposes of dispatching whatever hapless prey she chanced upon. By Cro-Magnon times, many women also carried large, flat bone knives up to nine inches long.[51]

These ancient women did not spend all of their time gathering, for foods had to be prepared and clothes had to be fashioned out of hides. Their duties may have included cleaning the hides with a scraper, drying and curing the skins over the smoke of a fire, using a knife or cutter to make the desired shape, and then punching holes in the skin through which leather straps or vines could be passed to create a garment that would keep out the cold. Women also wove, and by Cro-Magnon times, they were using needles to sew garments together. Although there is no way of knowing if women were the first tool makers, they certainly engaged regularly in these endeavors that required tools.

Women or men, or both, eventually developed the capacity to look at and feel a stone or bone and immediately assess its potential as a weapon, a tool, or an object of art. In this regard, they had to be attuned to the properties and grains of wood, stone, and bone, and to how each had to be struck and with what instrument for the desired effect. Making and using tools require not only have a hand capable of such work, but a brain that can control this hand and that can use foresight and planning in carrying out the work. Because of the survival and breeding of those who were successful in these activities, the left half of the brain, which controls the right hand, became increasingly adapted for the control of temporal sequencing, whether in tool making or in gathering, and hence for the temporal-sequential and grammatical aspects of what eventually became spoken language. As females engaged in gathering much longer than males, and as they may have been the first to manufacture and use tools, these changes may have been maximal in the brains of women, particularly in the motor areas controlling speech (e.g., Broca's area) and hand control.

Ontogeny follows phylogeny, and not only do modern human females demonstrate an earlier onset of language ability and less likelihood of suffering language disturbances, but our female ancestors may have fully developed the temporal-sequential aspects of language before men as well.[52] Perhaps this was the fruit of knowledge woman offered man only after having taken the first bite. In fact, we find in Genesis that the first gatherer, the first seeker of knowledge, and the first individual to hold a complete conversation (other than God and the serpent) was a woman.[52]

When we consider the many pressures acting on males to inhibit speech during the hunt, whereas females were allowed to talk quite freely while gathering (as they would have no fear of scaring off game), it certainly appears that these factors strongly promoted female linguistic development and a greater capacity to discuss topics unrelated to the hunt. Language was now being used for social bonding and only later became an instrument designed more and more, at least insofar as males are concerned, for the purposes of exchanging information related to business and sports (i.e., the hunt).

Hunting, Visual-Spatial Skills, and
Why Men Don't Share Their Feelings

Searching for large game animals required that the hunting band roam *quietly* over a huge expanse of varied terrain extending five hundred miles or more. Success required the following: a good directional sense, so that the hunters could wander and also find their way home without the aid of a street sign; good visual-spatial skills, such as depth and distance perception, so that they could walk and run without tripping and falling, as well as anticipate the trajectory of movement of a running prey; an excellent capacity to recognize, comprehend, and mimic animal, environmental, and nonspeech sounds; and a capacity to communicate nonverbally through gesture, body language, and, particularly, animal mimicry, so that game would not be scared off by the sounds of speech. These are all skills associated with the right brain of the modern-day human being, both male and female.

Indeed, the mimicry of certain animal sounds could communicate a host of meanings such as where something might be located (as those animals whose sound he is mimicking may be found only in a certain location), that a particular beast (which elicits cries from the creature being mimicked) is approaching or nearby, or a particular action that the group should take (fight or flight or freeze), and so on. These are all capacities at which the right brain of modern man and woman excels. However, as big-game hunting was an activity dominated by men, it should be no surprise that, after 300,000 years or more of engaging in these activities, males continue to demonstrate cognitive superiorities in these related abilities.

When males banded together for the big game hunt, they had to walk silently for long periods, and their communication was probably concerned only with facts and information about the hunt. If they had talked about anything else such as their family, they might have scared off the game. Thus, a premium was placed on men who could hold their tongues.

Also decreasing the need for talk was the fact that, as the hunters stalked their prey, they walked from twenty-five to one hundred yards apart. Two or men standing together were likely to warn a prey of their presence, whereas men standing apart had the advantage of chasing a beast into the waiting arms of a comrade.

Finally, it is likely that, once the hunt was complete and they returned to their base camp, men discussed, not their wives, children, feelings, or personal problems, but their own prowess as hunters, incidents related to the hunt, and whatever adventures they had had that would enhance their status among their fellows.

Gathering and Why Women Like to Talk and Share Their Feelings

In contrast to the silent, noncommunicative males, the women were able freely to chatter among themselves. Indeed, gathering fostered the development of language and, like hunting for men, served as a social activity, but one that was more physically close and socially intimate.

Our ancient female ancestors probably gathered in large groups of seven or more, as the typical size of a band was about twenty-five. Some women were pregnant or probably carried infants or were accompanied by children who would frolic about and play. Such gathering groups must have been noisy and very gay, filled with the talk of the women, the sounds of games and the yells of the children. Hence, unlike the men who had to remain quiet for long time periods in order to not scare off game, the women at work were free to talk to their heart's delight.

While they were in gathering groups, women probably did not wander far from the group because their smaller size might make them prey. Talking also served as a means of maintaining the location of the group so that if a gatherer or a child chanced to walk away, she (or her child) could always relocate the others by their hodgepodge of speech.

Talking thus became part of the gathering glue and served the purpose of keeping the group together and thus of bonding the group as a collective. The women talked about their children, their husbands, and each other and were thus allowed to expose their own feelings and thoughts to others who valued talking as much as they. Talking is a social bonding element for women even today.

When considering our evolutionary heritage and the fact that women have spent a good 100,000 years or more in female-dominated gathering groups, where socializing and freely expressing oneself about social and family matters were the norm, it should come as no surprise that modern-day females continue to respond similarly. It is thus little wonder that some women also feel a compulsion to shop and derive considerable enjoyment from making it into an all-day affair. Likewise, some modern men, bravely armed with high-powered weapons, also feel a compulsion to stalk and kill helpless animals or even other humans. It is hard to escape hundreds of thousands of years of evolutionary pressure.

In this regard, it perhaps should not be surprising that after almost half a million years of holding his tongue and jockeying for status and position amongst his fellows, modern-day man continues to respond similarly and have the same concerns, albeit translated and modified to some degree (i.e., sports, politics, business).

Of course, we are not ruled by our evolutionary history, nor is biology equated with destiny. Nevertheless, these influences, including those related to our personal history and the sexually differentiated manner in which we are raised, all continue to affect the manner in which we behave and how the sexes interact. Those forces that cannot modified should at least be understood.

II

PSYCHODYNAMICS

8

The Four Ego Personalities and the Unconscious Child and Parent Within

All living things are not only what they are, but what they were. An alcoholic is still an alcoholic although he has not had a drink in twenty years. Someone who has deliberately killed another human being is still a murderer although he has not taken a life in over a decade. A poor man might become rich, but the memory of these years of poverty never completely fades.

Metamorphosis is possible only for insects, as human beings have not yet learned to shed their skin. For mammal or human, what you were and what you have experienced greatly influence, if not determine what you will become. Growth is the nature of all living things, but complete transcendence occurs only with death.

The Central Core

As a tree grows, the young tree that it once was never disappears, rather, layer upon layer comes to be superimposed on its core. Deep inside, the baby tree that it once was is still alive. The way in which the young tree first took shape, the forces that acted on it, the twists, turns, bends, and breaks that have been caused by wind, rain, humans, or disease all determine the shape the tree assumes as it matures and ages. No matter how well it is cared for, it will never completely outgrow any neglect when it was just a sapling. If we were to cut down and examine the innermost portion of this tree, we would discover that the young tree that it once was continues to exist at its central core. It is still alive and forever retains its original form. What the young tree was, it will always be. What it was becomes the foundation for what it will be.

The adult tree retains this living core, having grown outward from it. If we were to rot out this central core, the tree would die, as the integrity of the entire tree depends on it. If the central core is weak and diseased, then no matter how expert the care, the adult tree will be as feeble as its foundation.

The Child Within

Just as the living tree retains its early core, within the core of each of us is the Child that we once were. This Child constitutes the foundation of what we have become, who we are, and what we will be.

Although as adults we have grown, matured, had new experiences, assumed new responsibilities, changed our minds over a thousand times, and done and said things we swore we never would, the Child at our central core remains the child it always was. This Child continues to harbor the same feelings, emotions, resentments, frustrations, and memories that were present during childhood. It is, in fact, a complex of the emotional and experiential associations that constituted a significant part of our early life. Although it is only a fragment of what we are, it remains the totality of what we were, what we experienced, how we were treated, and how we felt, with the same desires, fears, and hopes.

If this Child was rejected or abused, it continues to feel and act as if it were being rejected and abused long after we have attained adulthood. If it was predominantly loved, praised, encouraged, and treated with dignity and respect, it continues to expect such treatment long after we have become adults. How we were raised and the environmental stresses and parental pressures we experienced not only shaped the character of the Child but continue to exert formidable influences on how we behave and interact as adults. Indeed, sometimes, the Child never grows up and remains fixated in the character at a certain traumatic point in life.

The Unconscious Child Ego Personality

Young children are predominantly emotional beings, as language, thought, and the temporal-sequential aspects of consciousness are not yet well developed. They are mostly emotional and limbic in orientation, and their psychic functioning is governed by unconscious forces. Hence, our childlike central core is not to be found in the left but in the right brain and the limbic system. Indeed, the Child that we once were continues to exist in the central core of the unconscious in the form of a Child-like ego personality.

Broadly, the unconscious Child ego personality (hereafter referred to simply as the Child or Child ego) maintains the feelings of self-worth, the self-concept, the self-image, and all associated cognitions, memories, and emotions that were formed during childhood. It has the personality of a child, acts childish, and encompasses associated feelings and emotions aroused by others when we were young as well as the labels (e.g., *stupid, pretty, ugly, clever, retarded, talented, pig, generous, failure, successful*) that were repeatedly applied to us by other children, family members, and particularly our parents. Because these experiences and emotions are associated, they tend to be recalled and reexperienced together as a complex of interrelated feelings. This associated complex of unconscious feelings constitutes the Child ego. Because so many labels and so many experiences are linked so as to constitute the Child, even seemingly innocuous events may trigger, through association, an activation of the entire Child complex. The person may then behave in a very childish way.

Learning to Be "Not OK"

The Need for Physical and Intimate Association with Others

Physical, social, and emotional interaction during infancy are critically important in the neurological, intellectual, social, and emotional development of the child. In order to thrive, survive, and prosper and to withstand emotional stress and other negative onslaughts, children need all the love and physical stroking they can get. Children are so needy in this regard that they are biologically driven by their limbic systems to seek loving contact.

Abusive Parents and the Need for Love

Children can have parents who burn them with cigarettes and break their bones but will still cling to these parents in search of love. Like the clinging of Harlow's infant monkeys, this clinging is beyond their control, as they need love and are biologically dependent on their parents. Their parents are their world, the only world they have. Children are biologically predisposed to seek association and physical contact and once those bonds of physical contact are established, they are extremely difficult to break. So intense is the need for stimulation that sometimes even a bad parent is better than no parent. If mothering and the stimulation of physical contact are not provided regularly, the result may be death.

During the early 1900s, when the need for mothering and physical contact was not well recognized, the death rates of orphaned children under one year of age were more than 70 percent. Of 10,272 children admitted to the Dublin Foundling Home during one twenty-five-year period, 10,237 died.[1] Others who survived an infancy spent in institutions where mothering and contact comfort were minimal have been characterized by low intelligence, extreme passivity, apathy, severe attention deficits, and extreme difficulty in forming attachments.[2,3] In fact, so great is the need for physical and emotional stimulation that deprived infants will begin to stimulate themselves and sometimes develop bizarre and even self-abusive behavior. Their ability to form emotional attachments later in life is almost abolished.

In the extreme, consider the social behavior of adult monkeys that were deprived of mothering and were raised with terry-cloth surrogate mothers. As described by Harry Harlow, these monkeys, when they reached adulthood, would

> sit in their cages and stare fixedly into space, circle their cages in a repetitive stereotyped manner and clasp their heads in their hands or arms and rock for long periods of time. They often develop compulsive habits, such as pinch-

ing precisely the same patch of skin on their chest between the same fingers hundreds of times a day; occasionally such behavior may become punitive and the animal may chew and tear at its body until it bleeds.[4]

And why is this? Even painful, abusive stimulation is better than no stimulation at all. The alternative is death.

Feeling Rejected and Unloved

Infants and children who have neglectful, rejecting, or abusive parents are doubly damaged and deprived. In some cases where mothering and attention are minimal, the baby will fail to thrive physically, emotionally, and psychologically. Very young children who feel bad often conclude that they must have done something to trigger these bad feelings.

An abused child cannot conceive of the possibility that his mother or father is deranged or quite sick. Rather, emotionally he feels that if he is being called names, beaten, rejected, or ridiculed, it is because of something he has done, it is because he is bad. If he is unloved, it is because he is unlovable. For the abused child, there is nothing wrong with his parent, there is something wrong with him.

Divorce and Death

A child who is raised without a father or mother, the child who is adopted, and the child who loses a parent through death often have a common unconscious emotional reaction: the feeling of being rejected, abandoned, hurt, and angry. Unconsciously, they may be plagued by questions: "Why?" "Why me?" "What's wrong with me?" "Why don't Mommy and Daddy love me?" "If they loved me they wouldn't have left me." Consciously and unconsciously they are bothered by the suspicion "There must be something wrong with me." Since they feel bad they conclude that they are bad. It is this badness which must have driven mommy and daddy away.

Infants and children cannot reason like adults as their psyche is governed by the immediacy of emotion. Moreover, since the right hemisphere and limbic system are predominantly concerned with emotion and social interactions, they respond to their experiences in only an emotional and visceral manner and in accordance with its needs of being loved and accepted. Insofar as the limbic system is concerned, it wants love, and it wants it now. Reasons and explanations mean nothing.

We can explain to a child that his mommy or daddy loved him very much, and we can give eminently reasonable explanations for the death, divorce, or the necessity of his being left with relatives while mommy worked. However, children are emotional and are not capable of rationalizing and understanding these events; they take the absence of their parents in a profoundly personal manner. Even the most stable of adults may respond in a similar fashion when his parents or loved ones die—he feels miserable, frightened, upset, and guilty.

Unfortunately, children may not even be conscious of these feelings and may not express or describe them except indirectly.

We Are All Taught to Feel "Not OK"

We need not undergo the trauma of losing a parent or being adopted in order to feel that we are different and that there is something wrong with us. As so insightfully discussed by Eric Berne, Fred Harris, and W. H. Missildine,[5] each child is taught early in life that he or she is "not OK." The reason is that we must all shed characteristics which are not socially acceptable, and the limbic system must be tamed. Fortunately, many children receive "OK" messages as well. OK messages, however, are not as likely to make the same intense impressions, and "not OK" feelings often predominate and make a more lasting impact.

Because children are completely dependent on their parents, these God-like, all-powerful individuals provide their children with a mirror and the standards by which initially the self-image arises. There is a biological need for physical contact, love, and association, and it is through association with others that we discover who we are. To be accepted, to win praise, means to shed characteristics that our parents find objectionable. We learn what is undesirable by physical punishment, verbal scolding, admonishments, or the withholding of approval. We learn what is "OK" and what is "not OK." We learn that we must become other than what we are, and that what we are is in some way "not OK."

Even Good Parents Give "Not OK" Messages

The normal process of socialization and the steady stream of disapproval, punishment, scolding, spanking, deprivation and other admonishments from our parents as they attempt to teach us self-control create feelings of guilt and inadequacy in the child. Even "good children" who have very loving, accepting parents suffer punishment, the withdrawal of love, or looks of embarrassment. At one time or another, every child learns that something about her or him is "not OK."

Hence, a normal by-product of the early socializing process is negative feelings and emotions. Children do not like to feel "not OK" and also do not like to be thwarted in their desires. Even anger at the thwarting, disapproving parent may give rise to bad feelings because these emotions are so threatening. Even when angry, children need and want the love of their parents. Hence, even feelings of anger may give rise to guilt and the suspicion that they are "not OK."

While we are being socialized our parents or other authority figures impress on us what they consider good, bad, acceptable, objectionable, lovable, and unlovable behavior. They may do this by yelling, "No," or by hitting us or calling us names. When we are engaged in behavior they don't like, they may tell us, "Don't," "You can't," or "Don't you dare," or may give us looks of pain, embarrassment, anger, disappointment, rejection, shame, and so on, messages keenly observed by the right half of the brain. The constantly reinforced message is "I am not OK; there is something wrong with me; what I am is not acceptable; I must be different and must change in order to be accepted and loved by my parents."

Even parents who hold their tongue and try to distinguish between "being bad" and "behaving badly" may inadvertently instill this sense of inadequacy. Disapproval can be nonverbal and quite subtle: a frown, a sigh of disappointment, a look of sadness or exasperation, foot tapping, arms folded across the chest, a furrowed brow, pursed lips, head wagging, hands on the hips, pointing with the index finger, an irritated or unhappy tone of voice, or even a failure to acknowledge some success. However, the overall impact is similar, and all these messages are intently attended to and registered in the right cerebral hemisphere.

It is very important for children to receive these messages in order to learn self-control and how to set standards. Parents need to punish their children when they misbehave. However, more important, parents have to love their children all the more soon after punishing them, as well as when the children are well behaved or are just being children. Children need to know that they are loved, especially after being punished for behaving badly. Although punishment is not the best way to change behavior, it certainly is more effective in this regard when administered immediately after the act has occurred, by someone whom they know loves them. It makes a more meaningful impact when coming from someone you love and who loves you, and who you know will continue to love you again. It is in this manner that the limbic system comes under increased control, and the right and left halves of the brain learn the difference between being loved and being mistreated.

Feeling Unimportant

Few of us escape a sense of unimportance, which is instilled in us by our parents and society when we are little children. Our feelings are sometimes thought of lightly, or desires are thwarted, and when we are not being told to "clean up the mess," to "be quiet," to "go outside and play," or to "quit asking so many questions," we may be ignored, told to wait until we are older, or even punished for reasons that are not at all clear to us. We are forced to wear clothes and to eat foods that we do not like. Indeed, this feeling of being forced, of being told to wait, to be quiet, to go play outside, is yet another form of diminishment. In a sense, children often feel belittled, and even children who are "overprotected" are in fact being told, "You are inadequate."

Common messages sent even by well-meaning parents are "you're too young for that; you'll never be able to do that; if you do that you'll get hurt; you're not big enough; you're not strong enough; you're not smart enough." Children are told they are stupid; can't sing; can't catch and can't throw; lazy; absurd; naughty; shocking; foolish; idiotic; they are given a myriad of negative appraisals. Unless the child is also given a lot of love and has received positive feedback about his or her capabilities, these messages form the basis for both conscious and unconscious self-contempt in childhood and later, in adulthood.

Feeling Left Out

Childhood can be a happy time, and childhood can be a time of trauma. However, no matter how happy a childhood we had, other children responded to us, on occasion in a cruel, teasing, or hostile manner. Perhaps we were left out of some activity, were the last to be chosen as part of a team, were excluded from some group, or were otherwise made to feel inept, lonely, hurt, inadequate, rejected, and isolated. Indeed, messages of being "not OK" are frequently transmitted outside the family, and other children are sometimes very vicious and cruel.

Seeking to Maintain the Familiar

Feelings of being alone, left out, and unloved are always painful. Even if they are transitory and infrequent, they leave an indelible imprint, particularly with the mental system of the right half of the brain. That imprint, however, often remains unconscious, but although unconscious, it continues to influence the way we behave, think, and feel.

Because of these unconscious influences, some people seek out experiences or people who will create the same familiar "not OK" emotional atmosphere that was an earlier part of their existence. Even if unpleasant, the familiar is easier to accept than what is unfamiliar. By re-creating the past, they also get one more chance to "fix" that problem or to obtain the love they were denied. Unfortunately, if these early experiences and this sense of the familiar are maintained predominantly in the right brain, the left half of the adult cerebrum may have little or no knowledge of their presence and influence.

The Critical Parent Within

Our parents are the initial if not the only source of love and affection that is available to us during infancy and early childhood. We learn who and what we are by the way they touch us, look at us, and treat us; by the names they use to describe us; and through the way they make us feel. These experiences and feelings are linked, as all these messages are attended to by the right brain.

It is through our parents that our initial conception of self is formed and through whom we develop feelings of self-worth. This initial self-concept, in turn, becomes the foundation on which we continue to build our sense of self; what we are, what we are capable of, whether we are lovable, hateful, good, bad, ugly, or deserving of abuse.

Children learn who they are from how they are treated. They learn what they can be from the examples set by their parents. Parents are not only the child's world and universe, but for tiny children they are models of what they may become. Little girls want to be like their mommies, trying on their shoes, putting on their makeup, and little boys look to their fathers as models of what it means to be a man. Children often model these examples, incorporating what they see and hear, and engaging in similar behavior when they play. If their parents yell, fight, argue, and hit one another, children will yell, fight, argue, and turn the playground into a battlefield, where the wars observed at home are acted out again and again. If parents swear and curse, the same curses and oaths of anger will be repeated by their children, sometimes in the most innocent of circumstances.

For example, when I was very little I often spent several weeks, two or three times a year, staying with my grandparents on their farm. My grandparents had a very strict religion and did not drink, smoke, or even believe in dancing. My parents, however, often used very bad language, but not in the presence of my grandparents. Although raised in the city, I loved the farm and loved my grandparents dearly. One day, however, as I was riding in the back of my grandfather's pickup out among the fields of corn, I began to sing, "Goddamnsonofabitch, goddamnsonofabitch." Later, my grandmother said she could hear my three-year-old voice miles away at the farmhouse. My grandfather immediately stopped the truck, walked around to the side where I was gaily singing away, and slapped me across the face, thus cutting short my singing career. He had never hit me before and never did again, so this slap made quite an impression. Almost thirteen years passed before swearing again became part of my repertoire.

By example, our parents show us how to behave toward others, how loved ones should be treated, and how we should expect to be treated by those who supposedly love us and care for us. These early parental examples—of attitudes, emotions, speech patterns, tones of voice, facial gestures, nonverbal activities, and other behaviors—are etched into the fabric of the limbic system and the right half of the brain.

Sometimes the good and the bad are stored differently in memory. Positive experiences are more likely than the negative to receive a verbal label. Thus, the left brain may store and then be able to gain access to positive emotions and memories. Because negative experiences, along with all else that is emotionally significant, are stored in the right brain, the left brain may later be unable to gain access to past experiences that are still troubling. Later in life, if negative experiences are recalled by the right brain, and create emotional havoc, the left brain may be unable to gain access to the source of the difficulties and it will not know why it is upset, angry, and so on.

The Unconscious Parent

The mind of the child is malleable. Impressions are easily made and maintained. Indeed, just as an impression of someone's hand can be left in wet cement, the impressions made by our parents also make an impression in the very malleable, as yet unformed psyche of the child. These linked impressions stay with us forever in the form of an internalized Parent, which continues to exert influences similar to those exerted by our parents, but within the confines of our own psyche.

The shape of our mind affects the shape and form of our feelings and thoughts. The manner in which our mind was shaped will in turn influence the manner in which future thoughts and feelings take form. Hence, future thoughts, feelings, fears, hopes, and desires will echo the feelings, descriptions, pronouncements, and images repeatedly imposed on us by our parents. If we were repeatedly told we were failures, the parental voice will echo forever in the form of a Parent Ego personality (hereafter referred to simply as the *Parent* or *Parent Ego*) that is now inside us.

Just as the child that we once were continues to exist and to exert influence on our behavior and feelings, a verbal and nonverbal emotional image of our parents is also maintained, thanks to the interaction of the amygdala, the hippocampus, and the right half of the brain. The image is composed of a linked complex of feelings and experiences associated with how we were treated by those who ruled our early life, including, sometimes, older siblings and grandparents, in addition to parents. This complex of associated parental images and models of behavior constitutes our internalized Parent Ego, which comments on and criticizes us from within the confines of our own mind. For example, if our parents were emotionally abusive and treated us as if we were worthless and unlovable, the Parent Ego will treat us the same way. If they mixed the bad with the good, we will experience both from the Parent Ego.

The Parent Ego

The unconscious Parent Ego corresponds to our developmental experiences with authority figures, specifically our parents or their surrogates, and represents an internalization and incorporation—an imprint—of their values, admonishments, behavioral patterns, morals, judgments, and related traits, including how they treated us and made us feel about ourselves.

Indeed, as the child's psyche is in large part molded by the manner in which it is treated, in the final analysis the treatment shapes the mold so both become nearly identical and both exert their influences forever in the form of a parent image. The parent leaves his own image in the wet cement of the child's psyche in the form of a Parent Ego which forever remains as it was originally experienced.

That imprint often remains unconscious and continues to influence the way we behave, think, and feel. In fact, it often compels us to seek out experiences, situations, or people who will create the familiar parental emotional atmosphere that was a "normal" part of our upbringing. Again, as the good messages tend to be encoded by the left brain and both good and bad may be stored in the right brain, we may often feel confused, particularly if the bad far exceeded the good. The right half of the brain and the unconscious mind will seek out friends, lovers, teachers, or employers who will treat us as our parents did, or we will find friends, lovers, or employees whom our unconscious Parental voice can criticize. That is, we assume a Parental attitude, and the unconscious Parent Ego criticizes the actions of others, just as we were criticized when we were little.

The unconscious Parent will demand that we live up or down to its expectations and continue receiving the treatment we were told we deserved.

The Parent, Child, and Adult Ego Personalities

The Right Brain

The right cerebral hemisphere (including initially the limbic system) is concerned predominantly with social and emotional perception and expression, including the storage of emotional memories. It is the initial repository of all our childhood experiences, including our feelings and impressions. Because the ability of very young children to understand and reason is limited, and as their language abilities are also not well developed, most of these very early experiences are stored only within the right brain or in a code not accessible to language. Therefore, both the Child and Parent Egos are the result of visual and emotional images, associations and impressions stamped into the unconscious mental system maintained by the right cerebral hemisphere. The left brain may have little knowledge or understanding as to their presence, except in regard to positive memories which may be more accessible.

The right brain having selectively experienced and stored in memory the matrix of emotions and feelings which make up the Child and Parent, can at a later time in response to certain situations act on those memories. It can also reactivate the unconscious Child or Parent and all its attendant feelings and attitudes, much to the surprise, perplexity, or chagrin of the left half of the brain, which can only respond, "I don't know why I acted that way." As noted earlier, the language-dependent conscious mind and left brain cannot always gain access to memories stored in the right half of the brain, particularly if they are negative.

The Adult Ego Personality

Although the Child and Parent Ego matrix does not disappear as one ages, these aspects of psychic and emotional functioning are supplemented by the formation of a more mature ego: the Adult Ego personality. The Adult personality is similar to what we consciously recognize as our ideal Self, i.e., our Adult Self-Image (hereafter the terms *Adult Ego* and *Adult Self*-Image will be used interchangeably).

The Adult Self is more closely associated with consciousness and is therefore more easily recognized by the conscious mind, as it is formed largely after the development of language and the temporal-sequential aspects of consciousness. It is both an *ideal* of how an adult should behave and the actual more mature, responsible, and controlled aspects of our personality. We do not act as adults at all times and sometimes we can act very childish or parental.

Hence, whereas the Child and Parent are largely associated with the right brain and the unconscious regions of the mind, the Adult Ego is more or less maintained by the conscious aspect of our psyche. However, the Adult Self-Image is a composite of a variety of experiences and feelings. Although predominantly conscious, the Adult also transcends consciousness and occupies mental realms of conscious awareness maintained by both the left and right hemisphere.

Conscious Recognition of the Ego Personalities

Just as certain capabilities are maintained in both halves of the brain, so are certain aspects of the Adult, Child, and Parent. That is, our parents also spoke in words and sentences, and as we aged, so did we, and these linguistic experiences, be they associated with our Parent or our Child, are in turn stored in the left half of the brain. This is why we are not only aware of their presence, but can become conscious, with some effort, of this as well. However, what is most accessible to the conscious mind and the left brain, is the positive and our rationalizations and confabulations regarding the negative. What children are unconsciously aware of and what they are conscious of often comprise two different aspects of the same experience, not only because of the different way in which the right and left brains process information, but because of the immaturity of the corpus callosum. Thus, much of what children experience cannot be compared or become clearly associated. What is stored in memory are, in fact, two different experiences and two different memories, one of which is maintained in the right half and the other in the left half of the brain.

Indeed, neither the Adult, Parent, or Child Ego is wholly conscious or unconscious, as they have both verbal and nonverbal components. If they all were only conscious or unconscious, successful interactions among these different realms would be almost impossible; one would never know what the other side was doing. There would be no Self, but only fragments of a whole; a multiplicity of personalities and ego states each unrelated to the other. It would be as if there were two completely separate people existing but not communicating within the same head;

a condition that arises only after the corpus callosum has been completely severed.

The only aspect of the mind that is almost entirely unconscious is the aspect associated with the limbic system. This is the IDfant Ego personality, the most primitive aspect of the mind.

The IDfant/Limbic Ego Personality

The IDfant Ego personality (hereafter referred to as the IDfant) is a mental system that reflects the primitive interactions of the limbic system. It corresponds to the most primitive and the most inaccessible, infantile reflexive aspect of the unconscious and is somewhat similar to what Freud labeled the *id* almost a hundred years ago.

As mentioned before, the limbic system consists of several ancient and primitive nuclei buried within the depths of the brain. It is concerned with maintaining internal homeostasis, minimizing tensions, and regulating food and water intake as well as with the most rudimentary aspects of emotional functioning, such as pleasure, rage, feeding, fighting, fleeing, and sexual intercourse.

The limbic system—specifically, the hypothalamus—and its corresponding IDfant mental system are almost fully functional at birth, and although they can be modified through experience, they are intensely concerned with maximizing pleasure and minimizing unpleasant tensions. The IDfant serves the pleasure principle and our homeostatic needs.

If the infant is hungry, the limbic system causes it to cry and to engage in raging, screaming vocalizations if its needs are not met. Essentially, the IDfant Ego personality says, "I want it and I want it now!" It knows nothing of morals, has no conscience, and makes no judgments regarding the appropriateness of its desires.

The experiences and memories which make up the IDfant are generally not amenable to reflective scrutiny by right- or left-brain conscious awareness, as it is almost wholly unconscious. However, it can continue to influence and take control over behavior even among well-adjusted, moral, educated adults. When this occurs, the person is likely to behave in an impulsive manner, without thinking and without regard for the consequences. Such behavior is likely to be described as infantile, selfish, self-centered, and, in the extreme, sociopathic, psychopathic, and criminal. Nevertheless, in the guise of the Adult, Parent, or Child, we often pander to the needs and desires of the limbic system, and thus the IDfant portion of the psyche; hence comes our unending fascination with sex, food, and violence, which is exploited so skillfully by television and the movies.

Together, the unconscious aspects of the Child, Parent, and IDfant constitute an unconscious self-image. As a complex of unconscious interacting influences, they may be referred to collectively as the "Lesser Ego," for often (but not always) they exert influences that the conscious Adult Self finds undesirable or negative. The Lesser Ego is a source of personal doubt and feelings of inferiority. When activated, it often thwarts conscious goals and desires and is often the source of emotional conflicts and unhappiness. It is often active in seeking the familiar unhappiness of one's childhood.

The Conscious and Unconscious Self-Image

As we grow and develop, the limbic system and the IDfant mental system begin to lose their dominance over our behavior, as the neocortex (new brain) and the right and left cerebral hemispheres mature and gain control over these primitive systems.

The nonmotor and sensory regions of the right brain, however, appear to mature more quickly than the left brain, which is more involved early in life in developing motor control over the body. As a result, the more rapidly developing right brain, which is more involved in emotion and visual imagery, is the initial repository of all our childhood experiences, including our feelings about and impressions of how our parents treat us. The Child and Parent are the result of impressions stamped predominantly into the unconscious mental system maintained by the right half of the brain while we were growing up.

As we age, the conscious, language-dominant aspect of our mind and brain begins to play a greater role in controlling our behavior. Simultaneously a conscious self-image is fashioned, and over the ensuing years, this sense of Self continues to be modified and expanded long after we reach adulthood.

It is this conscious self-image that we identify with. It contains all the traits, tendencies, abilities, talents, goals, and fears that we are conscious of, and even some flaws and failings that we consciously view as comprising our essential character. However, just as the greater mass of an iceberg is hidden beneath the waves, a considerable portion of who and what we are falls beneath the horizon of the conscious mind. What we are consists of conscious and unconscious elements; we are both our right and our left brains.

The Two Self-Images

Just as we have a conscious and an unconscious mind, as well as a right and a left brain, we also have two self-images. One is consciously maintained, and the other is almost wholly unconscious. The conscious self-image is associated more or less with the left half of the brain in most people. However, this self-image is also subject to unconscious influences.

By contrast, the unconscious self-image is maintained within the right-brain mental system and is tremendously influenced by current and past experiences as well as by the unconscious Child and Parent. Together, the two self-images constitute the Ego.

Although the two self-images may be quite similar in some respects and may share many identical and overlapping qualities, in other ways they are exact opposites. They also interact. Indeed, sometimes the conscious self-image is fashioned in reaction to unconscious feelings, traumas, and feared inadequacies that the person does not want to possess but that, nevertheless, are unconsciously maintained. For example, a person with a "superiority complex" is often reacting to his or her own unconscious fears of inferiority. The person who is a "know-it-all" often unconsciously fears that she knows much less than she proclaims and is trying to hide her own ignorance. The man who over-emphasizes his masculinity may be trying to hide or compensate for a soft inner "feminine" core that he finds consciously unacceptable. People who see racist or sexist comments in even the most innocent of interactions may be responding to their own sexist biases or racism, or to their own feelings of unconscious inadequacy.

The unconscious self and the conscious self are in part a product of the "not OK" socialization process we all weather. There are aspects of ourselves that we had to hide, deny, suppress, or discard because our parents, friends, and teachers made it clear that to do otherwise was unacceptable. Slowly, a conscious self-image is fashioned as we learn the role, good or bad, that we are expected to play.

The fashioning of the self-image is thus the result of a compromise between the individual and society. As pointed out by Carl Jung, in some respects the conscious self-image is designed to make a particular impression on others, as well as to conceal our true nature from ourselves, our parents, and the culture at large. In this regard, the self-image hides as much as it reveals. In some respects, it can be considered a carefully constructed mask, or persona.

Indeed, because the self-image is fashioned both consciously and unconsciously, even the conscious self-image retains an unconscious stamp; so that why we are the way we are, and even who we are and what we truly want, often remains a mystery; that is, it remains unconscious. Of ourselves we tend not to be knowers.

Competition for Psychic Control

Just as there can be only one driver of a car, either the conscious or the unconscious half of the mind predominates in the control over our behavior at any given time. However, it is quite difficult for one aspect of the mind always to be in control. Sometimes the unconscious mind predominates, and sometimes the language-dependent regions of the psyche predominate, depending on the situation and our emotional state, or on whether we are, for example, reading a book, throwing a football, or performing arithmetical calculations.

Moreover, by nature of the very structure of our brain, sometimes the right brain and/or limbic system predominates, sometimes the left depending on our homeostatic condition and the actions we are engaged in. For example, an individual who has had too much to drink and is sexually aroused, may act on those desires, even with someone they may normally avoid, due to the driving power of the limbic system. The next day, when they reflect on their behavior, the left half of the brain may become active and generate considerable feelings of guilt. Conversely, an individual struggling with a physics problem may experience only minimal limbic interference.

Thus, depending on what we are involved in, one mental system and region of the brain will struggle for control. The left brain may resist acknowledging what the right brain is fully aware of, or the limbic system and all its unconscious processes may attempt to overwhelm the conscious psyche. Because of these competing influences, we sometimes behave completely contrary to our conscious self-image or in a way that we may later regret.

Struggles for psychic control do not occur just between the conscious and unconscious mind. Sometimes, conflicts and power struggles occur within a single mental system. For example, within the unconscious, the Parent and the Child may come into conflict, the Parent behaving abusively or very critically, and the Child feeling bad. Although the battle rages within the unconscious, the entire brain and psyche become upset, while at a conscious level, the person may not know why he or she feels so terrible.

The conscious Self is sometimes completely overwhelmed by the Parent, Child, or IDfant and the turmoil occurring within the unconscious mental system. As such he may respond in an irrational, destructive, violent, childish, impulsive, abusive, or loudly critical manner, or may just feel lousy, angry, upset, or moody and not know why.

When these conflicts become so intense that the conscious Self begins to be affected when confronted by unpleasant consequences, it often happens that this aspect of the conscious mind begins to employ various defense mechanisms so that the source and origins of its problems can remain unconscious. Common defense mechanisms include rationalization, projection, denial, self-deception, reaction formation, and compensation (described in Chapters 14–17).

The Adult Self as Mediator

Fortunately, the Adult Self and the conscious mind are able to recognize some aspects of the Child and Parent and can sometimes regulate the interactions between them so as to minimize their influence. That is, a person may recognize when he is about to act Childishly or as a critical Parent.

In this regard, when in the mode of the Adult (at least ideally), a person may attempt to mediate or appease the desires or may refuse to give in to the demands of the IDfant, the Child, or the Parent. Or the person may seek counseling or psychotherapy so as to get these aspects of the personality under control. Unfortunately, as the Adult (as described here) is somewhat of an ideal, and is subject to cultural and environmental influences, its influences may be quite meager, it may be poorly developed, or it may even join forces with these negative unconscious influences and then prey upon society, or whatever unfortunate victim comes its way.

9

The Unconscious Child Within

The Structure of the Mind

The human mind is multidimensional, multifaceted, and consists of several levels. The relatively easily accessible conscious mind resides in the left half of the brain. The right brain is associated with social-emotional, nonlinguistic unconscious awareness. The most primitive, relatively inaccessible, almost purely emotional aspect of the mind is referred to simply as the *unconscious* and is associated with the limbic system.

The Domination of One Aspect of the Mind over Others

All three mental domains are interlinked and ongoing and often process and respond to information simultaneously, each in its own fashion. In some instances, one aspect of the mind dominates and achieves expressive preeminence over the others. In other situations, the mental systems of the right and the left brains, or the right brain and the limbic system, work equally and harmoniously.

Although two or all three systems may interact harmoniously with the same goals and agendas, the three mental systems often act in opposition and independently, each according to its own memories, perceptions, likes, dislikes, goals, and desires. When one mental system's response is the opposite of the response of a different region of the mind, the result is confusion. People may act or feel "out of control," ambivalent, in conflict, self-destructive, "neurotic," impulsive, suspicious, "out of sorts," "depressed without reason," or even hungry or angry and may have no conscious rationale for their behavior.

Although the origins of these conflicts may seem a mystery, they are often due to the activation of childhood feelings, memories, and related experiences that helped form conscious and unconscious self-images, including the unconscious Child and Parent. In other cases, these conflicts are due to the perception of particular events that trigger emotional reactions or memories in either the right brain or the limbic system unbeknownst to the conscious left half of the brain:

While walking down the street on her way to work, Betty (i.e., her right brain) observed a man screwing up his face in a particular way and raising his hand as he hovered over a little boy. Her right brain observed the boy fearfully shaking his head up and down indicating "yes" as the man sternly hovered over him. Her left brain tried to mind its own business. As she continued down the street, she observed (with both halves of her brain) a whiskey bottle lying in the gutter. However, she paid it, like the little boy and the threatening man, almost no conscious attention because she was hurrying to work and had other things to think about.

By the time Betty sat down behind her desk she was feeling terribly anxious and upset. She couldn't keep her mind on her work and felt irritable. She had no idea what was bothering her. She snapped at her secretary for no apparent reason and decided to take an aspirin. It must be a headache. Right? Well, although body aches and pains are often a result of the inappropriate expression of emotions through the body (as both are linked within the right brain), this is not the case here.

When Betty was little, her mother was a closet alcoholic. As soon as her father left for work, her mother would begin drinking and by midmorning was usually quite drunk. Betty's mother was an angry drunk and would fly off the handle over little things. If Betty was noisy or made a mess, her mother would screw up her face in anger and raise her hand as if to strike Betty, who would always begin crying and promise to be good.

Although her parents had divorced by time Betty was five years old and she went to live with her father, these early memories—the drinking, the bottles of alcohol, the yelling, and the mother's screwing up her face and raising her hand in anger—were stored away in Betty's right brain.

When, as an adult, Betty walked down the street that morning, although her left brain noticed nothing of importance, the right brain attended to the man, his face, his hand raised in anger, the little boy, and the whiskey bottle. The unconscious memories were triggered, and all the hurt, fear, and anger came washing over her. By the time she sat down at her desk, she was terribly upset. But her left brain did not know why. Although her left brain could remember that her mother was an alcoholic, the emotional links between the bottle, the angry man, and the upset child were not apparent to it. The right brain, realizing these links, was upset, whereas the left brain had no idea what was going on.

The Ego Personalities

Spanning the bridge that separates yet interconnects the mental systems maintained by the left and right brain and limbic system are four ego states or rather ego personalities. These are the IDfant, Child, Parent (which together constitute the "Lesser Ego"), and the Adult Self. These ego states, in part, constitute the conscious and unconscious self-image, all of which constitute the Ego.

The Ego

The Ego is generally associated with our conscious subjective impression of who we are; that is, our self (as in "He has a fragile ego," or "Don't put your ego on the line," or "Don't be so egotistical"). However, as discussed by Freud so many years ago, the Ego may also be unconscious and thus represents part of the unconscious self-image. Hence, we are made up of multiple ego personalities. As defined here, the ego is a composite of the conscious and unconscious self-images and is maintained through the interactions of the frontal lobes.

The unconscious aspect of the Ego maintained by the right frontal lobe exerts inhibitory influences on both the right and the left halves of the brain and the limbic system. It greatly influences information reception and expression within the left brain and the conscious mind. The conscious aspect of the Ego maintained by the left frontal lobe is restricted to influences on the left half of the brain. Hence, the unconscious aspect of the Ego exerts considerably more control over personality functioning and the expression of impulses than does the conscious aspect of the Ego. When the frontal lobes are compromised, Ego functioning is severely affected and the personality becomes fragmented. Concern about consequences, the ability to establish and pursue long-term goals, and inhibitory restraint are destroyed, and the individual may act on every impulse or may simply cease to act at all.

Ego Personalities

As stated above, there is a multitude of selves, both conscious and unconscious. Although this multiplicity may give rise to numerous conflicts, it may also be beneficial and adaptive, allowing different aspects of our personality to come to the fore as they are needed on various activities. Different people and different situations bring out different aspects of our personality. Sometimes, it is OK to be childlike and playful, and certain people may stimulate us to behave this way, allowing our Child ego personality to come to the fore. However, if someone has hurt our feelings or made us feel insecure, a different aspect of the Child ego personality is activated, and we may fly into a childish rage.

In general, an ego personality may be considered a coherent matrix of interrelated experiences, feelings, memories, cognitions, and behavior patterns that were experienced during specific developmental periods, including infancy and childhood, encompassing the transitional period of adolescence, and continuing throughout adulthood.

Consider, for example, the formation of the body image, as it is maintained by the right brain. All sensory messages arising from the body's surface are transmitted to a zone of cells called the primary receiving area (in the parietal lobe). Each cell receives only a single piece of information from a single zone of the body. All the cells in the primary receiving area in turn transmit this information, through their nerve fibers (axons), to an adjacent row of cells. These secondary cells

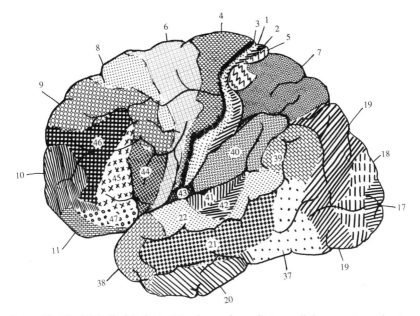

Figure 30. The left half of the brain. Numbers refer to distinct cellular groupings, that is, cells that are anatomically similar and presumably perform similar functions. Note Layers 3, 1, and 2, which are located in the primary receiving area for sensations arising from the body. Tactile and related impulses are received first in Layer 3, and then several cells from this region send converging messages to cells in Layer 1, so that Layer 1 receives more complex messages than does Layer 3. Many cells from Layer 1, in turn, transmit to cells in Layer 2, and the message becomes even more complex as these different forms of input are assimilated so that, for example, if we are holding something that is long, thin, and sharp, we may recognize it as a nail or a pin. From R. Joseph, *Neuropsychology, Neuropsychiatry, and Behavioral Neurology* (New York: Plenum Press, 1990).

receive messages from several different primary cells and thus have much more information. These secondary cells, in turn, transmit to a third zone of cells, each of which receives messages from many secondary cells. Finally, from this mosaic of overlapping and converging signals from the body, each of which is passed from cell to cell, a map of the entire body is formed. In this way, the body image—and by analogy, the self-image—is formed. That is, associations and experiences which are stored in various brain cells intercommunicate when activated such

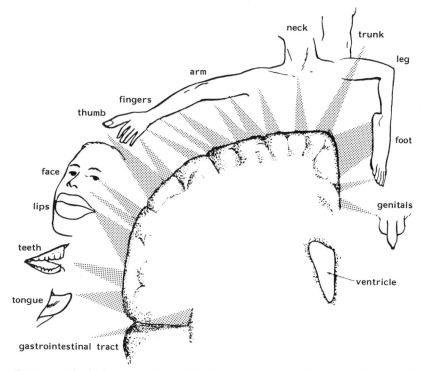

Figure 31. The body image. From P. M. Milner, *Physiological Psychology*. Courtesy of Holt, Rinehart, & Winston, New York, 1970.

that a mosaic of overlapping impressions are formed via their convergence. A self-image and ego personality are formed. It is in a similar manner that the angular gyrus (which sits at the junction of the auditory, visual, and somesthetic zones) is able to call forth associations to form ideas and concepts.

Some aspects of our personality are unconscious and correspond with how we were treated or felt as children. Other aspects are conscious and correspond to how we are treated as adults. Those aspects that were formed during childhood do not go away. Because our ego personalities tend to be age-associated and are formed during different developmental epochs, we form several of them.

The Ego versus the Adult Ego Personality

The Ego in its totality is affiliated with both the conscious and the unconscious mind; is based on subjective impressions, ideals, denials, selective memory, and cultural expectations; and is influenced by unconscious feelings. We may often confuse the Ego with the ideal of an adult self. Nevertheless, although we may wish to view our Ego or self as being adult, sometimes it acts instead like a child, or a parent, or an infant, or even a criminal. Being an adult and acting like an adult are not the same thing; many adults act like children or even infants.

Ego Activation

At any given moment, the ego personality of the IDfant, the Child, or the Parent, or any combination of the three, may dominate psychic and mental functioning and may determine our choices and the behavior we engage in. At any time, emotional reactions may be triggered, and the corresponding ego personality may be activated.

The IDfant. The memories and behavioral features mediated by the limbic system during infancy continue to be maintained and mediated by the limbic system throughout adulthood. When this constellation of associated traits is activated, a person may behave like an infant, displaying selfish, demanding, raging, narcissistic, egocentric, and impulsive behaviors, without any concern about the consequences. The IDfant says, "I want it now!"

The Child. Those experiences and memories stored during later infancy and early childhood in the memory banks of the limbic system and the right brain are maintained by these brain areas. When they are activated, a person may act like a child, feeling the same hurt, self-doubt, lack of confidence, unworthiness, neediness, deprivation, guilt, rejection, or abuse that characterized their self-image in childhood, and they may act accordingly and often inappropriately.

The Parent.   Our perceptions of our parents and how they treated each other and us while we were growing up exerted their greatest influence on the developing right brain. The reason is that much of our early learning was stored only in visual and emotional images. These impressions continue to be mediated by the right brain through adulthood.

The internal Parent is made up of our childhood experiences at the hands of our parents. When our internalized emotional representation of the complex associations that form the Parent is activated, it begins to treat us, or others, as we were treated as children and tries to tell us, or others, what to do and how we should feel about ourselves. In some respects, it is similar to what has been described as a conscience. However, it often likes to "guide" others as well, whether they like it or not.

The Adult.   The Adult Ego personality consists of idealized images of what society (and even television) informs us is admirable: truth, honor, justice, fairness, law, order, responsibility, and so on. The Adult Ego is predominantly conscious, but it includes unconscious elements as well. It is primarily our idealized image of ourselves.

The Ego.   The Ego is the Self in totality. It is sometimes overwhelmed by unconscious influences. However, it may also act as a mediator of the needs and desires of its composite elements. For example, a young man spies a very attractive young woman. His IDfant exclaims, "I want sex with her now!" The Parent says, "Are you kidding? You're a geek, a loser. You can't have her!" The Ego, acting as mediator, tries to meet the demands of both unconscious forces and its own conscious desires, depending on which are stronger. The Ego can accomplish this because it is both consciously and unconsciously maintained through the integrity of the right and left frontal lobes.

If the IDfant's desires correspond with conscious and other unconscious needs, so that the Ego is willing to act as mediator, the person may walk up to the woman, marshal all his right-brain social talents, and (although feeling Parental disapproval and thus some ambivalence and insecurity) try to start a conversation. If he already has a girlfriend, his left brain may rationalize or even deny his obvious intent; "I want her for a friend." On the other hand, if the man was raised in an environment where adultery and lying were commonplace, his left brain may feel no need to explain away what he is up to.

The Child Ego Personality

A child who is raised in a home where his parents yell and scream, hit each other, and hit their children will experience these outbursts as "normal." It is "normal" to the child because he or she has little or nothing to compare it with. As far as the child is concerned, this is the world and the only world he knows. As such, although unpleasant, it also becomes familiar; it is home. Later in life, this child may form attachments to people who behave similarly, for although they may consciously deny it, this is what they can "relate to."

Jerry was plagued by a terrible stutter almost from the moment he began to talk at age two. As he was growing up, he became increasingly aware that his parents were embarrassed by him and that his two brothers seemed to get all the positive attention. Neighbors sometimes commented on his stutter, and their children teased him unmercifully. Although not abused by his parents, he nevertheless felt rejected by them and by everyone else he knew.

Not surprisingly, he began to feel angry, resentful, and extremely sensitive about his "little handicap." Even when no one seemed to notice, he was sure they did and felt picked on wherever he went. However, being big for his age enabled him to act out his feelings. By the time he was five, he had become quite violent and had attacked several children who had laughed at and teased him. Fortunately, he began receiving speech therapy, and by time he had reached the fourth grade, his speech impediment had become a slight stammer, which disappeared almost completely over the next few years, except when he became upset.

Nevertheless, the unconscious Child at his core never stopped feeling upset, rejected, angry, and overly sensitive. Consequently, whenever he became upset and his inner Child became fully activated, not only were his childhood feelings and emotions expressed, but he acted like a child, speech impediment and all.

When he came into therapy with his wife, Donna, she complained of his violent temper:

Donna: It's like walking on eggshells. Everything sets him off. If I put too much in the garbage and a can falls on the floor, he's all over me, screaming and yelling that I did it on purpose. If he comes into the kitchen and I'm on my way out, he loses his temper and begins screaming about how I am trying to avoid him. No matter what I say or what I do, it's a put down or an insult. I can't even get up to go to the bathroom without his taking it personally.

Jerry: She is always nitpicking and insulting me. Nothing measures up to her standards. Everything I do is wrong. Everything I like she doesn't. I like to watch TV, and she doesn't. She knows I am trying to diet, and she makes big, lavish dinners.

Dr. J: Tell me about the incident with the garbage.

Jerry (becoming angry): It's goddamned filled to the top, and what does she do? She's got to put one more thing in it and the whole thing falls over. It's her way of trying to say I didn't take out the garbage. She did it to piss me off.

Donna: Why is that trying to piss you off? It just fell, that's all. It had nothing to do with you. I cleaned it up, didn't I? I swear, everything makes him mad. He comes home from work, throws his coat on the couch, and if I hang it up, he flies into a rage, accusing me of hiding his coat.

Jerry: You did it because you knew it would make me mad. Why couldn't you just leave it? I would have hung it up. It's just your way of trying to put me down, like you have to pick it up for me.

Donna: If I left it, then you would have been yelling about how I never clean up. I can't win with this man. Sometimes, it's just like living with a child.

Jerry: See what I mean? There she goes insulting me again.

Donna was correct, of course. Jerry did act like a child. And yet, Jerry was right about feeling picked on and insulted. However, it had nothing to do with his wife. He was still being picked on by the complex of feelings that made up his unconscious self-image.

Moreover, his unconscious Child was still hurting and still trying to strike out because of all the rejection he had experienced when he was young. Always feeling inadequate, his unconscious Child scrutinized every word and action for rejection and then struck back at every opportunity. By behaving in this way, he was ultimately inviting rejection, attempting to maintain a self-fulfilling prophecy. In any case Jerry was acting like a child because his Child was easily activated.

Indeed, this collection of childhood experiences never goes away; it is stored in the memory banks of the right brain where it continues to influence thoughts, feelings, and behavior forever. Hence, the inner Child often seeks to maintain the familiar and to avoid what is different or unusual.

John's father had been a selfish and violent man who often slapped and yelled at his children. From an early age, John could remember his father calling him "useless," a "sissy," a "lazy good-for-nothing," and a "selfish little pig." He would threaten to leave John at the city dump or give him to an orphanage. When John began to cry, his father sometimes slapped him across the face.

When John's father called him names, John took them as the literal truth about who he was and about his value. He could not see that his father was a cruel, hateful man who treated everyone with disdain and ill temper. To John, his father's words were like physical blows, which hurt just as much and which scarred him just as deeply as a physical beating.

Moreover, as a young child, not recognizing his father's shortcomings, John viewed his father's behavior as being the result of his own shortcomings. John believed he had done something wrong, was bad, and thus deserved abuse. Over time, being treated badly became "normal" to John—unpleasant, but normal.

Later, as an adult, John became involved with girl after girl, woman after woman, who treated him badly, picked fights, and generally walked all over him. Although he hated how he was being treated, as well as the arguments and name calling, he stuck with each relationship until the woman "dumped him." When this inevitably happened, he felt terrible and sometimes called the woman and begged her to come back.

At an unconscious level, he believed that he deserved the abuse he received because the complaints of his girlfriends always sounded familiar and made him feel guilty. He assumed that the fights and arguments were really his fault and were due to his own bad behavior, lack of understanding, selfishness, and insensitivity. Certainly, these were accusations that his girlfriends flung in his face. Their complaints seemed so familiar that he became convinced they were true and accurate descriptions of his character flaws. This poor treatment was only a continuation of what had long ago registered as normal. Insofar as his unconscious Child was concerned, John was a bad little boy. By then, he may have *become* a "bad boy" and had unconsciously provoked the annoyance of his girlfriends so that he would feel that everything was "normal."

When he got married, his wife criticized and attacked him in a very demeaning way, calling him "lazy, good-for-nothing, selfish . . ." She threatened to leave him and forced him back into the role of the bad little boy whom no one could love. In John's mind, fighting and being yelled at were not only familiar but the "price" he believed he had to pay to get love. Because he was so "bad," how could he expect to be treated better? Moreover, rather than fight back, he usually sulked and withdrew. In fact, he often panicked after one of these one-sided arguments and sometimes even began to cry and beg forgiveness.

Essentially, John was choosing women who corresponded to his own unconscious Parent; women who would criticize him, reject him, and make him feel bad, or women whom he could provoke into confirming his bad-boy unconscious self-image. This was the crux of the attraction, and in this regard, he and the women he chose were completely compatible.

In fact, as was pointed out by his wife during one of our sessions together, he often did things to provoke her. He would leave his clothes and underwear on the floor and expect her to pick them up, he would not clean up after himself when he used the kitchen or the bathroom, and he was always being forgetful and failing to do simple things that he had promised to attend to.

Thus, John's unconscious Child was not merely reliving its hurtful legacy. He did sometimes act like an immature, selfish child and behaved in accordance with his inner parental dictates. He was acting out a self-fulfilling prophecy and sabotaging his relationships so he would be treated badly and could relieve the hurt from the long ago.

Activation of the Child

When we are among adults, the Child still remains an active part of our psyche and can take control over or influence our behavior and feelings, particularly during times of emotional stress. When they are triggered by some memory or event, we replay these original memories and feelings of anger, guilt, depression, rejection, abandonment, and frustration as well as all of the associated behavior. Thus, we may react childishly, becoming upset and not knowing why. If someone complains that you are acting "childish," it may be because the inner core of your being (your unconscious Child) has come to the fore.

When they have had childish outbursts, I often ask my patients "how old" they were acting? If they say, for instance, "six years old," I then ask them what traumatic event they experienced around that age. Almost always they give a relevant answer. Sometimes getting in touch with that injured child enables the patient to gain control of or at least to understand what is disrupting her or his life. Why is that six-year-old still crying?

Self-Fulfilling Prophecies

Every person whose "not OK" unconscious Child has been sorely injured attempts periodically or continually to overcome the hurt or to achieve the love that was denied during childhood. In order to do this, she or he must find someone who can re-create the familiar hurt so that it may be overcome. This task is not difficult, as the unconscious Child is easily able to recognize individuals who will provide the familiar hurt; such people seem familiar (because they are) and are easy to relate to.

The unconscious Child changes very little. It attempts to live in accordance with the dictates of the unconscious Parent, so as to maintain its well-ingrained self-image. Regardless of the conscious Adult self-image, if that unconscious Child self-image is bad, the unconscious Child and Parent will attempt to maintain a self-fulfilling prophecy; it will fail in its endeavors, irritate others, find abusive mates, and so on. Many current situations are contaminated by this unfinished business. Indeed, some individuals who experience success (ardently strived for as a form of compensation) may do whatever is necessary to sabotage it so that the familiarity of the past can be reexperienced.

These original childhood experiences and their emotional hurt being maintained within the right brain can only be partially accessed by the language-dependent consciousness of the left half of the brain with some difficulty. The conscious half of the cerebrum may admit, "Yes, I was treated poorly," however, it will also likely declare: "But it didn't bother me that much," or "I got over that a long time ago."

The Unconscious Child Is Not Always Bad

Knowing that we have childlike qualities that must often be suppressed can be a source of embarrassment and potential tension, particularly if we must make a great effort to control this aspect of ourselves. However, as this is a part of us, we must treat it with tolerance, acceptance, and understanding. Rejecting it is like rejecting the fact that we have two legs.

No matter how hard we try, we can never completely erase or eradicate the unconscious Child. Good or bad, the child we once were is still part of us, and we all retain childlike qualities. Although these qualities may be controlled, "outgrown," or suppressed to varying degrees, it is very difficult to be "adult" at all times and in all circumstances.

The effort to be "adult" at all times can be a source of discomfort, for this requires that we deny who we are. Particularly among men, whenever any childlike qualities are expressed, be they good, bad, charming, or endearing, they are sometimes treated—these men treat themselves—as if they have behaved in an inappropriate fashion. This, in turn, engenders a sense of self-contempt. Even if these qualities come to the surface very seldom, we may still feel embarrassed and contemptible.

Is it not more contemptible to deny who and what we are? Is it not self-defeating and demeaning to suppress aspects of ourselves as if they are some major flaw? Even if we are flawed, suppressing, denying, ignoring our traits and limitations forces us to behave as if we were partially blind. How can we successfully navigate through life with our eyes half closed? Self-imposed blindness is a sure road to inadequacy. We cannot gain control over ourselves if we deny who we are, and we cannot overcome our weaknesses if we do not know what they are.

Having childlike qualities is no more a flaw than having two eyes and two legs. Everyone has them. In fact, the unconscious Child has many good qualities. The Child can also be inquisitive, charming, creative, playful, loving, open-minded, intuitive, inferential, and endearing. It is healthier to be able to cut loose and have fun or be silly at times than to be a perpetual straight arrow or uptight, disapproving Parent at all times.

People who are not in touch with the inner Child are truly inadequate, for they are not allowed to be who they truly are.

The Child Should Not Dominate or Control the Adult

For some women, the "little boy" that so obviously resides in their boyfriends or even husbands is a source of endearment. Although the man may deny that this "little boy" even exists, many women find this attractive and its presence sometimes makes it easier for them to love their mates and to express their own needs to behave in a caring, nurturing, and sometimes even motherly manner—attentions that this "little boy" seems at times to accept eagerly. This "little boy" enables the man to be less threatening, less distant, more caring, cuddly, sweet, sensitive, and vulnerable—qualities that appeal to the "maternal instincts" of some women—that is, until the "little boy" becomes childish, acts selfishly, or throws a tantrum or gets upset because he is no longer getting his way or because a son or daughter has been born and he can no longer be the "number one kid."

Similarly, many men find "the little girl" in their wives or girlfriends to be extremely attractive features. For some men, this childlike quality makes their mate more "feminine," innocent, and charming. For others, this makes their mate less threatening and insures a certain level of dependence and creates an illusion of being in control: "Me Tarzan, you baby!"

However, taken to the extreme, when this aspect of the Child dominates psychic interactions the person may not only act in an innocent, charming manner, but speak and interact like a child as well. This may take the form of excessive use of baby talk which, rather than endearing, can be quite irritating when prolonged. In others, this may take the form of reverting back to childlike behaviors or behaving in a helpless, dependent fashion at all times.

When in the mode of the Child (or even that of the IDfant), a person may behave in a selfish, self-centered manner, break things and throw a tantrum whenever he does not get his way, or engage in the continual seeking of attention, security, and thus reassurance. Indeed, some people never grow up and remain fixated at the level of the Child. They are a child most or all of the time.

A prime example is the man who although married and with children, feels a need to "be out with the boys" several nights a week, drinking, womanizing, and carrying on. "Boys will be boys" is fine for boys, but for men who make a career of being a "boy," there is certainly something inappropriate going on. They are demanding to be able to act like a child because basically the Child aspect of their ego predominates.

Of course, those who behave exclusively like children require other children as their mates, or else a perpetual Parent to take care of or rescue them. When they form relationships with those who take on the role of Parent, they will have not only someone to "baby them," but someone who will criticize or abuse them, or someone they can retaliate

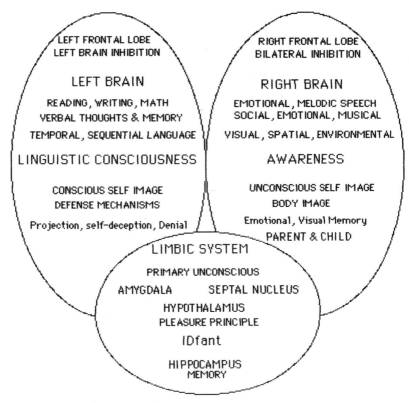

Figure 32. Schematic representation of the mind.

against for all the hurts that their own parents caused them, real or imagined. In some respects, this may seem rather ideal. However, by and large, it constitutes the formation of a potentially very damaging and emotionally upsetting relationship. The best relationships are those in which the Child and Parent are recognized and treated with understanding and acceptance by two adults who have decided to forge a life together as adults, and not as children or critical parents.

10

Unconscious Conflicts between Child, Parent, and Self

Dreaming, Creativity, and the Right Brain

The right half of the brain has a number of divergent capacities and properties, not all of which are directly related to emotional functioning. Visual and pictorial imagery, making inferences, visual closure, and deducing the whole from an abundance of seemingly unrelated parts, geometrical analysis of visual space, and even some aspects of dreaming are made possible by the functional integrity of the right brain.

Many capacities, such as reading and dreaming, require both halves of the brain. For example, in dreaming, the right brain provides the visual imagery and the emotional coloring, and the left brain provides the dialogue or narrative. Just as it takes both halves of the brain to make music (the right providing melody, the left rhythm), we also need both brains in order to dream.

Similarly, although the right brain appears to be the more creative aspect of the mind, for example, in the form of artistic expression, creativity is possibly a product of having two brains and two minds. Indeed, the creative process itself is perhaps made possible, at least in part, by the interpretation and guesswork that occur when the two different brains and regions of the mind tackle the same problem and then try to communicate with each other. In fact, the first evidence of prolific creative and artistic ability appeared at about the same time as fully developed language capabilities (with the exception of reading and writing) appeared on the scene and during the period when men and women became maximally physically and sexually dissimilar, that is, probably during early Cro-Magnon times. It seems that, when the two brains began to communicate and to respond to information differently—the right social-emotionally and visual-spatially, and the left temporal-sequentially, grammatically, and denotatively—people began leaving evidence of artistic and creative endeavors in caves and on cliffs ranging from South Africa to Russia and possibly to the Americas. Perhaps the creative spirit is due partially to miscommunication between the brain halves and the need for creative guesswork about what is going on in the other half of the brain. Perhaps the pushing of nonverbal capabilities out of the left brain and their further development in the right gave rise to these gifts. Or perhaps this blossoming of creativity, self-expression, and thus self-consciousness arose from recognition and reflection of oneself in the eyes of a loved one. Although there is no telling if Cro-Magnon man may have confused sex for love, there certainly is some suggestion that his first attempts at art may have been greatly influenced by his sexual feelings regarding the "fair" sex, as some scholars claim that the first paintings were of vaginas, although to others these particular paintings look like pictures of animal tracks and hoof marks.

Many authors, scholars, and artists have said that creative solutions to problems that they have long labored over have come to them suddenly during the course of a dream (and almost never during sex), or while they were engaged in activities involving minimal conscious or linguistic effort (e.g., shaving, showering, and so on). During maximal right-brain activity, creative solutions sometimes burst forth in dreamlike pictorial images, followed by a stream-of-thought explanatory commentary. Of course, the left brain also has creative capabilities, and both right and left often work together to solve problems.

Both halves of the brain engage in analyzing current experiences and conflicts, and some of these problems and experiences are then reexperienced and analyzed in dream imagery. For thousands of years, the importance of dreams in this regard has been well understood. Moreover, dreams sometimes warn us of troubles and dangers that lie

Figure 33. Some scholars have argued that these engravings represent the earliest forms of art and that they depict the female vulva. I suspect that they are engravings of animal hoofs or tracks and were probably instructional. Courtesy of Prehistoire de l'Art Occidental, Editions Citadelles & Mazenod, photos by Jean Vertut.

ahead: Jung believed that the unconscious sometimes anticipates the future by examining and drawing conclusions from facts that we are not consciously aware of, so that our fears or hopes regarding the future are played out in our dreams.

Neither the unconscious mind nor the right brain is ruled by our childhood experiences, and should not be considered synonymous with the unconscious Child or Parent. Nevertheless, our current experiences remain influenced by the past, so that the right-brain solutions we come up with are not always in our best interests.

The Parent Ego Personality

Developing simultaneously with the unconscious Child is the ego personality of the Parent. As stated earlier, the Parent corresponds to our developmental experiences with authority figures, specifically our parents or their surrogates, and represents an internalization and incorporation of their values, admonishments, behavioral patterns, morals, judgments, and related traits, including how they treated us and made us feel about ourselves. Just as the impression of someone's hand can be left in wet cement, the impressions made by the parents also make an impression in the very malleable psyche of the child. This impression stays with us forever and continues to exert influences similar to those exerted by our parents, but now within the privacy of our own thoughts.

As we've seen, if our parents were emotionally abusive and treated us as if we were worthless and unlovable, the unconscious Parent may treat us and others in the same way. If they treated us with love and respect, the unconscious Parent also has these capabilities.

The Critical Conscience

Loosely, the unconscious Parent is somewhat similar to what Freud referred to as the "superego" and to what others have called the *conscience*. It is more than a conscience, however, as it also engages in name calling and criticism, of others as well as of ourselves and the unconscious Child.

The Parent may represent an image of fear, hate, love, awe, and so on. If the actual parent was an unemployed alcoholic who acted like a helpless infant or child, the unconscious Parent will be helpless and childlike. If the actual parents often yelled, argued, and belittled each other and engaged in other inappropriate acts, the unconscious Parent will urge the person to act likewise or to accept similar treatment. It will say, "It is OK to treat others badly." Some children who grow up without a mother or a father or either may form a romanticized ideal of the Parent which then urges them to strive for similar ideas in themselves and in others.

Many major sources of early influence and authority, such as siblings and even television, may also become incorporated into the complex of Parental associations. This is increasingly true of television, which has become a surrogate parent, particularly when there is only one parent, who must work, or when both parents work. Unfortunately, television provides an unrealistic view of the world to those with minimal education, financial, or social opportunities. Moreover, television can exert formidable influences on the impressionable mind and

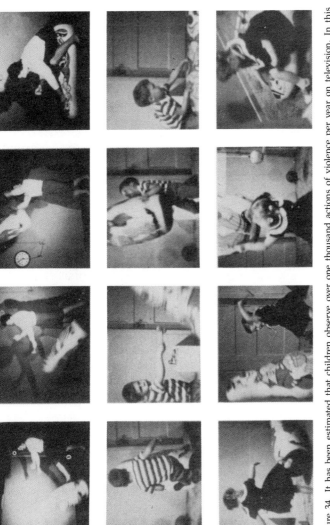

Figure 34. It has been estimated that children observe over one thousand actions of violence per year on television. In this experiment on learned aggression performed by Albert Bandura, children observed a film in which an adult model walked up to an adult-sized Bobo doll and, after making some demands that the doll did not obey, proceeded to punch it, kick it, and hit it with a mallet. Later, when each of the children was placed in a similar room with a Bobo doll and other toys, these children imitated most of the aggressive acts they had observed. From A. Bandura et al., "Imitation of Film-Mediated Aggressive Models," *Journal of Abnormal and Social Psychology* 66 (1963), 3–11. Courtesy of Albert Bandura.

brain of a child, and tends to pander to the limbic system, and thus to the lowest common but most arousing stimulant which links us all, an almost all-consuming interest in food, violence, and sex. Furthermore, television says it is OK to kill and harm others when you are frustrated or feel low in self-esteem; that it is OK to seek self-worth by using others as sexual objects or to engage in indiscriminate sex; and that the only route to happiness is to wear the right cosmetics or to drive the right car.

The Critical Parent

Although the Parent is an internalized emotional representation, its influences are not restricted to the privacy of our thoughts or feelings. Sometimes, the Parent criticizes not only us but others as well. That is, when others engage in behavior that our unconscious Parent finds objectionable or irritating, the Parent may become extremely critical, rejecting, or abusive.

Consider Jerry (the fellow with the speech impediment in the last chapter). Although his parents were not physically abusive and certainly never behaved in an obviously cruel or callous manner, he felt rejected, hurt, and ridiculed by them. The Parent that came to be formed in his mind was also hurtful and rejecting. Later, when he married Donna, she became associated with his unconscious Parent as she was someone from whom he wanted love. However, Donna was also taking the role of surrogate Parent in his mind, and his unconscious Child was able to strike back and accuse her of causing all the rejection and pain that he unconsciously believed his real parents, siblings, and classmates had caused him. At the same time, his critical Parent was able to treat her in the very same way that it was treating him. Whatever she did, he (i.e., his unconscious Parent) punished and criticized her.

Our thoughts and feelings can come to be controlled by the Parent, which then treats others as our own parents treated us. Sometimes, the unconscious Parent will seek out others whom it feels it can abuse in the way we were abused by our parents.

Most important, when we were children, our parents told us what to do, how to behave, and what they considered appropriate and acceptable. When the unconscious Parent continues this tradition, we may feel that we must obey, and when we don't, we may feel guilty or anxious, expecting disapproval and punishment. The unconscious Parent may thus be a source of tremendous conflict.

Becoming Our Parents

When the parental button has been pushed and gains control over behavior, we respond in a manner similar to that of one or both of our actual parents. We may use the criticisms, postures, gestures, and words that were used by our parents. In other words, we take on the state of mind that we perceived in one of our parents (or a parental substitute) and react as we feel they would react. Indeed, even when the Parental ego personality does not appear to be activated, we may find ourselves saying things our parents said or acting as our parents acted, even behaving in ways that we swore we would never mimic.

Contradictory Parental Messages

Often, we received conflicting messages from our parents. They may have yelled and screamed at each other and yelled and screamed at us but were absolutely frantic about what the "neighbors" might think and behaved only with the most decorous of manners and utmost restraint and tact when interacting with friends and acquaintances.

They may have admonished us for smoking, swearing, or drinking while engaging in these activities themselves. When confronted, they may have told us in essence, "Don't do as we do; do as we say." Although hypocritical and contradictory, these features have also left their imprint on our psyches, and our internalized Parent encompasses both extremes, contradictions and all. As a consequence, the Parent, when activated, not only acts to model and express the behaviors that we observed our parents engaging in but also expresses the disapproving Parental voice which told us how to behave. The Parent whispers, "It is OK to do this, but if you do, I will be upset." Within the unconscious this is a no-win situation, for whatever we do or desire, we may feel confused, anxious, guilty, and in conflict. We will get yelled at no matter what. For example, at home our parents argued, yelled, and screamed at one another; we were admonished to be polite and to demonstrate good manners when interacting with others. Now incorporated, these two seemingly opposing traits may both be expressed, so that we yell and scream at our spouse and children and then feel guilty and terrible about it.

On the other hand, our Parental voice may tell us it is OK to behave one way with one group of people and in a completely different way with others. We may treat our business associates, co-workers, and acquaintances with polite respect and then go home and scream at our kids for some minor transgression and feel this is "OK." We can live with ourselves and our bad behavior because it is acceptable to both the unconscious and the conscious self-images. In effect, we are modeling the contradictory behavior of our parents, and the message becomes "Do as I do, and do as I say; this is acceptable."

If the behavior were not acceptable to the unconscious self-image, both the conscious and the unconscious minds would work together to eliminate it. Such behavior would become overwhelmingly unacceptable and thus intolerable. We would not put up with it within ourselves or from others.

Not surprisingly, these parental messages, be they good or bad, may be passed down from generation to generation, conflicts and all.

The Good Parent

The unconscious Parent may be experienced as a source of internal punishment, criticism, confusion, and emotional pain. Nevertheless, it is not an evil personage altogether, although by and large it may be experienced as such.

The internal Parent is responsible for teaching us how to survive, how not to hurt ourselves, and how to stay out of harm's way ("Don't run out in the street; don't touch that because it will burn you!"). Of course, the emphasis was on the "Don't" or "stop that" or "no," and the concern for our well-being was not very obvious from the perspective of a child. What was perceived was the threat whose source was not the hot stove, but our parent, and the threat was that we would be spanked, yelled at, or punished in some other way. The potential for injury was perceived as arising not from our proposed actions but from our parents' disapproval. Our parents were often experienced as critical and disapproving even when they were good, caring parents. Still, by learning the "don'ts," we learned how to go through life without being injured.

Becoming Conscious of the Unconscious Parent

Although the unconscious Parent may be a source of confusion and emotional pain, most people are not aware of its presence or of the influence it exerts on their lives, even when the results of its influence are experienced consciously. We may hear ourselves being critical of others or may experience ourselves saying and doing things our parents said and did. However, rather than seeing the source of our critical or self-limiting comments as originating within ourselves, we (our Parent)—often blame others for them—"I am criticizing you because you need it and deserve it"—when the real conflict and need are internal: "I am criticizing you because I want to be critical. I want to treat you in the same way I was treated." Often, the people being criticized are merely serving as surrogates for our unconscious Child; we can now treat them as our parents treated us.

Examining and Confronting the Parent

It is sometimes very difficult to examine the Parent, not only because it is largely unconscious, but because the memories, inconsistencies, fears, and threats that it consists of are confusing and unpleasant, and it is difficult to face and analyze them. Most of us find it easier to avoid such confrontations, much less to analyze them in therapy. This is unfortunate because the Parent can exert tremendous disruptive influences on our lives and can make us and others quite miserable.

To confront the Parent is also to question its authority, and this is an act of defiance which triggers the wrath of the Parent. If the Parent then attacks or becomes critical this activates the unconscious Child, which will feel bad. When this occurs, the Parent may begin to scold, berate, threaten, criticize, or abuse the Child, which may then cause some miserable aspect of childhood to be reexperienced. As a consequence, the person feels considerable emotional turmoil and confusion. Hence, most people consciously avoid attempts to scrutinize the Parent (and the Child).

Parent–Child Unconscious Interactions

The creation of the internal Parent occurs at the same time as the creation of the internal Child, and they are linked. Together, they help to constitute the unconscious ego and self-image. However, as noted, these personality features probably should not be described as a "superego," for rather than serving up an ideal, they often exert negative influences on the self-image and the ability to interact with others in a psychologically healthy manner. In this regard, these unconscious personality features could be called the "Lesser Ego," particularly in that they are often expressed together. However, the Parent does not only express negative qualities (at least in most cases) since it can have its loving, nurturing side.

When the Parent is activated, the Child is often triggered as well. The Parent criticizes and the Child is criticized. Or the Parent praises and affirms, and the Child feels secure and loved. This is how the relationship of these two ego personalities was first formed, and this pattern merely repeats itself.

How much of the positive or the negative predominates depends on the entirety of our early and subsequent experiences. If we habitually experienced the negative, the unconscious Parent may criticize the unconscious Child for no other reason than to just do it. It is a habit. The Parent sometimes criticizes the Child because we are about to do something it does not approve of. For example, if your parents frequently scolded you for being "no good," "worthless," and "a failure," and you act otherwise by becoming happy, successful, or involved in a promising healthy relationship, the Parent will do its utmost to sabotage you so as to maintain the familiar. It will try to anticipate and shape the future based on past expectations and bad needs. What is familiar is comfortable or at least tolerable. What is unfamiliar is uncomfortable and may be avoided. If what is familiar is to be hurt, used, rejected, neglected, and to feel badly or worthless, experiences which do not conform to these well-ingrained expectations will be rejected as well. When success is suddenly within grasp, the unconscious Parent and the Child as well may insist, "No, you don't deserve it. You will fail. Stop. Go back. Don't do it. This is not familiar," and you will begin to feel depressed, irritable, unhappy, and blue, for "no particular reason," when consciously you know you should be happy.

If you are subject to unhappy feelings that seem inappropriate or incongruent with how you consciously know you should be feeling, it may be useful to ask, "Why is my Parent yelling at my Child?" "What are my Child and Parent trying to sabotage?"

Derick didn't look like the stereotypical physician. He had piercing dark eyes, was over six feet tall and thickly muscled, and had a menacing air. He came in for counseling because he had recently been threatened with the loss of his hospital privileges after he got into a shouting match with the head of his department, whom he had threatened with physical violence.

This was not the first time he had been in trouble. By his own admission, he had been arrested twice for drunk driving and had been threatened with expulsion while in medical school for drinking and once for behaving in a threatening and hostile manner with a fellow student, whom he had thrown against a wall. In fact, although he denied being an alcoholic, Derick had an obvious drinking problem.

Paradoxically, Derick was also considered brilliant, and after completing medical school and his residency, he had been a candidate for a teaching position at one of the better known American universities. The offer was withdrawn, however, after he got drunk at a faculty reception, where he had gone to meet with members of the department following a more formal get-together earlier in the day. He insulted one of the female instructors by calling her a lesbian:

> She was really coming on to me. But I was not at all responding. So she got irritated and began laying all this feminist propaganda on me, as if *I* had been coming on to *her*. What was I supposed to do, just stand there and be insulted? So I said, "Honey, I'm not interested in you. I thought you were a lesbian."

Derick had a history of destroying successful possibilities. He had been drinking heavily since high school and had been raised to believe he would be a failure. Both his parents were alcoholics. His mother was a physician, and his father was a highly successful attorney who was also extremely abusive:

> My earliest memory of my dad is when I was about two. I guess I was afraid of the toilet, as I would usually crap on the floor. Anyway, my father was probably extremely drunk at the time, but he must have decided that the way to cure me was to beat me and to rub my own shit in my mouth. It worked. In fact, it is one of my earliest memories.
>
> Usually, my father's solution to a problem was to drink until he was almost senseless. All my early memories are of him storming around the house like Frankenstein on the loose, drunk and angry, and slapping or kicking me for any reason. It wasn't just me he would yell at. My mother was

very scared of him, and she had her own drinking problem as well. It was my misfortune that, when she was mad, she usually took it out on me, too.

It seems my entire childhood was a blur of being punched, kicked, and called the most horrible of names. Everyday, it was "failure," "stupid queer," "good for nothing bastard," "retarded asshole," and on and on.

It was probably also my misfortune to be extremely brilliant. I never had a problem with school. The teachers were always stupid, and I could get an A simply by reviewing an assignment the night before the test. You would think that would make my parents happy, but it didn't. I learned quickly not to show them my report card. I remember one time, after my father was drunk and screaming about me being "a retarded queer," that I responded by proudly producing my report card. That must of been in the third grade, as it was the first time I had got letter grades. He took one look and then glared at me as if he was going to kill me. He then slapped me across the face, grabbed me by the back of the hair, and started crumpling the card into my face and yelling, "A is for asshole."

After that, I decided not to tell my parents much about anything I was doing, and I spent as much time away from home as possible. I also started drinking and hanging around with a pretty rough crowd, but even then, I still got all A's, and for two years in a row, I played on the varsity football squad. Even that got ruined, though.

One day I started drinking early in the morning and then showed up at school completely wasted. I had never done that before, at least not so anyone would notice. I guess I made a spectacle of myself and got into a fight with one of the teachers, who I shoved, and then I got in a screaming match with my football coach, who threw me off the team. Man, it has been like that ever since.

Rose was a beautiful, willowy brunette who had worked as a professional model for almost a year. When she came to see me, her career and love life were in shambles. After our first four sessions, she admitted that she was almost addicted to cocaine. As her use and dependence increased, she began showing up late or not at all and missed several assignments or came in so strung out that she was sent home. She had not worked in two years.

Like Derick, she came from a home where she had been abused. However, rather than being screamed at or hit, she had been repeatedly sexually molested by her adopted father and was later abandoned by her family. Her real father had been killed in an auto accident before she was born, and her mother had married his best friend before Rose was a year old, so to all intents and purposes he was her father:

> My father had a drinking problem, and my mother, I think, was popping pills. Despite his drinking and her pills, it seemed to me that everything was real happy around our house until I was about five. My father, who was a police officer, liked to wrestle and tickle me, and we were always playing, probably more than my mother liked because I can remember her always telling him to not play so rough.
>
> And then everything seemed to change. My mom had left to go shopping, and my dad and I were sitting on the couch. I remember him telling me he had a little kitty and asking if I wanted to see it. I did, and then he took out his penis. I had never seen one before, and it scared me. But then he told me to pet the kitty, and I did, and then he told me to kiss the kitty, and I did that, too. After he got me to lick on it he told me that the kitty was our little secret and never to tell Mommy. I remember feeling scared the way he said it, the way he looked at me and held onto my arm. For the next several months, every time my mother would leave the house he would put his penis in my mouth and make me suck on it until he came. It was awful because I was scared, and sometimes I would gag, which would make him laugh.
>
> I never said anything, though I knew somehow that what we were doing was bad, and I always felt afraid. Then, one day, he was sitting on the couch drinking and, like many times before, had me sit on my knees in front of him and suck. All at once, my mother is standing there yelling and screaming, and I can hear him yelling back, saying he was passed out drunk, and that I was a little whore and a slut. And I'm not even seven years old.

My whole life changed after that. My mother almost never talked to me, and my dad completely ignored me. They didn't let me sit at the table with them and made me stay in my room when they were watching TV. I felt so bad I didn't know what to do. Somehow, I felt like everything was my fault, and I tried to make up for it by cleaning and vacuuming and washing the dishes and setting the table without anyone asking. But nothing ever changed. It was almost as if I didn't exist.

Everything continued like that for almost the next six years, and then one evening, after I had come home from a girlfriend's house, where I practically lived, I went to take a bath. All at once my father walked in; he must have picked the lock or something, and I could tell he was drunk. He was smiling as if he was mad at the same time, and I just froze. He started accusing me of being on drugs and telling me he was going to have me arrested and put in jail. I was sitting naked in the tub, and I didn't know what to do. I was thirteen years old, and since he was a captain with the police department, I was really scared, so I started to cry, telling him I was not on drugs. Well, he said he knew I had drugs on me, and he was going to search me, and if he found any, he was going to take me to jail. He then started feeling between my legs, and then, after he had been doing this for a while, he told me he'd found some drugs and was going to arrest me or maybe just shoot me. I was so scared I started pleading with him, telling him I'd do anything he wanted. Well, what he wanted was the same thing he'd wanted seven years before, and now it started all over again, sometimes twice a day.

I don't know why I let him do that to me. I felt like the most awful whore. But then it got worse, and he started taking pictures of me naked. That's when I decided to tell, and that's when all hell broke loose. Nobody believed me. My mother kept saying I was crazy, and then she slapped me across the face and they kicked me out. I went to live with my real father's parents, who were really, really good people.

By then I was fifteen and had started high school. It's weird because everybody thought I was funny and the happiest person around. Nobody would have guessed that I hated myself and was always thinking about suicide. When I turned sixteen, my grandparents bought me a car. They were really wonderful to me. But by then, I had started drinking on the sly and would do really crazy things like drink too much and then speed like crazy, almost getting into wrecks.

In my junior year, I tried out for the cheerleader squad, mostly to please my grandmother, but also because some of the girls I knew kept telling me to. When I got picked, I was so excited I didn't know what to do. So I got really drunk and almost crashed my car. But then everything started going so wonderfully for me. Somehow, one of the most popular boys at school, and a great athlete and everything, became my boyfriend. Everyone seemed to like me. People kept telling me how pretty I was, even the teachers. But deep inside, I was still depressed. And then I did something really crazy. Some of the other cheerleaders and I went over to the local college to a fraternity party. I drank way too much, and the next thing I knew, I was in one of the bedrooms with all these guys and I was so drunk I just let them do what they wanted. It made me feel sick. I hated it. I just don't understand how I could have got so drunk and let them do that to me.

A couple days later, it was all over school, and I felt like such a total disgusting whore that I just couldn't face anybody. I dropped out, and because I was too young to do anything else, I went to beauty college. It was there that one of the teachers started telling me I should be a model. She made these arrangements with this photographer, and the next thing I knew, I start getting these little jobs and then bigger jobs, and pretty soon I was being flown to Los Angeles and San Diego, and I was making way over a thousand dollars a week. I was just nineteen years old. That's when I started getting into cocaine, which I really, really liked. Actually, it was my boyfriend at the time who got me started, and pretty soon we were freebasing and then I was so strung out that I was angry and depressed at the same time. I started screwing up, showing up late and sometimes not at all for my modeling jobs. All at once, I had no more assignments, no more money, and my boyfriend, who was living off me, couldn't support us either. That's when I really hit bottom. I was addicted to crack, and when I couldn't afford it, I began selling some of my clothes, and then I just started screwing guys for whatever. Fortunately, my grandparents came to my rescue.

You know, it's almost as if I wanted to screw up. It was as if I couldn't believe what was happening, and I knew it was going to end. It was almost as if I was thinking, "What the hell, I'll end it, since it's going to end anyway."

Abuse and Self-Destructiveness

There are different forms of abuse. A child may be constantly beaten, screamed at, belittled, sexually molested, neglected, ridiculed, rejected, or made to feel worthless. Regardless of how the conscious aspect of the personality comes to view itself, the unconscious self-image is fashioned in response to these parental messages and to the child's feelings about how she or he is being treated.

Sometimes, people respond to unconscious failure messages with reaction formations and compensations (see Chapter 14). They may become overachievers or Type A personalities and may do what they can to overcome their unconscious feelings of inferiority and inadequacy. In some cases, they merely develop a superiority complex to mask an inferiority complex. Others may be driven to succeed at all costs so as to prove the Parent and Child wrong. Some people are so gifted that they achieve despite themselves.

Individuals who had bad childhood experiences and still succeed in life may be plagued by unconscious feelings of doom and failure. Their conscious and unconscious self-images may be in conflict, and if the person actually succeeds, the internal Parent and Child may do whatever they can to sabotage this person's ability to enjoy or even to maintain her or his success. The Parent and Child will attempt to impose the familiar. If the familiar is being "no good," "a whore," "a failure," any form of success will feel unfamiliar and thus uncomfortable. Various tensions will arise, coupled with feelings of depression, and before the person knows what happened, he or she will have done something foolish, made a stupid comment, failed to show up for an appointment on time, insulted someone inadvertently, or got drunk, become involved in drugs, or formed a bad relationship, so that everything is destroyed or at least constantly put in peril. In conjunction with their drive to succeed, such people have an equally strong desire to fail, and to destroy everything, even themselves:

Norma's mother had been married and divorced twice and had had several affairs before Norma was born. Apparently, the mother had been so promiscuous that she had no idea who Norma's real father was.

Norma's childhood was a nightmare of neglect, abandonment, and rejection. Her mother did not like being a mother and was very unstable, as was her grandmother, who tried to smother Norma to death when the child was about two years of age. Her mother also suffered violent fits of rage and depression and was able to care for Norma only very haphazardly, often leaving her baby with friends or relatives for days and weeks at a time. Curiously, although she had a promiscuous past, Norma's mother was religiously superstitious and was consumed with fears regarding sins, some of which she claimed little Norma harbored in her soul.

When her mother became more mentally unstable and indifferent, she put Norma in a foster home so that she would be freed of the burden of having a child around. Then, after several months, she took her back. Over the next several years, this pattern was repeated, Norma being left at various foster homes or placed with various caretakers until, by the time she was fifteen, she had been placed in ten different foster homes, had spent two years in the Los Angeles Orphan's Home, and had lived several years with a guardian. The guardian's husband had sexually molested Norma, and a neighbor raped her and made her pregnant. Norma was allowed to bear the baby, and then it was taken away from her soon after birth. She never saw her son again.

When she was still only fifteen years old, her current guardian was planning to move to another state and did not want to take Norma with her. She presented the girl with the option of marrying the son of a neighbor, whom Norma knew only vaguely, or going back to the orphanage. Norma chose marriage. After a few years, she was divorced.

Over the next several years, she worked briefly as a call girl; got employment as a model (she was very beautiful); had numerous simultaneous affairs, usually with older men; became repeatedly pregnant; and although seemingly a vivacious young woman, was frequently overcome with insecurity and severe depression. Indeed, over the course of the next twenty-one years she underwent over twelve abortions, was in and out of psychiatric treatment, married and divorced several very prominent and not so prominent men, made repeated suicide attempts, and abused alcohol and drugs, but managed to forge a career. However, as her fame grew so did her depression and her frequency of suicide attempts. She finally committed suicide (or possibly was killed) at the age of thirty-six. They buried her under the name of Marilyn Monroe.

Seeking the Love You Do Not Deserve

Children need to feel that they are loved and protected by their parents. Parents who are rejecting, overly critical, frequently absent, mentally ill, drugged, or intoxicated, or who withhold love, are in effect punishing their children for having normal needs. The children of these parents are taught that they are not worthy of love, and even that their needs for love and support are abnormal. Because they are made to feel bad, they begin to suspect that even their desire for love is only another indication of badness: They want something they do not deserve.

Some children who are rejected and neglected become extremely needy, dependent, and/or rejecting, filled with self-doubt and even self-hate. These extremely needy individuals develop an unconscious attitude that seems to cry out, "My needs are so deep and so painful, I will do whatever is necessary to meet them. I will accept whatever abuse is dished out, as long as love is promised in return."

Others do not cry out in this way. They scream, yell, criticize, and threaten all those who represent the rejecting parent: "I hate you. I hate what you did to me. I hate what you might do to me again!" Consequently, they fight with, reject, and abuse those who offer the promise of love and those who in any way represent the rejecting parent: a teacher, an employer, a spouse, or even a position of responsibility. The relive again and again the painful familiarity of the past, and in relationship after relationship, they reexperience or provoke the same terrible emotions and deprivations, causing themselves to fail or to engage in some act that destroys them.

Creating Rejection and Self-Destruction

However, sometimes such people respond to their feelings of inadequacy by driving themselves to overcome any and all deficiencies so as to achieve the love they were denied as children, so that they can prove they are, in fact, worthwhile. They may become great athletes, movie stars, politicians, business leaders, or even high-priced prostitutes. They seek greatness, superiority, fame, or fortune or the acclaim, applause, admiration, or desire of others in order to overcome their sense of being unworthy, inferior, unwanted, inadequate, and unloved.

Fortunately, some people are able to overcome these early handicaps. Others may never convince themselves that they are worthwhile and continue to fear rejection or failure as they are unable to accept themselves. When success does not deliver the love they sought and they continue to feel depressed and insecure, they self-destruct as the internal Child or Parent takes over. Sometimes the Parent or Child only threatens destruction, so that the person feels overcome with depression or insecurity even as he or she reaches new heights.

Indeed, many people with horrible childhoods live seemingly successful lives and may even achieve fame and fortune. In some cases, however, their ability to enjoy their success is sabotaged, as their unconscious Parent may tell them they do not deserve the success and their Child also tells them it is undeserved and won't last. They may feel that the success is illusory and that the real "me" is the person who goes home sad and depressed and who takes drugs or drinks herself or himself into oblivion, or they are just miserable all the time. They may be plagued by feelings of insecurity, unhappiness, dissatisfaction, emptiness, and depression; feelings that seem to grow worse as their life becomes outwardly more successful. Regrettably, even when offered love and true acclaim, they may remain miserable because the unloved unconscious Child within them can never get over its terrible hurt and feelings of being unloved and unwanted.

Neglect

Children who have parents who are always too busy and thus tend to neglect and ignore them are in an environment that is just as abusive as those homes where the children are screamed at and beaten. Worse, because the critical social, emotional, and physical interactions and intimacy are inadequate, the child becomes impaired. Many children raised in homes where they are neglected become extremely needy, yet they have difficulty feeling close to others, giving or receiving love, and feeling worthwhile.

The Busy Parent

Some parents are neglectful without being aware of it. They may be preoccupied with earning enough to maintain a beautiful home for their children, or they may be preoccupied with social responsibilities, such as PTA and the Girl Scouts, which they feel make them good and caring parents. Some of these parental behaviors are unconsciously designed to ward off closeness, which makes the parent uncomfortable. Being at PTA may be a convenient excuse for not being home hugging and loving their children.

This behavior may be confusing to children. At an unconscious level, they may feel neglected; simultaneously, they consciously believe or have been convinced that their parents must be good because they participate in these activities. That is, the left and right brains receive completely different messages. Very young children, however, are presented with a void because they cannot consciously appreciate the "good" intentions of their parents: Mom and Dad are busy or not home, and this does not feel good. Not surprisingly they may feel not only hurt and rejected, but very angry and vengeful.

Children who express these feelings are sometimes told they are being selfish or that their feelings are totally unjustified. Hence, they feel not only neglected and bad, but guilty as well. This does not stop them from expressing their neediness and anger later in life.

Neglect and feelings of rejection may also occur when the parents are home but busy with other things. They may communicate to their children, both verbally and nonverbally, that they don't want the child's presence and need to be left alone. Even if the rejection is not intentional and there is no conscious desire to make the child feel unwanted, rejection may be implied by a look, a sigh, a grimace, a head or body movement, and a tone of voice—all of which are attended to by the right half of the brain. Thus, the child feels bad, guilty, and needy.

Even if the parent is truly, justifiably very busy, the children may infer that they are unimportant and that what they say and feel is of no consequence. These feelings are amplified further when parents fail to interact with their children, fail to answer their questions, ask them to be quiet, or constantly interrupt them when they are trying to share something.

Although unconsciously feeling bad, or "not OK," the children of such parents may not consciously understand that they were neglected or what they are feeling bad about. It is the limbic system and the right half of the brain that suffers from a lack of contact comfort. This lack may give rise to considerable feelings of neediness and unhappiness, the reasons for which are not clear to the left brain and the conscious mind. It is very difficult for children (or even adults) to consciously recognize the absence of something that they have never been exposed to. However, as there is a physical and biological need for love and affection, this absence is fully felt even when it is not understood.

However, as they unconsciously feel deprived of the love that all children need and require, this sometimes merely motivates them to try all the harder to get that elusive love and attention. Nonetheless, if their needs continue to be neglected, they may become extremely needy and hungry for love and affection throughout their life and may in fact inappropriately demand love from or offer it to anyone who shows them even the barest amounts of affection or attention.

The Unconscious Parent and Child as a Single Hurtful Unity

The boundaries between the unconscious Parent and the unconscious Child are somewhat indistinct, as both are formed during the same period and in response to one another. They may have similar desires and similar goals and may be expressed simultaneously. The internal Parent and Child may act in concert either to assist the person in achieving success or to overwhelm the competing desires of the conscious self; for example, by sabotaging success or healthy relationships or by engaging in self-destructive behaviors.

Repetition Compulsions and Reactivating Hurtful Legacies

The unconscious Parent and Child continue to influence the individual long after he or she has reached adulthood. Often the unconscious Parent and Child act as a force that drives the person to engage in certain behaviors or to seek out certain individuals or situations that allow the reenactment of familiar experiences, good or bad. Both the right and the left brains sometimes feel compelled to provide the same familiar solutions to the same problems, with the same consequences. Although familiar, and therefore comfortable, or at least tolerable, the consequences of these unconscious strivings may be unpleasant, or supportive and encouraging:

Christy's mother and father fought bitterly before and after she was born. Throughout her young years, she witnessed frequent name calling, yelling, screaming, crying, and threats of physical violence induced by drinking and suspected adulteries. Although her parents never hit one another, she was frequently terrified, and as she grew older, she constantly feared they would divorce. A week did not go by when she did not feel the threat of impending doom, a constant fear that the two people on whom she was most dependent, who made up her world, were about to destroy one another. Moreover, as they were so busy fighting and being unhappy, they seldom expressed any love or tenderness for her.

Because Christy was also incorporating the turmoil surrounding her, she also felt considerable guilt and inadequacy. Because she felt bad, she felt that she had done something bad and that this was why her parents fought and were seldom nice or loving to her. She wanted her parents to get along and tried in her own childlike way to make this happen. She became overly good by helping without being asked, and by doing whatever she could to make everyone else feel OK, to "fix" her parents. When this didn't happen and the battles continued, within her emotional right brain she continued to feel threatened and felt it was in some way her fault; hence, her guilt and inadequacy: "I am not OK." She expended enormous energy trying to do better, to be good, to make things work, which, of course, had no effect. Unfortunately, the continued fighting and her father's frequent absences and drinking simply confirmed that she was just "not good enough."

In Christy's right brain, her parents' admonishments and punishments were stored, along with their patterns of interacting, their neglect of her needs, the unloving manner in which she was treated, and her feelings and reactions to what she was experiencing. Because she was a child when she had these experiences, the reasons for her parents' behavior (his alcoholism and philandering) were not recorded. Nor were they understood.

The unconscious Parent is comprised of emotional experiences, not of interpretations or explanations of that experience. Christy's internalized Parent was thus a source of anger, rage, rejection, interpersonal hostility, and anxiety and served as a model of how couples and families *should* interact as well as how she should be treated by the people she loved, that is, badly.

When Christy turned eighteen she married a man several years her senior to escape her parents and to find the love she had always wanted. Although seeking security, she married a man who was frequently unemployed because, so he had her believe, he had not yet found a job equal to his talents. Eventually, she obtained steady employment as a secretary in a law firm and began making a good salary, so that she was able to tide them over when he had no income. She was being very good. She was still trying to fix things. She accepted this role because her unconscious mind knew that she had to go the extra mile to get love. How else would anyone love her or put up with this bad, unlovable little girl?

Because he was ostensibly looking for a good job, her husband was often gone most of the afternoon, and to her chagrin, he would often stay out late at night with his friends, "relaxing" and having "a drink or two." When he came home drunk several times, she became increasingly upset, and they began to argue frequently.

When his drinking and unemployment had continued for a particularly long time, she confronted him. A terrible row ensued. He told her the reason he was staying out was that she was "such a bitch," "always complaining about money," "always tearing him down," "always selfishly thinking of her own needs. What about his needs?" In other words, it was her fault, and if she didn't lighten up, he would get a divorce. He took on the role of the threatening, rejecting Parent.

Christy immediately felt terribly guilty and frightened of losing him. Her unconscious Child felt responsible, because he had taken on the role of her unconscious Parent and activated her own unconscious Parent as well—the same Parent that had always neglected her and made her feel guilty and inadequate. Unconsciously, she believed not only that she deserved what was happening, but that it was familiar and thus tolerable. She resolved to be more understanding and to be a better wife, that is, to be good and to "fix" things, this time for good.

When his behavior deteriorated and he failed to come home at all on several occasions, she again became extremely upset. One night, a woman called and asked for him and then laughed and hung up when Christy asked who she was. She became terribly angry and again confronted him.

Sticking to his previous mode of interacting (which had proved successful), he again blamed Christy for his behavior. He also taunted her with the possibility that he was indeed seeing another woman. He told her that, if that were the case, it was because she was such a lousy wife and lousy lover.

Christy again backed down, primarily because his accusations rang familiar bells. However, the bells were from the past, and what was familiar was only the fact that she had had similar experiences and feelings when a child. Besides, husbands and wives were not supposed to get along; misery and turmoil was "normal." Husbands were supposed to drink and fool around. This was the way she had been brought up.

Rather than realize that she had married a man like her father (who had also been an alcoholic and who had engaged in extramarital affairs) and that this man was making her feel the same feelings she had had as a child, Christy instead reactivated her Child ego personality, accepted the blame, and again tried to be good so that this man would love her. She was condemned to repeat her early experiences again and again, so as not only to feel and be treated badly, but to someday fix what had originally gone wrong so that she could get the love that she had always been denied.

Why did she put up with this? In effect, she was being given, or so she unconsciously believed, a second chance to win the love she had never been able to attain. Besides, she knew at an unconscious level that she did not deserve to be treated with love or respect. This was familiar and normal and thus tolerable. That is, it was tolerable to her unconscious self-image. If her conscious and unconscious self-images had responded similarly to this mistreatment and had rejected it as unfamiliar and undeserved, she would have walked away from the abuse. Nevertheless, although consciously undesirable, this treatment rang familiar unconscious bells.

What Christy had observed during childhood had registered as normal. Fighting, rejection, drinking, and having affairs were her only form of reality. She had not had other models for comparison.

These early experiences form the core of our being and the foundations on which we build our lives and our identity. We cannot destroy this foundation and survive. Consequently, it is replayed throughout our lives. Fortunately, this internal tape recording can be modified and may even be replaced with a new tape. Although we cannot erase the recording, we can dampen its amplitude and change the meaning that one applies, cease to take personally and feel responsible for treatment that more than likely had absolutely nothing to do with one, as any child born in another child's place would have been treated exactly the same, and thus fight it appropriately. It can be controlled and through the application of the above, a new tape may be inserted in its place.

Failure to Become Conscious of the Parent and Child Ego Personalities

Although the Parent and Child are often a source of confusion and emotional pain, most of us are not aware of their presence or the unconscious influence they exert on our lives. This is true even when we consciously experience the consequences of their influence.

When questioned, many people will erroneously state that their childhood was good and that they had good parents, even if their parents beat them or abused them verbally. It is difficult for many adults to acknowledge that their parents treated them badly because such an acknowledgement not only seems disloyal but is a reflection on their own self-worth. This is why when we were children it bothered us so much for someone to say that "My dad is better/stronger/smarter than your dad," for what this implies is that your dad is worthless, and you are worthless because he is your dad.

It is even more difficult to confront these unconscious elements because many people are frightened by the intensity of their feelings toward their parents. They are also afraid to allow hurt and resentments that have been bottled up for years to emerge, for these emotions may be extremely destructive and violent.

Recognizing When Mom or Dad Are "Not OK"

Although some children may at an early age begin to sense that perhaps their parents are also "not OK," this is frequently a function of resentment and frustration due to irritation at being punished and thwarted in one's desires. In this regard the parent is viewed as an obstacle. In fact, parents may be viewed as an obstacle even when they spoil their children, for once they refuse to indulge the child's whims, the child is forced to confront an uncommon source of restraint and they may resent it, and later in life even feel unconsciously cheated for not being completely catered to.

Some unfortunates who have had truly bad parents not only realize as they grow older, that the parents were "not OK" but begin to hate them. This occurrence does not extinguish the internal Parent. Rather the parents and all associated authority figures (e.g., teachers, employers, the police, and one's spouse) come to be hated and resented, and the person must confront each substitute parent in turn, creating constant turmoil. However, the battles between the unconscious Child and Parent may rage on even in the minds of those with good parents. The battles only become horrible wars if the childhood experiences were horrible or in some way terribly limiting and depreciatory.

However, for those who had truly "not OK" parents, the struggle goes on consciously as well as unconsciously and they may feel forever compelled to destroy the "evil parents" wherever they reside. The battle never ends because the war is never won.

When we attain adulthood and are able to analyze and confront how our parents raised us, our recognition may have little effect on the unconscious Parent. The reason is that the unconscious Parent is not synonymous with our actual parents, and our unconscious perceptions do not always correspond with what we experienced consciously. The unconscious Parent is an emotional composite that includes unconscious feelings, perceptions, fears, and even misperceptions and may be based on a variety of authority figures, such as an older brother or sister, our grandparents, our teachers, other children, and even the police.

Nevertheless, although the internal Parent cannot be completely erased, those who confront its existence and influences can recognize that it is wrong if it tries to treat them badly. They can assert that they are "good," "worthwhile" individuals who have limitations and weaknesses but who nevertheless deserve happiness and success. Those who can confront and begin to control the influence of the internal Parent, even if they have difficulty accepting or feeling good about themselves, will increasingly discover self-acceptance until finally, one day, with conviction, they can proclaim, "I am OK."

11

Unconscious Parent and Child Repetition Compulsions

Seeking to Maintain the Familiarity of the Past

Many people blame failure, lost jobs, lost opportunities, and arguments on bad luck, bad timing, bad people, fate, destiny, and sometimes even on themselves (i.e., bad judgment). However, in many cases there is a repetitive familiarity to it all. There seems to be a recurrence of situations in which we lose control over our emotions, make people angry at us, become upset only to regret it later, or destroy some promising situation or relationship.

If we were to examine the circumstances surrounding these occurrences, including our feelings and the events that seemed to provoke them, it sometimes becomes clear that a pattern has been in motion for a number of years. The same bad things seem to happen again and again.

Seeking the Familiar

I have had patients confide that this is the third marriage or relationship in which the husband or boyfriend has turned out to be physically and emotionally abusive, the fourth marriage to a man who is an alcoholic or had a drug addiction, the third marriage to or relationship with a woman who cheats on them, the second marriage to a woman who has turned out to be mentally deranged (e.g., schizophrenic).

In one case, a woman patient complained that she was divorcing her husband because she was shocked to learn he was homosexual. She then confided that the last two men she had dated were also homosexual or bisexual, and that, like the situation with her present husband, she had not realized their predisposition until several months into the relationship.

Tanya complained that her current boyfriend, who was unemployed, at first borrowed money from her incessantly and then started trying to control all her finances, telling her she was spending too much. He was also doling out an allowance to her: "Money that was mine to begin with!" She then admitted that her last boyfriend, who was also unemployed "because he was in trade school," had also borrowed money from her.

In addition, her current boyfriend often tried to control her by telling her whom she could see and for how long. He would take her car keys away and not let her drive her own car—even when he wasn't borrowing it. However, she said she could understand this behavior because "he was really badly hurt by his last girlfriend, who went out on him. He has trouble trusting women."

Moreover, she claimed that this one was an angel compared to her first real boyfriend, whom she had married: "That relationship was a mistake from the very start." He liked to do "speed" and would go through violent mood swings. He sometimes beat her because he was "insanely jealous," or if she did anything to displease him.

Jackie confessed that she had maintained a long-term relationship with a man who was impotent. When they broke up, she began an intense relationship with another fellow and within weeks moved into his apartment. She then discovered he was a cross dresser and preferred sex when dressed in women's underpants.

After breaking up with him, Jackie thought she had finally found a good man, the vice president of her company, whose home base was in another city. She carried on a relationship with him for two years, although the sex was never satisfying to her, hoping for marriage, which he repeatedly promised but kept putting off for one reason or another. One week after they finally became engaged (after she had given him an ultimatum), Jackie heard at the office that he had been arrested for molesting a little boy.

When Jackie came to me for counseling, she wanted to know why she kept having such "bad luck." She was adamant in the belief that she had never had a clue to why she kept getting mixed up with "screwed-up" people.

She is not alone. In each instance, these and other people steadfastly claim that they hadn't realized that the spouse or boyfriend was that kind of person when they met. It was only after several months that what was obvious to their friends suddenly became undeniably apparent to them. However, if that were true, how was it that they repeatedly ended up with similar kinds of people? Bad luck? Not likely.

What is more likely is that, when they met that "special" person, they unconsciously detected certain familiar cues that alerted them to the possibility that this person was a liar, cheater, beater, or alcoholic. Their unconscious radar led them to that person, whom they then found "attractive." They ended up chasing after a number of these people who displayed familiar characteristics that cued them to their abusive/alcoholic/philandering potential. They simultaneously rejected those who did not meet their particular criteria for abuse, hurt, or rejection.

What were the cues? And why did they keep forming relationships with people who were bad for them, who made them miserable?

The Familiarity of the Past

When asked why they were drawn to this or that particular person, some people respond that they felt an almost immediate feeling of closeness and a comfortable sense of familiarity or a feeling of being "needed," or, conversely, intense excitement and in the extreme, love at first sight: "It was easy to relate to him"; "I felt he really needed me"; "We were crazy about each other from the start"; "I felt as if I had known her for years—we were that close."

More often than not, what is familiar is the similarity to relationships they have maintained in the past with an alcoholic, a philanderer, an abuser (perhaps their mother, father, or first boyfriend). Upon meeting a new person, these people unconsciously recognize personality and behavioral traits that tip them off to that person's potential for destructiveness. The forces that make these choices and exert these influences reside in the unconscious Child and/or Parent ego personality, and in the right half of the brain, which has a keenly developed ability to read body language and related nuances.

Meeting the Expectations of the Unconscious Parent and Child

All too often, even in good relationships, we seek partners who meet the expectations of our unconscious Child or Parent. However, depending on how in touch we are with our feelings and our past, and on how much we successfully rely on our right-brain capabilities, we may also be led to the best person, someone who will be supportive and encouraging. How much these forces actually influence choice and behavior is determined by personal experience and a variety of factors.

Often we may be attracted to individuals who remind us of our mother or father or even our own unconscious Child, providing us with the opportunity to re-create painfully familiar scenarios from childhood. If we know who we are, what we truly enjoy, and what our true and growth-promoting desires may be (e.g., hiking, dancing, reading, music, dogs, physical fitness, freedom, movies, and personal exploration and growth), and can recognize the unpleasant familiar when encountered, then we can utilize the right (as well as the left) brain and these same emotional representations, to meet and bond with someone who possesses the same growth-promoting potential.

Of course, many people seek mates who may seem completely different from their parents. They do everything they can to avoid reexperiencing what was so painful. Nevertheless, the unconscious Child and Parent may exert their influence, and the conscious aspect of the personality may do all it can to oppose these forces.

In their attempts to escape parental influences, some people seek mates who offer qualities that were missing in their early lives. However, rather than being directed by their unconscious Parent, they are being influenced by their unconscious Child as much as by their conscious mind, which says, "Enough! I've had it. I won't take it anymore!" They are repelled by those who remind them of their parents. Of course, the factors involved in the formation of relationships are more complex than this, as we are not completely at the mercy of the unconscious. Nevertheless, some people are attracted to individuals who represent their own unconscious Child or Parent, or who can activate these psychic forces.

Seeking a Parent or the Unconscious Child in a Partner

Some people seek a mate who is like their father or mother in some way in order to strike back at him or her. Or, they seek someone who corresponds to their own unconscious Child and treat him or her accordingly; their unconscious Parent will now mistreat this new victim. That is, they will find someone who will tolerate being treated as they were treated when they were a child. In these cases they are not seeking a parent substitute but someone who reminds them of themselves when they were children. This enables them to escape their own sense of inadequacy by employing a substitute. This person takes the place of their own unconscious Child and is treated accordingly. Of course, this requires that their mate possesses some flaw or other deficiency (either real or imagined) which can then be attacked.

When Billy reached the age of two, his mother went back to work; he felt very upset and rejected. He was also the youngest and always quite small for his age. Whereas his brothers were athletic, he was quite clumsy, and they often teased him about his clumsiness and his small stature. Moreover, his brothers purposely left him out of all their activities. His father (who drank much too much) sometimes joined in the teasing and sometimes laughingly called Billy disparaging names. However, the father also often tried to help him overcome his lack of coordination by playing catch with him when he was small, but still sometimes laughing at his son when he would trip or fall down, or if the ball bounced off his head. Unfortunately, he also became extremely critical and yelled at Billy when he kept making the same mistakes.

Billy did not appreciate his father's frequent "help" because he wanted to be "big" and to do everything on his own. Moreover, he (i.e., his right brain) could tell from his father's constant "helpful" criticism that he was embarrassed by his clumsy son. Hearing his father's helpful advice and the occasional yelling only made Billy feel worse, and his mother's absence compounded the problem. Before he was even five, he felt terribly inadequate, stupid, and clumsy, and as if he were an embarrassing failure whom everyone made fun of or disliked.

By high school, however, Billy had had a growth spurt, and although he never became a true athlete, his performance improved markedly. He also became extremely handsome. He tried out for baseball and managed to make the team, his first real accomplishment in life, and one he took great pride in. However, before the first game, he dumped his motorcycle and broke his right arm in two places. This accident came as a terrible blow and shattered a lot more than his arm, such as the little self-confidence he had achieved. Nevertheless, being so very handsome and sexy, he made the most of his looks by "screwing his brains out."

After graduation, he got a job as a welder's assistant and soon after met and married Charlene. Together, they made a very stunning couple, but although Charlene was stunning to look at and very sexy, he did not stop his "screwing around."

Charlene never met her real father, as he had abandoned the family soon after she was born. Her mother did not want to keep her baby but was pressured to do so by her own mother and her oldest stepsister, who were very intolerant and strictly religious. Charlene's mother resented her because of the restrictions having a baby placed on her and because it reminded her of her former husband. Having to work to support herself and the baby, and then having a new husband who resented the child, this mother began leaving Charlene with the stepsister and sometimes her grandmother during the week and picked her up only on the weekends. Sometimes, she left Charlene at the grandmother's for weeks at a time.

Charlene told me that her earliest memories were of sitting in her aunt's house and not being allowed to do anything and being yelled at if she made the slightest noise or even the smallest mess. Her grandmother was the same way. If she took some toys out to play with, her grandmother was there in an instant, picking things up and putting them away, commenting on what a messy child she was:

> The worst part was when my mother would pick me up on Friday evening. Every Friday, it was the same thing: my grandmother or aunt complaining about the "messy," "dirty," "disobedient" child and then my mother glaring at me. When we left the house and walked to the car, my mom would start cursing me and shoving me, and sometimes, she

would kick me over and over again on my butt as we walked to the car.

This pattern pretty much repeated itself until she was seven years old, when her grandmother died. She then began living with her mother and her stepfather full time:

It was like being transferred from one prison to another. No matter what I did, my mother would be right behind me telling me what a dirty, spoiled little bitch I was, and then she would take my things and put them away and tell me to go to my room. I mean, she wouldn't even let me play. And she would never let me do anything. She would even tell me I was too stupid to clean up my own room, and she would clean it. She would make me stand there while she even made my bed, all the time cursing. It was almost as if she wanted me to be messy so that she could be mad. She was always mad and always yelling and always telling me to do this or that. Then, she would immediately interfere and say, "Never mind. I'll do it," and then start cursing under her breath and glaring at me. When she got married for the third time, everything just got worse. Now I wasn't just stupid, messy, and an embarrassment, I had to be invisible.

When Charlene met Billy she was willing to do just about anything to escape her home, so when he asked her to marry him, she was terribly flattered and immediately accepted. However, although they were a beautiful couple to look at, their marriage began to fall apart almost immediately:

Billy: She doesn't give a damn about anything, and she doesn't even bother trying to do things right. We get married and right away she says she can't cook. No matter, I say, I'll do the cooking; you do the dishes. But then she leaves the dishes all day and when she finally cleans them, they are still dirty. And then the house is always a mess. I come home from work and the place is filthy. And what is she doing? Sitting on her lazy butt watching TV. Everything is half-assed. It makes me embarrassed to bring anybody around. And now she's packing on weight. We've only been married four months, and already she's waddling around like a pig. She doesn't want to work, she doesn't want to go to school, she doesn't want to improve herself, she doesn't want to grow, she doesn't want to do anything.

Charlene: You knew I couldn't cook before we got married, and you're exaggerating about the house. Sure, I'm not the neatest person in the world, but you're just using that for an excuse to pick on me. No matter what I do, it's not right, it's not good enough, and he is always telling me how great he is, and how much better and smarter than me he is, and how no-good I am at anything. No matter what I say, it's "stupid," or he laughs at me and calls me "ignorant" and a "dumb cunt." I can't even have an opinion. And when I say something about getting a job or going to school, he has a fit and starts in on how I'm too stupid to go to school.

Billy: No. Not too stupid. Too lazy. I brought you home an application, and all it did was sit there and finally got thrown away or lost or something. You don't want to grow, that's all there is to it, and no matter how much I try to help, you just give up or won't even try. You act like a baby. Why don't you grow up?

Charlene (starting to weep): Why should I try? All you want to do is pick on me, and then you don't even come home until after midnight. You say you don't come home because it's dirty, but that's not the real reason. You're out screwing around. I know it. You don't even care about me. You like teasing me, putting me down, and criticizing me. If you loved me, you'd try to help me.

Belinda was the last of four children, all of whom were in their late teens or twenties when she was born. Her parents were in their late forties, and both worked. She had been unplanned for and was seen as a burden by her parents. Although they never beat or chastised her, they tended to ignore her and took little interest in her accomplishments. Her parents also tended to argue, and on more than one occasion, she could recall her mother's crying and her father standing over her yelling or ordering her around. Belinda sometimes cried when this happened or would withdraw to her room and get into bed.

Belinda responded to this childhood experience not by becoming independent, but by developing a sense of helplessness and dependency, through which she attempted to elicit her parents' reactions and love. Her parents did nothing to alter this merging self-concept of being helpless. In the morning, her mother would quickly dress her, make her lunch, and send her to school. She would then quickly make the child's bed and quickly clean up the room before she went to work. When Belinda tried to help, her mother would smile indulgently and tell her that she was in a hurry and it was faster if she did it herself. Her father simply ignored Belinda and took no interest in her whatever.

Belinda grew up to be an exceptionally intelligent and attractive young woman and scored so highly on the SATs and other college entrance exams that she was offered several scholarships. Despite her mother's misgivings, she went away to college. Five years later, she managed to graduate with a C average. That she graduated at all was a wonder, as she spent most of her time procrastinating and lying in her dorm room daydreaming and depressed. Three years into college, she became so distraught that she sought out counseling. Her major complaints were that she felt paralyzed and indecisive and just couldn't seem to get started on any of her tasks, all of which seemed to overwhelm her. Instead, she would withdraw to her bed and lie there depressed.

However, near the end of her fourth year, she met a graduate student, Brent, who had an intense need to control and who was extremely attracted to her as well. Essentially, he took over her life. He would show up in the morning and make her get up. Later, in the evening, he would grill and chastise her if she had not done her homework. Most of all, he would continually criticize and belittle her and often made her cry. Sometimes, he would even spank her if she had not done what he had instructed. Although she often felt humiliated, particularly by his spankings (after which he usually made her have intercourse with him), she felt she loved him and needed him.

Essentially, this graduate student became her surrogate Parent and exercised the control that was a familiar part of her upbringing. Although her own parents had never belittled, criticized, or even spanked her, in this way he also served the needs of her unconscious Child, which felt worthless, unlovable, and bad. He was able to verify her feelings of being bad or worthless by spanking her, calling her names, and controlling her.

On graduating, Belinda agreed to marry Brent. Unfortunately, her unconscious Child continued to rule even after they had two children. Brent would come home from school (where he was now an assistant professor) and with shock and dismay would daily criticize her for the messy condition of the house. "Vandals," he would cry. "It looks like vandals ransacked this house." At which point she would then begin to clean in a haphazard manner with him standing over her giving directions, threatening to spank her, and sometimes making her cry.

When Opposites Attract

A person's unconscious Child may search out a man or woman who will take over the role of the unconscious Parent. This is what Charlene did. She found in Billy another critical, rejecting Parent.

The converse also occurs. The unconscious Parent may be attracted to someone who is similar to his or her own unconscious Child and will then abuse or love him in accordance with these unconscious needs. This is what occurred with Billy. He took on the role of the critical and sometimes absent parent. Instead of criticizing or trying to "help" the "Child Billy," his unconscious Parent tried to "help" "the Child Charlene" and mistreated her in the way Bill felt mistreated.

At a conscious level, Billy and Charlene had little or nothing in common except their good looks. The same could be said of Belinda and Brent. However, at an unconscious level, they were a perfect match. How did they manage to find each other?

Making Unconscious Choices

Jennifer, who was raised by an abusive alcoholic father, may unconsciously detect (via her right brain) that a certain man at an office party shares certain mannerisms or other qualities that unconsciously remind her of her father or of an alcoholic or of someone who is potentially abusive. Because this new man has these qualities and has given off these cues and clues, he seems comfortable, familiar, and she is able to relate to him and thus feels attracted.

At a conscious level there is no knowledge of why she is attracted, only that she feels attracted. Jennifer then makes a number of conscious interpretations based on that feeling. Not consciously knowing the real reason she is drawn to this person, she misinterprets the sensations of tension and familiarity as something else: "I like this person. He is interesting and exciting. I feel comfortable interacting with him. I want to see him again."

Three months later, much to her dismay Jennifer discovers she is again in a relationship with a "total bastard" and an abusive alcoholic and wonders, "Why does this keep happening to me?"

If, on the other hand, the man she meets seems kind, sincere, stable, and supportive, her unconscious Child and Parent will find him unfamiliar, and she will feel uncomfortable and thus uninterested in him. Consequently, she will discard him, labeling him inappropriately as a chump or nerd.

An attractive young woman with a background almost identical to Jennifer's goes to a nightclub with her girlfriends. A good-looking man walks up and she immediately scans his nice attire, his comfortable movements, and his look of confidence. He asks her to dance, and by his tone of voice and choice of words, coupled with his style of dress, she is able to infer that he has a good job, is educated, and seems fairly stable. She wrinkles up her face and sarcastically replies, "What? With you?" As he walks away she turns to her girlfriends and laughs, "What a geek!"

A little while later, a rather physically attractive man walks up. He has not shaved, is wearing an old shirt opened halfway down his hairy chest, and is staring at her smugly. He asks her to dance, and she accepts. They dance several more times, and she gives him her phone number.

Later in the evening, her girlfriends ask her about this man, and she replies, "Oh, he is really a nice guy. He's so easy to talk to, and he's really interesting."

"What kind of work does he do?" asks one of her friends.

"Oh, I don't know. I think he said something about being between jobs."

A few weeks later, after he has borrowed money, has stood her up twice, has failed to call her several times as he has promised, and, after an argument has slapped her face, she begins to wonder, "How come I can't ever meet any nice guys?"

Indeed, many people who are rejected or abused or neglected in childhood are attracted to people who are losers and "takers," and who will use and then reject them, who will treat them with contempt and constantly belittle them. Their unconscious Child expects to be treated in this way, and the unconscious Parent tells them they deserve nothing better.

Of course, in many cases the individual may finally decide that he has had enough and will end or at least seriously consider ending the relationship. Often, however, one kind word is enough to keep that person tied to such an individual. Hope springs eternal. Or, they will cling to the relationship in the hope that their boy or girlfriend will somehow change.

Seeking a Second Chance and Reliving the Family Script

Why do some people stay with lovers who mistreat them? Usually, it is because the painful familiarity of the past is being re-created and they are continually attempting to obtain the love and affection that was not only denied, but that deep down they do not believe they deserve. Hence, they feel that the cruel partner gives them the opportunity finally to win the parental love they were denied; that is, if they keep trying (or so they unconsciously believe).

However, knowing (unconsciously) that they do not really deserve love, they are reluctant to end this relationship because they believe that they do not deserve and cannot get anything better. Their unconscious Parent tells them, "You deserve rejection and abuse." Their unconscious Child tells them, "I am not OK. I deserve rejection and abuse. I am lucky this person even puts up with me."

In addition, some individuals, as we've seen, seek the comfortable familiarity of the past in their relationships because, in part, they are trying to rewrite the family script, to achieve finally what they were always denied so that they may live happily ever after. If Daddy was an alcoholic who was never loving, she will find a man who is an alcoholic and make him love her.

The feelings and emotional reactions that characterized our childhood may linger forever. The struggles of childhood all too often become the struggles of adulthood in a disguised form. Fortunately, if we are sufficiently willing to engage in some painful self-scrutiny, these forces can be diminished and controlled.

Nevertheless, because these experiences during childhood serve as the foundation on which our adult lives are built, we sometimes seek to maintain our lives in the way that is *best* fit for that foundation, like searching for the right key to open a lock. "Best," however, is not necessarily "good."

Activating the Unconscious Parent in One's Spouse

In part, it is the unconscious Parent and Child that attempt to *impose* this road leading to the familiar, treating either us or our mate in a manner that we observed and experienced during childhood. The Parent and Child also re-create certain feelings that plagued us when we were young. The unconscious Child invites our partners (or even strangers) to treat us as we were treated by our parents, siblings, or other authority figures, seeking approval, conflict, rejection, or abuse. Often, in fact, the unconscious Child attempts to elicit the unconscious Parent who resides in the spouse by provoking him or her.

In general, although conflicts between partners often seem to occur at a completely conscious level, they may be occurring at an unconscious level as well. Indeed, in simmering and long-standing conflicts, the actual battles may be taking place completely within the unconscious of each partner. Although the partners may consciously search for the reasons that they are upset and may confabulate a variety of explanations to justify the conflict, its true origins may be completely unconscious and may not even have anything to do with the partner. When relationships become wars, the battles may be between each partner's Child (Child-Child), one spouse's Child and the other's Parent (Child-Parent), or between their Parents (Parent-Parent), all of which is fought upon the battlefield of the conscious and unconscious mind.

When one considers the number of factors involved, it is little wonder that many relationships are troubled. When two people interact they are interacting on a conscious level and, in addition, bring their Parent and Child into the relationship. Thus, the marriage takes place among six ego personalities: two Children, two Parents, and two Adults. Each ego personality may have its own agenda, causing disturbances intrapsychically as well as interpersonally.

Communication Breakdown

When interactions are taking place on an unconscious as well as a conscious level, one aspect of the mind may espouse one set of needs and desires, and another aspect of the mind may have needs that are in all respects completely opposite. The conscious self may be looking for love and acceptance, whereas the unconscious Child and Parent are looking for abuse or the opportunity to behave abusively. This opposition, in turn, not only creates difficulties in regard to internal homeostasis and balance but may disrupt communication between two individuals as well.

If Bob says, "I want a woman who can hold down a job and be responsible," whereas his unconscious Child fears rejection, then Gail, who holds a job and is independent, will be perceived as a potential

Figure 35. Within each human there are an unconscious Child and Parent, which keenly observe our interactions and the behavior of others. Often when we feel depressed or angry "without reason," it is because of turmoil occurring between the internal Child and the internal Parent. Frequently, arguments between couples are unconsciously mediated and serve the bad needs of the unconscious Child and Parent residing in both of them.

source of rejection. Bob's unconscious Child may need someone who is very dependent, whereas his unconscious Parent may need someone who is independent. Therefore, Gail and Bob will probably have countless arguments, all of which are based on unconscious conflicts and a failure in unconscious communication. However, this failure will probably not be apparent to either partner. She will not know what to believe when he states what he wants. Moreover, he will be unhappy no matter what she does.

Which aspect of her or his own mind should a potential spouse believe? This depends on her or his abilities to perceive, interpret, and respond accurately to unconscious right-brain cues. A person who relies solely on left-hemisphere modes of interacting, will fail to pick up cues or to understand what is being communicated by a partner's unconscious right-hemisphere Child or Parent. In fact, a person who remains in a left-brain mode may fail to attend to cues emanating from her or his own right brain as well.

For example, if Fred relies predominantly on his left brain (and his Parent and Child are largely submerged), whereas Tammy often interacts in the right-hemisphere mode of her Child, they will have difficulty interacting and communicating successfully. They will not understand one another and will often be in conflict.

Meeting Unconscious Needs

Parent–Child Communication

Communication breakdown also occurs when both partners interact in a predominantly right-hemisphere mode. One individual may communicate in the mode of the Child, and the other in the mode of the Parent.

For example, Tammy, having baked Fred's favorite dish, sits across from him expectantly waiting for acknowledgments and compliments. He takes a bite, grimaces slightly, and takes a sip of water:

Tammy (Child): You don't like it, do you? There's something wrong with it, right?

Fred (Parent): You know I like lots of salt. You didn't put any salt in it. What's the matter with you?

Although the above transaction could be interpreted in a number of ways, the result is a conflict between Parent and Child. In the case of Tammy and Fred, this conflict represents not only a communication breakdown but unconscious game playing. That is, for this couple, the above exchange is a ritual, and each partner has a role that has been dictated by unconscious needs and past experience. What is really being communicated is this:

Tammy (Child): You don't like me. There's something wrong with me. Why can't you love and appreciate me?

Fred (Parent): Because you're stupid, and this irritates me because you reflect poorly on me as a wife.

Another way in which the above could be translated is as follows:

Tammy (Child): I'm worthless and stupid, and I need you to tell me that, no matter how hard I try to please you, I'm still not good enough. For that reason I purposely (unconsciously) didn't put enough salt in the food because I knew it would provoke you.

Fred (Parent): I married you because of your poor self-concept and your need to be victimized and because you remind me of myself when I was a child, when my parents criticized me for my shortcomings. Now my parents can criticize you instead.

Although seemingly in conflict, we can now see that Tammy and Fred are in fact responding to one another's unconscious needs. Although unpleasant, the conflict satisfies these destructive needs. If the discussion were indeed about the food and there were no unconscious needs to be fulfilled, Tammy and Fred would interact as adults:

Tammy: How is everything?
Fred: Good. But it needs a little salt.
Tammy (reaching for the salt shaker): Here you go.
Fred: Thanks. I'll get it.

The Need to Be Hurt and Rejected

Does Tammy need to be victimized? The answer, unfortunately, is yes. Does she deserve to be victimized? Absolutely not! What one wants and what one deserves or needs are not always the same thing. Consciously, she hates what is happening. Unconsciously, she is merely attempting to maintain the familiar; once abused, always abused.

Just because people need and desire something does not mean that it is good for them. In Tammy's case, what we would like to believe is that she really needs love, affection, appreciation, and emotional support so that she can develop her self-image and improve her feelings of self-worth. Consciously, this is exactly what she is after, and this is what she truly deserves.

However, Tammy was always told by her older brothers and sisters that she was stupid and ugly. Her mother and father, who both worked, never had time for her because of their own busy schedules. Moreover, because her mother was so busy, rather than give Tammy the encouragement to do and accomplish things on her own, she would impatiently intervene with the admonishment: "You're not doing that right." In consequence, Tammy's child learned that she was helpless, inadequate, of little importance and not worthy of love and positive attention. This is how she was raised and what she was familiar with.

In marrying Fred, Tammy found a man who would re-create the familiarity of her own childhood so that her internalized Child and Parent would be comfortable. He, in effect took the place of her unconscious Parent and satisfied the bad needs of her unconscious Child. If she had married a man who was nice and loving, her Parent would have protested that she was not worthy, and her Child would have complained that she was not being treated in the way with which she was most familiar. Indeed, she would never have married such a man, because she would have avoided any man not corresponding to her own unconscious Parent.

Maintaining the Familiar

In many bad relationships, one partner tries to control or continually mistreats and criticizes the other partner for supposed shortcomings when, in fact, he or she is merely reliving and expressing some hurtful aspect of childhood. This aspect was part of the dynamics in the relationships of Brent and Belinda, Billy and Charlene, and Tammy and Fred. All were attempting to maintain the familiar of long ago, and all were attempting to fulfill bad needs.

These particular types of relationships require that one spouse remain forever inadequate. The "good," "helpful," "critical" spouse does not really want the mate to change. If the unconscious Parent did not have this victim to mistreat, it might turn the attack back on him or her. Consciously, both mates may want their relationship to change for the better. However, there are also unconscious forces at work that will do their utmost to maintain the familiar.

As I do with many couples, I saw Billy and Charlene together and separately. Billy voiced the same concerns and disappointment alone as he did when they were together. I discussed the possibility of Charlene's going to college and he seemed all for it. When I brought up the field of nursing, which she had expressed a keen interest in, Billy spontaneously commented on how much nurses make and how it might bother some guys that their wives made more than they did, but that it wouldn't bother him at all. I spent the rest of that session trying to nudge him to see what was really going on.

Over the next several sessions with Charlene, as we worked through some of the self-imposed restraints that were holding her back, she finally told me she had made up her mind and was going to enroll at a local community college and get her degree in nursing. She seemed extremely pleased with herself and was obviously quite happy and excited at the prospect of what lay ahead.

The very next session, I saw Billy and Charlene together. Immediately, the topic of Charlene's going to school came up:

Billy: Yeah. I really think it's great she is going to go to school. But, you know, we both kind of thought that maybe going straight into a full-time program might be too much for her right off the bat. I mean, she's got housework and stuff, and someday we'd like to have kids. Anyway, I thought maybe she ought to just take a course or two to begin with, something she's really interested in.
Charlene: Billy thought that, since I've been saying for years that I would like to learn how to draw, maybe I should take a drawing class.
Dr. J: Is that what you're going to do?
Charlene: Either that or a cooking class.

Billy then proceeded to thank me for the help I had provided and indicated that, as things were going so well, they had decided to terminate counseling. I made them both promise me that she would actually take one of the classes. Billy looked at me doubtfully.

In effect, although truly not conscious of his unconscious intentions, Billy nevertheless required that Charlene always remain inadequate and was threatened by the possibility that she might make something of herself and, worse, make more money than he did and thus make him the more inadequate member of their partnership. So, he simply seemingly sabotaged her chances for success by talking her into doing something irrelevant and maybe even getting her pregnant so that school would become at best a remote and far-off possibility. Unconsciously, he never wanted her to succeed, as he needed someone whom his unconscious Parent could criticize and belittle.

This characteristic is seen of many similar relationships. One partner has a vested interest in the other's "not changing," as change would be unfamiliar and intolerable. Fortunately, sometimes both partners can be persuaded to accept change within limits.

Charlene, too, obviously had a say in all this and in fact seemingly sabotaged herself. Essentially, her unconscious Child informed her that being a nurse was a role she could never play, and her unconscious Parent conspired with Billy to see to it that she would never be successful. Being too successful, even trying to take too much control over her own life, was very threatening to her. Being unfamiliar, it was easy to reject.

She never became a nurse, but fortunately, as Billy was not opposed, she took the cooking and several drawing and art classes, which, it turned out, she was quite good at. Think she is happier? You bet. Feeling more confident of her abilities and meeting many new and supportive friends in school, she began not only to feel better about herself, but to keep the house cleaner. Even Billy was pleased although, according to Charlene, he often made fun of her paintings. The praise at school, however, made his comments meaningless.

Of course, not all people involved in relationships such as these require that their spouse, boyfriend, or girlfriend remain inadequate. Many truly want their mate to change for the better, if only as in Charlene and Billy's case, within limits.

On the other hand, many relationships and marriages are entirely based on one spouse's being identified as the "sick," "helpless," and even "hopeless" half of the partnership, and the other spouse's taking on the role of provider and caretaker. Not infrequently, both will try to maintain this status at all costs.

"I Will Reject You before You Reject Me"

When disappointment is constantly experienced, when their desires are thwarted and their needs go unmet, when they are faced with rejection or when they are made to feel powerless by seemingly unreasonable authority figures, children sometimes react with extreme anger and resentment toward all those who are in a position to exert control over their environment.

One way of reacting is by rebelling. Sometimes children rebel by turning their backs on those who are important to them. In effect, they are rejecting something they really consider highly desirable, for example, a hug, or a toy advertised ad nauseam on television. Adults sometimes act in the same childish manner, as if they don't want what they in fact intensely desire, for example, forgiveness, love, a date with an attractive person, a high-paying job, or a promotion.

"I will reject you before you reject me," "I didn't want it anyway," and "Who needs it" become their mottoes in life. This attitude protects them from further hurt and rejection. If they pretend something means nothing to them, it loses its power to make them feel bad. Of course one must feel bad in the first place to go to these extreme of self-protection.

Indeed, it is natural for a child to mistake disappointment for catastrophe and to strike out in anger and rage and even to destroy what was desired simply because something is wrong or missing. Ultimately, these kinds of behaviors are self-destructive and self-negating. The internal Child "blows up over anything," and even trivial incidents lead to fights, arguments, and tantrums. This is sometimes referred to as "cutting off one's own nose to spite one's own face." However, a person who cuts off his own nose also feels that he has nothing to lose and that he deserves to be punished. In this case, the punishment is self-inflicted. Unfortunately, much of the pain generated by the unconscious Child and Parent is exactly that: inflicted on the entire self and the right and left half of the brain.

Looking for Rejection

More often than not it is the internalized Parent who exerts these self-negating influences. Unconsciously, the person feels genuinely undeserving of having his or her desires or needs met and is convinced that, given the chance, others will reject or thwart him or her, so as to administer punishment for some blemish in his or her personality. A person who feels this way not only expects rejection but is always on the lookout for it and is always prepared to deal with it, even if inappropriately. When this occurs, the internal Child is activated and disrupts his or her life through its unconscious influences.

Due to these unconsciously occurring turmoils a person may inadvertently insult or put off someone whom she or he would like to have as a friend or a lover and may thus alienate that person. Or in the early stages of a promising relationship, the person may fly off the handle at some real or imagined slight and destroy all future possibilities of closeness—only to deeply regret what he or she has done or, worse, to be completely in the dark about why another relationship has ended badly.

On a less significant level, such a person may blush or feel ill at ease when complimented, may feel unhappy or depressed about some event that he or she has been looking forward to, may distrust those who offer friendship, may feel embittered or deeply annoyed if someone makes light of a gift or a compliment, or may simply behave in a self-protective arrogant manner so that no one will ever be able to get close enough to hurt him or her.

By their very manner, such people may so alienate others that they inevitably become hurt in return. Nevertheless, others' alienation only reinforces the arrogant behavior that increasingly serves as a protective wall.

Essentially, these individuals are impelled by unconscious forces that tell them, "You do not deserve to be treated well"; "Things should not go well for you"; "No one can love you"; "You are no good." Of course, one who is "no good" has nothing to lose by behaving badly and a vicious circle is created.

Tommy was a very strong-willed little boy. He tended to rule over his mother but feared his very controlling, domineering father. His father, being in the military, was at home infrequently and very inconsistently for the first three years of Tommy's life. Because of his mother's indulgence, Tommy was at best semiobedient and frequently relied on himself. At an early age he began to feel a degree of self-mastery and a sense of independence.

When Tommy reached age four, his father was discharged. He was not pleased by his son's willful disobedience and began to exert control, ordering his son about and administering daily punishment in a militaristic fashion. Tommy did not easily acquiesce to this new pattern and became resistant. He was frequently beaten into submission when he refused his father's demands. For the next thirteen years, he and his father were in frequent conflict. For most of his young life he felt humiliated by and extremely inferior to his powerful father. His father's incessant bullying manner made Tommy feel considerable self-doubt and even a tinge of self-hate as he felt he had less and less control over his life.

As Tommy grew older, his father often seemed to make arbitrary, unreasonable demands or simply refused to allow him to engage in some activity "because I said so." The intensity of Tommy's anger and feelings of resentment increased dramatically. He had long concluded that this Parental figure was "not OK." However, he also felt very bad about himself not only for being treated this way but for taking it in the first place. In nursing his hatred, his major solace was the thought "Just wait until I'm big."

Once Tommy was an adult, his resentment never dissipated, and although he was able to go through college and land several good jobs, he was often in conflict with his bosses and supervisors. He was seen as having a chip on his shoulder and, in consequence, was frequently fired or quit. For him, every supervisor represented his unconscious Parent, and when someone told him, "You must" or "You will," his unconscious Child automatically responded, "I'll do what I want to. No one can tell me what to do. I'm big now." In a sense, his strong-willed Child, now having the conscious adult Self as an unknowing ally, not only fought with but was able to win battles against the hated internal Parent.

However, the Parent was not defeated. It began to focus on others as well. As a result, his unconscious Parent and Child almost became allies. The Child hated being pushed around, and the Parent began to act like a bully with others.

Because Tommy was so angry and resentful, he often made others feel nervous or even threatened when around him. Strangers, acquaintances, business associates, and secretaries seemed to treat him with disrespect or in a sullen irritated manner. "Everyone seems to have it in for me!" he complained. "And I don't even have to do anything. I just walk into a room and bang, I'm getting these bad vibes. Even the cashier at the Burger King gets an attitude when I walk up."

He was, in effect, "rubbing them the wrong way," not so much by what he said or what he did, but through his arrogant manner, the way he said and did things, including his tone of voice, facial expression, and gestures. Essentially, he was behaving nonverbally in a very threatening, intimidating manner. He communicated his intense anger (via the right brain and the limbic system) without having to say a thing.

When he walked into a Burger King and gave his order, he (unconsciously) half-expected that it would be goofed up (thwarted by his Parent). This expectation was also displayed (via the right brain) by his body language, his tone of voice, and the intensity and anger in his expression. Often, the clerks behind the counter would unconsciously respond to these cues; that is, their right brain would respond to the messages being sent by his right brain. In consequence, they would feel intimidated and anxious, and they would perform their duties haphazardly.

In other cases, he was perceived as looking down on and being critical of those who served him or who were merely doing their job. In this way, he would inadvertently activate the unconscious Child and/or Parent residing in these other people, and they would lash out at him. That is, by his actions, tone of voice, and movements, he would unconsciously remind the people trying to help him of their own critical Parent, who would respond with both conscious and unconscious resentment of Tommy's treatment of them. As a result, these people would often unconsciously (and thus unintentionally) retaliate by being difficult, making him wait, goofing up his order, ignoring him, or making some inappropriate comment.

Right-Brain Intuitiveness

The right brain is the repository of early feelings and emotions, including the impressions and feelings of the Parent and Child. It also controls vocal-emotional functioning and the perception and expression of body language, facial gestures, and other forms of nonverbal communication and comprehension. The right brain listens to how someone says something, not necessarily what she or he says.

Tommy often behaved in a confrontational way when entering a store or dealing with a service representative, although he did not consciously intend to be rude. Nevertheless, his manner, body language, facial expression, and tone of voice offered a challenge as well as a display of contempt.

Essentially, Tom saw his father in every person who had the power to thwart his desires or who exercised some form of authority or control over his environment, whether it was a cashier in a department store or the manager of his office. Being no longer small but "big," his Child ego personality could now display its anger, irritation, and resentment in all its childish fury with no fear of being beaten or bullied. All such people allowed him the opportunity finally to beat the Parent who had tried so hard to control his life and who had made him feel so inferior when he was small.

However, he was forced to go to war again and again because that battle was never won. It was continually being fought between the Parent and the Child inside him. Essentially, Tom was at war with himself, and all innocent bystanders were drawn into battle.

12

Repetition and Rejection
Dreaming, Self-Fulfilling Prophecies, and the Seeking of Failure

Learning to Be Who We Are: The Power of Names

When a child first learns his or her name he or she also acquires the ability to think and to conceptualize himself or herself in symbolic terms, as an abstract entity. By use of this linguistic symbol both the child and others are able to refer to and to think about him or her. Indeed, as the name is repeatedly applied, the child begins to conceptualize that he or she and this name are one and the same, much as the word *mommy* and the child's mother are one and the same, at least in the abstract.

When the child hears and begins to use his or her own name, the image of the self becomes more firmly established, and specific characteristics, also in the abstract, begin to be realized and applied to the developing self-concept. The child thus develops not only a visual, tactile, and emotional image of the self, but a verbal one as well.

Most of us were probably told at one time or another that "Sticks and stones will break our bones but names will never hurt us." Most of us also knew that this just wasn't true. Children learn that certain names applied to them by their parents or siblings, or by neighborhood children can indeed hurt and leave tremendous scars on the psyche. *Stupid, idiot, pig, ugly, fatso, asshole, clumsy, good-for-nothing,* and *failure* all leave their imprint.

Children learn who they are from the names that are applied to them and from how they are treated. Through others' interactions with us, we begin to learn who we are and whether what we are is good or bad, inferior, or superior.

We all require association with others to discover ourselves. We apply the names that others apply to us and begin to see ourselves as others describe us. Even in the privacy of our rooms, we can easily visualize how we may appear to outside observers. Our parents, our siblings, other children, and then, later, our adult friends and lovers become mirrors that reflect us. As we ponder ourselves in their eyes, we learn to see who we are. If Mother tells us that our performance is "good" and "wonderful," then we are good and wonderful. If at three years old we are repeatedly told that we are strong and brave, we believe it and begin to live up to what we have been told, even if, in fact, we are not strong and brave, but merely like every other three-year-old.

We also learn who we are by comparison with others. If in the first grade during gym we can run faster than our classmates (and our speed is verified by the comments of others), we add more to our knowledge of who we are and how good we are (i.e., the fastest or slowest kid in the first grade), and this knowledge may become a source of pride or shame. One cannot feel shame or pride except in relation to others.

Take, for instance, an adolescent boy masturbating in the privacy of his bedroom. He is neither thinking about what he is doing nor judging it (although he may be imagining that he is doing something to someone else). At that moment, he does not feel shame or guilt, just the pleasure of his self-manipulation. Suddenly, his bedroom door swings open, and his mother walks in. A look of disgust crosses her face. He has been seen touching his own body and enjoying it. Good heavens! How could he do such a thing? What a disappointment! What a pervert! How completely abnormal! Immediately, this boy feels guilt and shame, for now he sees himself in the eyes of another. Recognizing what his mother is probably thinking, he now applies these labels to himself; feels shame, guilt, and embarrassment; and later promises some deity (or whomever) that he will never touch his own body again.

Vicki is riding her bicycle across the campus quad when, turning a corner, she loses control and falls to the ground. Immediately, she feels like a fool and is overcome with embarrassment. Sheepishly she gets up and discovers, to her surprise, that there is no one around. No one saw her fall. Immediately her feelings of self-reproach vanish. She rides away vowing to be more careful and no longer feels foolish or embarrassed because no one saw her.

Nevertheless, if Vicki in fact had a poor self-image to begin with and has been told by significant others that she is a "klutz" or "foolish," she will probably continue to feel foolish for falling and will probably visualize how her mother or father, or brother or boyfriend would have viewed her on the ground. Moreover, needing to verify this self-image, Vicki will probably feel compelled to tell her friends, boyfriend, siblings, or parents about how she "stupidly" fell off her bike. She will do this because she will need to have her self-image updated and maintained by the reaction of others, who will verify her poor self-concept.

Maintaining an Inaccurate Self-Image

Trying to win the good opinion of others so that a good and healthy self-image may be established or maintained seems rather natural. Surprisingly, some people have a bad self-image and also seek to have it verified, reconfirmed, and maintained. This is true even if the bad self-image is inaccurate. Most people require not an accurate self-image, but an image that is familiar, one with which they feel comfortable.

If Vicki, having been raised to believe she is foolish and clumsy, begins to get feedback that does not fit this image, she will be surprised, unbelieving, and uncomfortable, because she *knows* (i.e., believes) that this positive feedback is inaccurate. Indeed, if her friends begin to comment on how well she rides a bike, or on how quickly she has solved a problem and thus how smart she is, she may feel compelled to set them straight by arguing with them: "No. It was only luck." If they persist in complimenting her, she may even begin to avoid them and instead try to form friendships with others with whom she feels more comfortable. That is, if Vicki feels bad about herself, then she needs to associate with people who will treat her badly or, at a minimum, will allow her to maintain her poor self-image without challenge. If she was rejected or made to feel guilty or inadequate as a child, she will later seek out friends and boyfriends who will make her feel the same way. How they treat her will enable her to maintain her self-image, even if it is inaccurate.

Rejection

The Conscious and Unconscious Self-Images

Very young children view their parents as powerful sources of love and security, and as mirrors of their true self. If Mommy says I am "bad," then I must be bad. If Mommy and Daddy treat me badly, I must deserve it. If I feel bad, then I must be bad. If being treated badly becomes a normal part of life, children will unconsciously learn to expect to be treated badly by others, even though consciously they may absolutely hate what is happening. Being bad and being treated as bad are part of their self-image, which they need to maintain. Even when treated well, they may "read between the lines" and see badness where there is none.

Conversely, those who are given love and respect as children tend to believe they deserve love and will seek it out. If as adults they are treated poorly, this treatment will seem unfamiliar, and they will avoid those who behave this way. It will not fit with either their conscious or their unconscious self-image. People who have a healthy unconscious self-image are able to marshal both conscious and unconscious resources so as to avoid bad treatment or misfortunes. Those with an unhealthy unconscious or conscious self-image are less likely to resist, as they "halfway" expect things to go wrong and relationships to sour, and to be treated badly. When this occurs their unconscious, in the form of the Child and the Parent, actually conspires against them. Unconsciously, they do not resist because what happens to them is perceived as familiar, normal, and thus tolerable.

Blaming Oneself and Seeing Rejection Where There Is None

Rejection can take many forms, and people who are severely rejected in childhood by their parents, siblings or other children, or whose parents frequently fight (even if they don't lash out at the child) may, in addition to feeling bad, also feel bitter, hostile, and angry, feelings that they may turn back onto themselves. They become very sensitive and may be easily hurt. Sometimes, they turn the anger on themselves and become depressed.

Some individuals are plagued by an unconscious feeling that they are not liked, that others see them as flawed or undesirable, and that, given the chance, others will reject and hurt them. Indeed, these feelings are so pervasive that they may scrutinize almost every interaction, every comment, for slights, rebuffs, and rejection. They may interpret the comments and actions of others as rejecting whether they are or not; they will see racism or sexism where there is none.

Brenda was often ridiculed by her parents, particularly her mother, who frequently withheld love and approval, and who called her daughter terrible names if she disappointed her in some way, particularly in manners and "good taste." Brenda's mother, being a very vain and beautiful woman, also expected her daughter to look beautiful and often supervised and criticized her daughter if she did not dress appropriately or look her best.

Brenda's mother was also extremely competitive, and Brenda always came up short. This fact was communicated either verbally or by looks, gestures, or tone of voice. Of course, Brenda was always hurt by this treatment.

Later, as an adult, Brenda was also easily hurt. If a friend gave her a compliment she would reject it off-handedly. However, if the same friend made a helpful suggestion, Brenda would take it as rejection and would become terribly hurt and then angry.

Later, after she married, one of her husband's main complaints was that living with her was "like being on pins and needles. You have to watch everything you say because you never know what's going to set her off."

As in Brenda's case, the unconscious Child not only suspiciously searches for rejection but demands rejection and creates it even where none exists. The internal Child may do this by being overly demanding of love and reassurance, or by behaving promiscuously, self-destructively, in a rejecting way, cynically, hostilely, irritably, accusatorially or downright meanly and nastily. Sometimes, however, people treat themselves this way and then feel very depressed. In any case, their relationships suffer and their marriages tend to be rocky and unstable.

The Scapegoat

Being unable to recognize the unconscious origins of their unhappiness and their own responsibility for it, the members of a dysfunctional family often choose the weakest member of the family as a scapegoat. This is usually one of the children, or it may be one of the parents if he or she has an undeniable flaw. Nevertheless, family members often come to believe that their problems are caused by this particular individual and then treat him or her miserably.

Children raised by rejecting, abusive, or neglectful parents tend to feel awful about themselves and may even blame themselves for the treatment they received or the fighting they observed (even if they themselves were not actually victimized). However, rejection during childhood and infancy often has nothing to do with the child, and has everything to do with the mother and father. Of course, the parents may have also been treated badly when they were small. Who is to blame?

The mother is of primary importance in the nurturance and development of a small child because a tremendous biological and emotional bond has been in formation since conception. To sever or attempt to weaken this bond can lead only to feelings of insecurity and confusion in the infant. Although dads and others can make good surrogates, the bond with the mother remains primary because it was there first.

Because of this bond, which becomes strengthened during the first year of life, mothers exert a tremendous influence on an infant, and feelings of rejection and inadequacy often stem from her reactions. For example, some women, although they have desperately wanted a baby, begin to resent it for preventing them from sleeping at night, from going out and having fun or for any ruinous effects that the pregnancy and birth had on their bodies. The baby may also be seen as a burden because of the care that it requires. Some of these are realistic and quite normal feelings, as long as they do not dominate the mother's interaction with her child. Although a mother may not put these feelings into words, they are nevertheless expressed via body language and vocal tone, particularly if she often feels this way. In some cases, a mother who (naturally) wanted a beautiful baby and instead got a plain or even ugly child may unconsciously reject it. In other cases, a mother who dislikes her husband may unconsciously transfer these feelings to the child, who is a product of him and who reminds her of him.

Some women (like some men) do not like to be touched, hugged, or cuddled. Although not intending to be rejecting (and although they may actually love their child very much), this feeling of aversion is communicated via the right half of the brain. The child responds accordingly because of its own right-brain perceptions.

Some children, as well, do not like to be touched, and some have suffered an abnormality in the limbic system and cannot respond appropriately to physical contact. Unfortunately, if such a child is paired with a noncuddler mother, it will receive minimal touching and social stimulation, and its social abilities will wither. As an adult, such a person may be friendless and withdrawn and, in extreme cases, may be abnormal emotionally because of the atrophy of certain limbic nuclei as a consequence of deprivation.

Fathers, too, exert a tremendous influence on the infant and young child; A man may resent the birth of a baby because, before it, he was the number one kid and he now has a competitor. If he has a son and the son is not particularly athletic or bright, the father may see this as a reflection of himself (of his own inadequacies) and may torment, ridicule, or otherwise emotionally abuse the child. Moreover, his own fun may now be curtailed, and he may resent the financial burden the child places on the family. Consciously or unconsciously, verbally or nonverbally, these feelings are conveyed to the child, who then feels rejected.

However, in assessing blame, one needs to look at a variety of situational variables. These parents may have been abused by their parents, who may also have been abused. Or these parents may have suffered a severe illness, financial disaster, been fired by a boss who was truly an "asshole" and be unemployed for a long time, and so forth, all of which cause an extreme amount of stress affecting everyone in the family. Although it is not right, often the weakest member of a family becomes a scapegoat.

On the other hand, sometimes the strongest child willingly becomes the scapegoat in order to fix the family and because he perceives that no one else is able, willing, or deserving of this role. He may become overly good and try to attend to the family needs, or may be overly bad so as to take the pressure off the family and thus diminish the uncontrolled uproar of the household. Such children feel if they do not become overly good or overly bad, the entire family will fall apart, and this possibility is more horrible than their own abuse, neglect, punishment, and rejection.

As we have seen, when subjected to abuse by parents or a family group, victims often assume the role assigned and begin to act as if and believe that they are what they have been labeled; the Child ego personality accepts and lives in accordance with the self-image provided. Hence, we may act bad and not care, because we know we are bad. If our self-image is that of a failure, we have nothing to lose by failing. We are only maintaining the image of ourselves that has been fashioned by the labels and treatment applied by others.

Even if parents do not directly communicate their difficulties, children sometimes perceive their troubles and conflicts and then personalize them. The children feel responsible, guilty, and bad even though no one has done anything bad to them.

Stephanie was severely rejected by both her parents as she was unwanted; the pregnancy had been unplanned. In addition, she had been a truly ugly baby, who at an early age began to put on weight. Both her parents were alcoholics, and when not neglecting her, criticized her and used her as a scapegoat for their own problems.

Her father told her she was retarded, worthless, and ugly and would be better off dead. Anything she accomplished or showed interest or talent in, no matter how great or trivial, was met with derision or contempt. In effect, she was told not only that she was worthless and unworthy of love, but that she would never amount to anything and would never succeed.

Stephanie had two major talents—math and computer programming—which she excelled in while in high school. After graduating, however, she did not go to college but began drinking quite heavily (i.e., imitating her parents) and drifted from one low-level job to another (living up to their descriptions of her).

When she finally was referred to my office (for counseling after an industrial accident), she underwent a battery of tests which revealed her extremely high aptitude for math and programming. When I encouraged her to take some classes in these fields in college or to pursue a related job, she responded adamantly that, although she loved doing math and programming, she just wasn't any good and would never be able to get a job in either of these fields. In fact, she seemed incredulous that I should make these suggestions. They did not fit in with her self-image.

Similarly, when we talked about college, she was convinced that she would not be able to do well. When I pressed the point, she began to make up excuses about why she would not be able to attend (e.g., it cost too much, it would take up too much time, and she wouldn't be able to work). When I persisted, she began to challenge the accuracy of my tests and perceptions.

In other words, the possibilities I presented and the abilities I told her she had seemed so unfamiliar that she became extremely uncomfortable and had to invent a number of reasons to maintain her poor self-image. Her left brain began to confabulate. When this didn't seem to be working, she simply negated my test results and opinions. She was a failure and that was that, and I was wrong to think otherwise.

Nevertheless, several sessions later, after discussing with her the concepts of the unconscious Parent and Child, I asked her, "Who is telling us you can't do this? You or your father (i.e., unconscious Parent) or the Child within?" In truth it was both. Her Parent was saying, "You can't succeed because you are worthless and stupid," and the unconscious Child stated exactly the same thing.

Of course, her difficulties were deeper than just this. Failure was a way of life for her. It was familiar, safe, comfortable, and predictable and therefore completely normal. Attempting to change her life was frightening simply because it would be abnormal for her. There is little security in traveling uncharted waters. She already knew who she was, although that self-image was inaccurate, and although she was unhappy, she was comfortable with that image.

For Stephanie, just the possibility of success was so unfamiliar that she expended considerable effort avoiding it. Her unconscious Parent and Child conspired to ensure that the comfortable familiarity of her early self-concept would be maintained and that any attempt at positive growth would be derailed.

Who Is in Charge: The Child, the Parent, or the Unconscious Self?

A major question, particularly for individuals who are unhappy and who suffer from frequent emotional and social difficulties, is: "Who is in charge? Who is making the decisions? Who is controlling my behavior and emotions in this situation? My Parent, Child, IDfant, or the Adult rational aspects of my conscious and unconscious mind?" When this question can be asked, the first major step toward gaining positive control over one's life can be made.

How can we tell if the Parent or the Child has been activated? First, by the intensity of our emotional reactions. The more intensely we feel, or the more out of control, or the more we act childish or critical, the more likely it is that unconscious forces have come to the fore. However, there may be just and rational causes for emotional arousal or even outbursts. If your house is on fire, both halves of your brain are going to be upset.

It is when irrational emotions come to the fore that unconscious parental or childish influences should be suspected. If our feelings and responses to others are nonlogical, irrational, or exaggerated, their origin is usually unconscious. For example, if we are feeling depressed "for no particular reason" or are suffering from destructive, pessimistic thoughts, it is usually because the unconscious Parent is belittling the unconscious Child. We should ask, "Why?" and then be alerted that something good (e.g., a job promotion or a date with an attractive partner) may be about to be sabotaged.

Another clue may be feeling extremely hurt, insulted, or rejected by an insignificant statement or action by a loved one or even by someone of little significance to us. If a man asks a woman to dance and she says "No" and this is a crushing blow, he is probably responding to the pain of childhood, which has very little to do with this woman. If a man who has taken a woman out fails to call again as promised and this causes her terrible feelings of anger and depression, she is probably responding to the pain of the past, which has very little to do with this man. If a simple comment from a loved one sends a man into despair and he is overcome by feelings of rejection and anger, his internal Child has reached out and interfered with his life.

Every person whose unconscious Child has been sorely injured has difficulties with love and success simply because the Child is still hurting and easily stirred. That is, the person expends energy to meet the needs of the unconscious Child, often in the form of a repetition compulsion. While feeling undeserving of love, the person continually attempts to overcome the hurt or to achieve the love that was denied when she or he was a child.

From Generation to Generation

Children who are abused often become abusers when they grow up. Children who were molested sometimes molest other children when they become adults. These conditions are sometimes passed from generation to generation. Children neglected or rejected or abused by their parents may, in turn, allow their internalized unconscious Parent to treat their own children the same way. Their unconscious Parent tells them it is OK for parents to mistreat their children. People learn by example.

Sometimes, these bad early experiences and their effects on future generations are not so straightforward. If the unconscious Child feels rejected, it may feel that others cannot love it. When this person has children, these feelings contaminate his or her ability to be a good and loving parent or to accept affection from his or her children. Indeed, such people may erroneously feel rejected by their children and then increase their own inappropriate responses. Individuals who dislike and reject their own parents may, in turn, unconsciously reject their own children and then have disturbing dreams in which their children are hurt or killed. A mother who hated her father may reject her son. A father who was rejected by his mother may unconsciously reject his daughter.

Repetition

Learning who we are requires not only association with others, but repetition. We are called the same name over and over, treated in the same manner, and engage in the same activities, day after day, and slowly an image of the self is formed.

When a name is applied to us over and over, we learn that that is our name. When we find we can perform a certain action over and over, we gain confidence in our ability and define ourselves accordingly. Practice makes perfect, because through practice (repetition) an ability or capacity becomes easier to perform. If we do or say something enough times, it becomes second nature, like a reflex.

In children, the process of repetition is abundantly obvious and sometimes a source of irritation for parents. However, the childish need to repeat things over and over is a natural part of being alive, like having two legs and using them for walking. Children like to repeat what they see, hear, experience, or observe. They may ask the same question over and over, sing the same song, say the same rhyme, want to hear the same story, or watch the same movie two or three times or more at one sitting and then want to see it again and again the next day.

Children like to learn. Through learning and repetition, children begin to gain mastery over themselves and their environment. They learn who they are and what their capabilities are.

Infants who find that a sock can be removed, whereas a foot cannot, discover the boundaries where the self begins and ends. They also take delight in this new ability to perform some new action, to affect and control their environment. They then begin to do the same thing over and over. Babies may take their socks off as often as their mothers are willing to put the socks back on; they throw a cup on the floor each time it is put back on the high-chair tray; they throw a bottle out of the crib over and over again. By doing this, they learn self-mastery and how to control and influence their world. Repetition is the essence of learning, and it is through repetition that we learn who we are.

The Brain, Learning, and Repetition

Traumatic Repetition and Nerve Cell Activation

Children can often be observed reenacting incidents that were emotionally troubling to them. They may make two dolls fight, draw pictures, tell stories, or engage in certain violent play activities, or they may start fights with other children or behave in an inappropriately sexual manner. All of these actions contain and repeat the theme of the original trauma, such as having been beaten or molested, or having viewed or experienced some terrible event.

The need to repeat and reexperience may be due to unresolved unconscious forces repeatedly demanding entry into consciousness. In young children, it is also related to the immaturity of the frontal lobes, which are unable to inhibit these urges. Repetition is also partly due to an attempt to achieve mastery over the event, by controlling it. If the experience is troubling, the child (and later, the adult) lessens its significance by reducing the tension associated with it, for example, by talking about it and thus placing it under the control of language.

However, we are also compelled to repeat because, at the very level of our brain, the nerve cells involved in that particular experience have become more complex and more sensitive and are thus more easily activated. Like a bad habit that we cannot extinguish, it thus becomes easier to think about the bad event, repeatedly.

Axons and Dendrites: Learning and Memory

Different regions of our brain communicate through nerve cells called *neurons*. The neuron (or nerve cell) transmits messages to other nerve cells through a nerve fiber called an *axon*. Neurons receive messages through their dendrites.[1]

Each neuron contains an axon and several dendrites. Axons can be quite long and sometimes travel considerable distances to other cells. The corpus callosum is made up of millions of axons. Dendrites, although very short in length, are quite bushy and form what have been referred to as *dendritic trees* with many spines and branches.

At the synaptic junction, axons make contact with the dendrites of other cells. Dendrites, because of their extensive branches, can receive messages from many axons, as they have very extensive receptor surfaces. Axons, however, transmit impulses only to a single dendrite.

Axons fire in response to an "all-or-none principle." That is, the axon either becomes sufficiently activated and fires or it does not, depending on if its firing threshold has been surpassed. For this to occur, the "resting potential" (i.e., its base level of activity) must be increased in order to surpass this threshold.

When the axon is triggered and fires, it releases a chemical neuro-transmitter. Each dendrite has a receptor surface that receives chemical neurotransmitters released by the axon of a different nerve cell. If sufficient transmitter is released, the dendrite becomes activated and allows the information to be transmitted to its cell body, which, through its own axon, may relay the message to the dendrite of the next cell. However, although a dendrite (and its cell) may become stimulated, if the threshold for axonal activation is not surpassed, the axon leading from that cell to the next dendrite will not fire. If the axon is sufficiently activated, it will release its neurotransmitters.

The chemical transmitter acts on the dendrite, the cell body, and its axon by increasing activity until the axon's firing threshold is reached, at which point it ignites and discharges. If the receiving dendrite, cell body, and its axon are insufficiently activated, the impulse received fades such that information transmission stops with that cell. However, for a brief time period, the cell remains slightly activated so that it remains close to firing.

Due to this principle, although a single impulse may not cause the cell to fire, since the dendrite and neuron have been partially activated (such that it is well above its resting potential), if a second or third low-level impulse is also received (such as from the same axon or a different axon making synaptic contact at a different branch of the dendrite), there is an additive effect and the threshold will be easily surpassed and it will fire.

When brain cells are insufficiently utilized, they tend to die and drop out. However, those which are repeatedly exercised often grow larger and the dendrites they possess become more extensive. Moreover, new axons may grow connections to the new dendritic branches of these very active cells. This is, in part, how learning and memory is made possible. As more and more information is assimilated and learned, more and more cells become associated and interlinked, thus enabling mental processing to become more complex and elaborate.

In addition, an axon that is repeatedly used for information transmission will increase its supply of neurotransmitters. Conversely, a dendrite that is repeatedly stimulated not only becomes more complex, but each individual receptor surface (at the synaptic junction) may become more extensive so as to take advantage of the increased amount of neurotransmitter available.

An axon that is repeatedly stimulated also becomes "potentiated"; that is, it remains activated (although below firing threshold) for long time periods, even when there is no activity. Therefore, when a stimulus is received, the axon is more likely to fire.

Due to these many changes, although the firing thresholds remain the same, when the receiving cell is already partially active and now much more sensitive, its threshold can be surpassed more easily. Moreover, due to potentiation, and since there is so much neurotransmitter available, a highly active neuron can now fire with minimal stimulation. The cells involved can now become activated with ease due to the physical and chemical changes which have occurred between them.

Such conditions can be produced artificially as well, such as via the administration of various drugs such as narcotics. That is, the dendrites may become more extensive and the receptor surface will increase as well. However, since an axon is not producing these chemicals, this requires that the drug be administered more often and at higher doses. It is in this manner that some people become addicted to drugs so that they soon need more and more. There is an increase in the nerve cell receptors which respond to that drug, which now has to be periodically replenished in greater amounts.

As noted, in learning situations there often results increased intracellular complexity. Other cells may grow nerve fibers into this area and all cells involved may become more complex as they undergo physical changes to accommodate the increase in activity occurring. That is, a dendrite will attract the axons of other cells to it, which then make contact so that even more information can be exchanged and processed.

With repeated use and with the alterations in dendritic growth, the receptor's surface, the amount of neurotransmitter available, and the number of axons contacting a particular dendrite, it now takes very little to cause one cell to activate a second cell and so on, so that the same message becomes more complex as well as repeated again and again. There is an increase in the number of synapses, the number of different cells making contact, the amount of transmitter available, as well as an increase in the resting potential of the nerve, all of which requires that fewer or less powerful signals be transmitted in order to cause activation. By repetition and practice it becomes easier to perform a certain action until it finally becomes like a reflex as it takes very little to trigger a response. Indeed, this is how some habits are formed. Practice makes perfect.

However, eventually, with rapid repeated firing, a depletion of the amount of available neurotransmitter occurs, cellular activation ceases, and the response or activity ceases or becomes haphazard. At a behavioral level, we could say that fatigue sets in.

The same thing happens in regard to certain thoughts and even complex actions. The more often we produce the same thoughts and entertain certain feelings, the more often will they be experienced in the future as it now takes very little to set off the whole chain of events. Eventually, however, we reach a point where we can think about the subject no more, at least for a while, only to discover later that we are again dwelling on the same subject matter.

Nevertheless, when certain cells repeatedly communicate, a circuit of experience is created. That is, an assembly of cells becomes associated via their interconnections (which become stronger with use) and mutually lowered firing threshold such that now complex actions can be initiated in an effortless and routine fashion. For instance, in using the toilet, a whole sequence of associated actions takes place that are so well learned that one need not even think about the different steps involved (e.g., unfastening clothes, etc.). The entire circuit of experience is activated and occurs almost in reflex fashion.

It was this circuit which was apparently triggered in the seizure patient already described, who would expose himself and urinate. This individual always performed essentially the same act because the seizure was repeatedly activating the same assembly of nerve cells and thus the same circuit of experience; that which controlled the unzipping of his pants, the grasping of his penis, and urination.

Thoughts, Visualization, Repetition, and the Talking Cure

Via repetition we not only learn, we gain control over our behavior and the environment. By repeating certain thoughts and by visualizing, we can gain control over our feelings and past experiences. If that which we wish to subdue is not amenable to persuasion and physical force or has already occurred and is beyond our ability to alter, then our only recourse is to think, visualize, and talk about it, like talking about the fish that got away.

Unfortunately, some people engage instead in repression or denial and choose not to discuss what has occurred. Essentially, the frontal lobes, in attempting to protect these people, engage in massive inhibition.

Nevertheless, whenever most of us have been traumatized or have had a tremendously enjoyable experience, we want to talk about it, to reexperience it, and possibly analyze its significance and implications. In the case of trauma, it is especially important to talk out one's feelings, for this eventually lessens their impact—like blowing off steam—and lessens the tension associated. This is why psychotherapy (the "talking cure") can be so important and helpful. The mind is now made conscious of the unconscious forces demanding attention.

In fact, by talking about what has occurred we now change the circuit of experience, be it based on something from the past, or some current trauma. This may occur by focusing on different aspects or possible meanings of the experience, by tying in past events or future hopes and desires. In this manner, that which was so troubling can be experienced in a different fashion and thus a new assembly of interlinked ideas, feelings, thoughts, meanings, and thus nerve cells can be formed. We have taken control by changing the experience, and by repeatedly therapeutically visualizing or talking about what has occurred. The firing threshold of this new circuit of experience is thus lowered such that now we can think new thoughts and have different feelings about what has happened or will happen.

This is also why it is important to visualize (as well as think about) what one wants to occur or the goals one wishes to achieve and avoid negative thinking patterns. The cells involved become linked and easier to activate, and our ability to efficiently and effortlessly perform that action, be it harmful or advantageous, is maximized.

Sometimes, however, a trauma, or emotional pain, never becomes consciously expressed except indirectly. This again is sometimes a consequence of the limitations involved in transfer between the two halves of the cerebrum as well as the frontal lobe (the right frontal lobe in particular and thus the unconscious Ego) acting to prevent information reception or processing in the left, conscious half of the brain. Rather, the pain and bad feelings continue on an unconscious level with a life of their own at the central core of our being. These feelings continue to be replayed over and over, unconsciously, creating havoc with our lives and relationships.

However, in this regard the "talking cure" and the use of "free association" can also be helpful. This is because events occurring within the unconscious half of the brain can trigger the formation of thoughts in the language-dependent regions of the mind through information exchange via the corpus callosum. Although these left-brain thoughts may be off base or only indirectly provide hints as to what is really occurring, they are nevertheless linked to the central core which has stimulated their production. By following these associational links and hints, a skilled therapist can then interpret a person's thoughts so that the underlying core can be extracted and made conscious. Indeed, the stimulation of thoughts and words by unconscious processes is what is presumed to take place when administering an "ink blot" or "word association" test.

The Complex

In that these troubling experiences occurred during a particular time period at the hands of particular people, they are associated and emotionally interlinked to a centrally disturbed core—like spokes to the hub of a wheel—what Jung has referred to as a *complex*. Consequently, any future experience which in some manner resembles or reminds us of one of these bad early experiences can inadvertently trigger one of these emotional spokes and activate the entire complex. Indeed, it is possible for anything remotely resembling hurt or emotional pain to activate the entire circuit of experience. Like a bad habit, the unmet needs and associated pain will be experienced again and again.

The formation of these complexes is not limited to childhood experiences but can also occur during adulthood as well. For example, consider "Penny," who after much inner turmoil and confusion decides to undergo an abortion. Afterwards she feels sad, cries for a few days, and then slowly forgets about it. Weeks and even months later she finds that toys in a store window, little children playing, the sound of a child's voice, make her feel uneasy and even a little bit depressed. They do so because through association they are linked to a complex. The children, the toys, all remind her (at least unconsciously) of the baby she had aborted, and the originally experienced turmoil, guilt, confusion, and depression are activated and reexperienced, to varying degrees.

The associational linkage of external experiences can activate memories from the past or even seemingly unrelated ideas (in fact, they are related). Lucy mentions to Paul that her roommate just got a new kitten. He immediately remembers that he has to buy dog food and some milk. *Dog* and *kitten* are associated because both are animals. *Milk* and *kitten* are associated because kittens drink milk. The notions of *milk* and *dog food* are conjured up because, although forgotten, they remained active needs and were easily triggered.

Prior to being triggered these memories were based on impressions that had been pushed out of consciousness. In this regard they are unconscious. Unconscious verbal memories, however, are not stored in the right brain, they are stored via the hippocampus of the limbic system and within that area we have described as the primary unconscious. However, since it is a verbal memory, it is stored in the left hippocampus which is located in the left temporal lobe of the brain. Similarly, visual, emotional memories are stored in the unconscious mental system maintained by the hippocampus and limbic system but in the right half of the brain. Hence, verbal associations can call forth unconscious verbal and even emotional memories since the right and left limbic system are subcortically linked and communicate together.

When Lucy mentions to Penny the new baby kitten, she becomes depressed and upset. Penny does not think of milk or dogs, she is reminded of her aborted baby, because a kitten is a baby. She then feels and recalls her emotional pain. In fact, children, bicycles, school buses, teachers, nurses, and so on may be incorporated within the single "abortion complex."

A complex can act like a magnet that draws associations. The stronger and more powerful the complex, the more associations it will attract, so that seemingly divergent and quite unrelated events and experiences cause emotional upset. For example, a few weeks after her abortion, Penny was going through a drawer, saw a piece of chalk, and soon felt depressed. She associated the chalk with school, which is associated with children, and so on. Related memories and feelings are easily triggered through association.

Arnie was madly in love with Roberta. It was a whirlwind romance, and he had been swept off his feet, as had been so many men before him. And then, a few months later, she suddenly dumped him and refused his calls. He was devastated and could not stop thinking about her.

Moreover, everything reminded him of Roberta. If he saw a red car or a convertible, he would remember her in her red Ford Mustang convertible. Just the color red sometimes created the same domino effect in his mind. If someone mentioned the make of any automobile, he would immediately think of "Ford" and her Ford Mustang. In fact, if someone said, "Let's go for a ride," the entire string of associations would again lead to Roberta.

If he heard a certain song on the radio, he would remember that he had heard it in her company. If someone mentioned going to a concert, Arnie would suddenly remember that Roberta loved listening to rock and roll. Everything reminded him of her, and everything made him feel sad and depressed.

Essentially, everything reminded him of her because, in some tangential manner, it was linked to a "Roberta complex." Anything associated with Roberta extended throughout his psyche like the arms of an octopus or the spokes to the hub of a wheel. Because this hurt was still so intense, anything associated with one of these spokes triggered and activated the entire complex (i.e., "Chevy"–"Ford"–"Ford Mustang"–"Roberta's Ford Mustang"–"Roberta") and he would become depressed. In this regard, it is important to note that complexes and other neurotic emotional disturbances are not necessarily rooted in childhood.

As pertaining to complexes formed during childhood, all are linked to the central core we have referred to as the Child ego personality. If the Child is hurting, then its influences can be quite powerful as all related associations go on alert and stay in that manner for years. Its memories and feelings can sometimes be easily activated via associations experienced during adulthood.

Sometimes, however, traumas are so deeply buried and the actual memory and the disturbed core of the complex is so deeply submerged and hidden that it is extremely difficult to actually regurgitate the original incidents which have led to a lifetime of hurt, anger, or insecurity. We may feel hurt and in pain but don't know why; a pain that continues to wreak havoc in our lives and relationships. Search and analyze as we may, the source of our problems cannot be given up by the memory banks of the right brain and limbic system. Like the roots or core of the tree, their overall existence is buried within our depths and it is truly hard to look that far inward. In some instances these complexes were formed so early that they are wholly associated with the primary unconscious mental system maintained by the limbic system.

Often these original traumas that continue to influence our lives are stored in the memory banks of limbic nuclei located in the right half of the brain, memories to which the left brain of the adult cannot gain access. Because adults utilize a set of codes which tend to be temporal-sequential and language-oriented, it is quite difficult to access these early memories. The key to the code has been replaced by a new key and a new code. As we've noted, those keys and codes utilized by adults cannot open those which were utilized by children, and these early memories remain well-kept secrets.

This is why it is particularly difficult to recall a series of connected memories of events and experiences which occurred before age four, although fragments certainly can be accessed and reflected upon. Thus, much of this material, although present, remains hidden within the memory banks of the limbic system and the right brain, and unrecognized by the conscious aspect of the left half of the cerebrum even though the whole brain may hurt.

The Royal Road to the Unconscious? The Origin of Dreams

Frequently, traumas are repeated not only in our thoughts but in our dreams. Dream imagery is predominantly a product of the right hemisphere, although it may be stimulated by thoughts and concerns of the left.[2] Similarly, complex visual hallucinations are also reported more often with right-brain irritative lesions, which result in abnormal activation.[3,4] Even daydreams reflect certain peak levels of right-brain activity, coupled with reduced levels of left-brain activation.[5] This is why the conscious half of the brain seems to be completely overwhelmed when a daydream is experienced. The left brain is at a low level of arousal and is flooded by right-brain-generated feelings and imagery. It is this same difference in dream production and brain activity which accounts for the fact that most dreams are very difficult to recall. This is because they are locked away within the right half of the brain. It is the left brain which cannot recall the dream.

Dreaming and Hemispheric Oscillation

Up to five stages of sleep have been identified in humans. However, dreaming largely reflects the oscillation of two distinct sleep states. Dreaming occurs during a sleep stage referred to as *paradoxical sleep*. It is called *paradoxical*, for electrophysiologically the brain seems quite active and alert, similar to its condition during waking. This indicates that a considerable degree of mental activity is taking place. However, the body musculature is paralyzed, and the ability to perceive outside sensory events is greatly attenuated. If it were not for this paralysis, we might walk in our sleep and try to act out our dreams.

As is well known, the formation of vivid, hallucinatory dreams is also associated with rapid eye movements such that the eyes seem to dart about as if looking at and following something. Rapid eye movements (REMs) occur during paradoxical sleep.

By contrast, periods associated with lack of dream activity, or the production of nonvisual, verbal, thoughtlike dreams occur during a stage referred to as *slow-wave* or *synchronized sleep*.[6,7] This has also been referred to as *non-REM sleep* (N-REM) as the eyes remain fixed.

Most individuals awakened during REM sleep report having dreams approximately 80% of the time. When awakened during the N-REM period, dreams are reported approximately 20% of the time if at all.[8-10] However, as noted, the type of dreaming that occurs during REM vs. N-REM sleep is quite different. For example, N-REM dreams (when they occur) are often similar to thinking and speech (i.e., linguistic thought), such that a kind of rambling verbal monologue is experienced in the absence of visual imagery. Such activity probably should not be referred to as a "dream." It is also during N-REM sleep that an individual is most likely to talk in his or her sleep.[11] In contrast, REM dreams involve a considerable degree of visual imagery and emotion and tend to be distorted and implausible.

The production of REM sleep is correlated with high levels of activity within the brain stem, the occipital lobes (where visual information is first analyzed), and other select brain regions. Electrophysiologically, the right brain becomes highly active during REM, whereas the left brain becomes more active during N-REM sleep. Similarly, measurements of cerebral blood flow have shown an increase in the right temporal and right parietal regions during REM sleep in subjects who, on wakening, report having dreams. These results strongly suggest a specific complementary relationship between REM sleep and right-brain activity.[12-15]

Forgotten Dreams

Since most dreaming is associated with right hemisphere activation and low-level left hemisphere arousal, it thus occurs during a period when the left brain is more asleep and less aroused as compared to the right, which *paradoxically* behaves *as if* it were wide awake. In some respects this is similar to the experimental condition when the left vs. right brain has been anesthetized in preparation for neurosurgery.

As we all know, it is sometimes very difficult to remember our dreams in the morning, although some innocuous event later in the day might trigger their recollection. In this regard, it is interesting to note that if awakened during various sleep stages, it is easier to recall one's dreams if they are still in a right-brain mode of high activity. Conversely, it becomes progressively more difficult to recall one's dreams as one spends time in or awakens during N-REM, which is associated with high left-brain and low right-brain activation. That is, as the left brain becomes more active and the right decreases in arousal, the left brain is unable to access the dream dreamt by the right brain. Thus, are dreams really forgotten or are they locked away in a code that is not accessible to the speaking left hemisphere during waking?

Daydreams, Night Dreams, and Hemispheric Oscillation

There is some evidence to suggest that, during the course of the day and night, the right and left brains oscillate in activity every 90 to 100 minutes and are 180 degrees out of phase—a cycle that corresponds to changes in cognitive and intellectual efficiency, the appearance of daydreams, REM (night dreams) sleep, and conversely, non-REM sleep.[16–20] That is, like two pistons, it appears that when the right brain is functionally at its peak of activity, the left brain is correspondingly at its nadir and that these shifts occur throughout the day. This would suggest that during the course of every day, there are time periods in which one is best able to utilize their right vs. left brain.

In fact, shifts in cognitive abilities associated with the right and left hemisphere have been found during these cyclic changes during the day and after awakenings from REM and N-REM sleep. That is, by measuring brain activity during sleep, and by waking people during certain sleep phases and testing them, correlations between the type of test (right vs. left brain) and brain sleep and dreaming have been established. It has been found that performance across a number of tasks associated with left-brain cognitive efficiency is maximal after N-REM awakenings, whereas right-brain performance (e.g., point localization, shape identification, and orientation in space) is maximal after REM awakenings.[21,22] Moreover, left-hand motor dexterity was superior to the right when awakened during REM and that the opposite relationship was found during N-REM, i.e., right-hand superiority. This indicates that the right brain is more aroused and more capable of efficiently processing information during periods when dreams occur, whereas the left brain (and right hand) dominates during N-REM periods.[23]

Some investigators, however, have claimed that the left, not the right brain is responsible for the production of dreams. In support of this notion some have pointed out that when the left brain has been damaged, particularly the posterior portions (i.e., aphasic patients), the ability to verbally report and recall dreams also is greatly attenuated. Of course, aphasics have difficulty describing much of anything, let alone their dreams. Indeed, evidence presented in support of this supposition is not very convincing.

Nevertheless, it certainly seems likely that the left brain contributes to the production of dreams. That is, just as the right brain is responsible for the visual, pictorial, emotional, and hypnagogic aspects of dreaming, the left brain too mediates a peculiar form of dream activity; the dream is in words and thoughts. Just as the left brain mediates language and the production of verbal thoughts, it is also responsible for producing much of the verbal commentary that accompanies visual dream imagery. If as part of a dream, we are speaking to another person and we hear that person talk and understand what they say, this is made possible by the left brain. The right brain cannot understand complex conversations, not even those which occur during a dream. Hence, the left brain, although at a low level of arousal, is still able to function to a limited degree and thus provide the speech we hear and speak while dreaming. Moreover, like the egocentric child, the left brain participates, comments, and explains the activity initiated and maintained by the right half of the brain during sleep as well as during waking. Otherwise we would not understand what people say as part of our dreams. The verbal monologues produced by the left brain as an accompaniment to the visual dream produced by the right brain may, in fact, trigger more dream material.

Some, including Freud, have argued that dreaming is a form of wish fulfillment. As some wishes and desires are unacceptable to the conscious self-image (and the left half of the brain), some impulses can be presented only in disguise. Presumably, this is why some dreams are so difficult to comprehend or even to remember. They represent hidden and disguised impulses.

As previously noted, the frontal lobes inhibit the transmission and processing of information within the rest of the cerebrum. The frontal lobes are also associated with the conscious and unconscious aspects of the Ego. The Ego often acts as a mediator, so that impulses can be expressed and fulfilled in a way that is acceptable to both halves of the brain and the limbic system. One presumes that the frontal lobes, and thus the Ego (at least as conceptualized here), also play a role in the formation of dreams. It is noteworthy that frontal lobe damage can result not only in loss of restraint, so that the person acts on impulses without regard for the consequences, but in the cessation of dreaming. Is it possible that the removal of the Ego also removes the need to disguise impulses during sleep in the form of dreams?

Creative Problem Solving and Right- and Left-Brain Miscommunication

Just as verbal thoughts may be stimulated by the right hemisphere during waking and dreaming, the left brain can stimulate the production of right-brain dreams and thoughts. That is, the right brain and the unconscious mind often attempt not only to interpret and analyze what is occurring socially and emotionally but also to put into action ideas and thoughts generated by the left brain. It is because of these forces that creative solutions suddenly burst into the conscious mind, or problems are probed or solved by dream imagery.

However, just as the left brain must sometimes guess in interpreting right-brain-generated feelings and desires, the right brain must sometimes guess at what is in the left. This is most apparent during waking. For example, you are sitting in the living room, engrossed in what you are reading, when it suddenly occurs to you that you need a marking pen. You put the book down and get up. A few moments later, you find yourself standing in the kitchen staring into the refrigerator. You can't remember why you got up, but you know you're not hungry, so what happened?

The left brain thought of obtaining a pen for writing and then completely abdicated dominance to the right brain, which then proceeded to maneuver your body through space, all the while trying to guess what the left brain wanted, as this information was not available to it. It decided to go to the refrigerator and then psychically nudged the left brain to take what it needed. Coming back on-line, the left brain surveyed the scene and wondered what was going on. Both halves of the brain were now confused. You then sat down again, picked up your book, and remembered the pen.

Dream Stimulants

Attempts by the unconscious mind and the right brain to analyze their surroundings are often most apparent during sleep and dreaming, although these attempts go on all day as well. For example, the right brain may respond to a sensation experienced during sleep by creating a dream to explain it. When this occurs, we sometimes dream backwards.

As an example, you are dream you are walking in San Francisco lugging large bags of gifts. Feeling tired, you set them down on the sidewalk. You look for a bus and see a cable car coming toward you. As it pulls up, the conductor begins to ring its bell. The sound of the bell grows louder and jolts you awake. Someone is ringing your doorbell. Hearing the bell seemed to be a natural part of the dream, and it was. What seems paradoxical, however, is that the dream seemed to lead up to the bell, so that its ringing made sense in the context of the dream.

The dream did not lead up to the bell, however; the bell initiated the dream. The dream was produced to explain the sound of the bell in the unique language of the right brain during sleep. The bell was heard, and the dream was instantly produced in explanation and association. Because most dreams last only a few seconds (although they may seem to take place over the course of hours) and because the right brain does not analyze in temporal and linear units, the sequence of events is not very important to the right half of the brain.

This is another reason why dreams often do not make sense to the left half of the brain and the conscious mind. The nonsequential, simultaneous, and parallel arrangement of the dream is not comprehensible to the left brain, which depends on temporal-sequential language for understanding. Indeed, the nontemporal, often gestalt, nature of dreams requires that we consciously scrutinize them from multiple angles in order to discern their meaning, for the last may be first, and what is missing may be just as significant as what is there.

Fortunately, backward dreams are the most easily comprehended because the left brain recalls the dream from its ending forward and then, like a reflection in a mirror, reverses all that it perceived so that the dream makes temporal-sequential sense.

One noted dreamer dreamed that he was in eighteenth-century France in the midst of the French Revolution. After a trial in which he was found guilty, he was being led down a street lined with yelling and cursing French men and women. At the end of the street was the guillotine, where he could see the heads of various political criminals being chopped off at the neck. Mad with fright, he felt and saw himself being led up the stairs and his head being placed in the yoke of the chopping block. The executioner gave the signal, the crowd screamed its approval, and he could hear and sense the blade falling. With a loud crack, it struck him across the neck, with such a jolt that he awoke to find that his poster bed had broken and that a railing had fallen and struck him across the side and back of his neck.

Dream Patterns

Crucial

Dreams often reflect something significant about the mental and emotional life of the dreamer. Experimental studies have shown that, when subjects are awakened repeatedly during REM sleep over the course of several days and report their dreams, an evolving thematic pattern, like an unfolding story, can often be discerned. These patterns sometimes reflect mental-emotional activity concerned with the solution of particular problems the subjects are currently facing.

One experimental subject said that

> after being woken many times and seeing three or four dreams a night, I could realize there was a certain problem being worked out, like coping with responsibilities that were thrust upon me, but that weren't necessarily my own but I took on anyway. It was working out the feelings of resentment of taking somebody else's responsibility, but I met them well in my dreams. A good thing about spending time in the sleep lab was you could relate a common bond to some of the dreams.[24]

Similar bonds and patterns were recognized by Freud and Jung many years ago.

Dreams and Repetition

It is not surprising that emotional traumas, because of their right-brain association, are often replayed and analyzed by the unconscious mind in right-hemisphere dream imagery. However, because the right brain uses a "language" that the left brain does not speak, dream imagery is often incomprehensible to the left half of the brain, even when the left brain is providing the accompanying narrative or dialogue.

For example, Sara's parents fought, screamed, yelled and argued almost nonstop, even before she was born. There was pushing, shoving, slapping, and constant threats of divorce. However, both her parents treated her fairly well. Nevertheless, being brought up in the midst of this battle was extremely traumatic, as Sara felt she had no control over what was happening around her. Her world was being turned upside down and was in complete chaos.

Sara seemed to be a wonderfully good little six-year-old girl, except that she often woke up at night screaming about the creek, which she dreamed about almost nightly. According to Sara's mother, the little girl was afraid of a very big creek that ran near their house because she had been told that hoboes lived under the bridge; she had been told never to go there without her mother or father. I had Sara tell me her dream:

> I'm walking on the sidewalk near the big creek and go to the edge and stare down at the big rocks at the bottom. All at once, the whole world starts to shake, like it's turning upside down. It's trying to throw me into the creek. I get scared and start to be afraid and start grabbing at the trees and bushes to keep from falling into the creek and onto the big rocks. Sometimes, I see this hole and I crawl in. Then everything is OK. Sometimes, I fall. Mostly, I fall and fall and fall and I can see the big rocks, and I know I'm going to fall on them. When I fall on them, I wake up because it hurts.

Sara had another troubling dream in which she went riding on her bike; when she came back to her street, her house was gone. Every house on the street was the same, including the neighbors', but when she asked about her family, no one knew what she was talking about and no one recognized her.

Although the symbolic content of these dreams was not apparent to Sara or her mother, one need not be a psychologist to decipher their meaning. Sara's emotional world was being turned upside down and was in chaos because of her parents' horrible fighting. She was terrified of losing her home and of the catastrophe she perceived as befalling her family. Her very identity and functional integrity as a person were at stake, for if she lost her family, she lost her self.

The imagery involving the hole that she climbs into is also interesting; it suggests the desire to return to the safety and security of the womb. However, to re-create a womb symbolically by dreaming of a large, safe hole would require one to have some memories of being in the womb that the dreaming right brain taps into when it experiences certain traumas. Certainly, this little girl did not consciously know what a womb was, nor was she (or her mother) able to make sense of this part of her dream. Either this prebirth experience was registered in memory (presumably by the hippocampus of the limbic system), or the hole in the earth that Sara climbed into was merely that, a hole. The latter is possible because dreams often mean exactly what they seem to mean, and there is no hidden meaning.

It is important to emphasize that the mental activity we observe in the form of a dream does not cease on our waking. Right-brain and unconscious mental processing, including all significant memories and feelings, continue to influence daytime functioning as well. However, because the body is not paralyzed as it is in sleep, the waking person can act on her or his dreams as well as dream them.

Repetition and Therapy for the Child Within

Children frequently experience the same dream over and over. Like much of their experience it represents the need to repeat and to attempt to gain mastery of what is occurring in their young lives as well as the influences of unconscious forces repeatedly demanding entry into consciousness.

Just like the child that we were, the Child that continues to live within us all, within the confines of the unconscious mind, has the desire to repeat. If the unconscious Child is hurting it usually repeats experiences, feelings, and emotions and recalls experiences that were highly significant and traumatic. These traumas, and any feelings of neglect, rejection, or worthlessness, continue to reverberate throughout the unconscious mind in the form of a compulsion to repeat childhood experiences, which is experienced consciously as depression, mood swings, and cycles of misfortune.

Such cycles continue only until one deals with the trauma and the associated feelings both consciously and unconsciously. To accomplish this, a person must learn to talk about his or her feelings and what is causing him or her difficulties, unhappiness, anger, or depression, and to feel these feelings and allow freely associated thoughts, visual images, and emotions to be expressed. By allowing one's mind to wander unchecked among one's feelings and visual images, one can become an active participant in influencing or counteracting these emotions rather than a passive victim of their unconscious influence. When these feelings are allowed expression, without inhibition and internal censorship, such as by free associating in words or images, lost thoughts, feelings, images, and memories not only are recalled but are allowed to emerge into conscious and unconscious awareness, where they can be recognized for what they are.

By thinking, talking, visualizing, and free associating with the aid of a skilled therapist, we can recognize and alter the various complexes and their associations, and we can assign new meanings to what affects us from so long ago. Then, we can create new circuits of experience. A bad experience ceases to be something that "happened to me" and is no longer viewed as reflecting something unpleasant about us.

Many things that happened to us as children have nothing to do with us personally and everything to do with the people who caused the unpleasantness. If given the chance, these people might have treated anyone similarly. By taking responsibility only for our own contribution to the circumstances and not feeling bad because someone else acted badly, we can manipulate, reinterpret, and analyze these experiences, and bring out into the open the associated thoughts, feelings, and images. This changes the experience and changes the circuit in the brain that has been formed around the experience and allows it to be dealt with by both halves of the brain and mind. The circuit now includes our interpretations, analysis, and the new meaning we have. Although we cannot change the past, we can change what it means to us. The original experience thus becomes altered and increasingly under control.

In this manner, memories and bad feelings, including the complex core we have identified as the Parent and Child, need not remain well-kept secrets of the right brain. They can be recalled and dealt with by both halves of the cerebrum. It is in this fashion we can gain mastery over ourselves, establishing inner harmony, and thus gain control over our own lives. However, control, harmony, and self-mastery do not come spontaneously. One should think positive thoughts, visualize positive experiences, over and over again in order to overcome and change the meaning of the past and extinguish the need to repeat and reexperience the pain of childhood.

13

Love, Criticism, Sex, and Abuse

Criticism and Rejection

Criticism often has a more profound and lasting impact than praise. Praise, many of us believe, is often "phony" and may be motivated by a desire to manipulate, to win some concession, or merely to maintain the flow of conversation in a socially acceptable manner. "She just said that to be nice." "I bet you say that to all the girls."

By contrast, although we may well realize that the person who has criticized us may have "just been trying to be mean," we don't assume that they probably "say that to all the girls." Rather we tend to think that they believe their nasty comments, and that there may in fact be some truth behind their evil words. That is, we tend to take criticism more personally and may even think we are in some way responsible for what is being said.

Having been criticized as children, similar treatment later in life has a familiar ring to it. Although we do get praise as children, even at a young age some of us begin to suspect that praise may not be truthful, so we tend to feel critical words much more deeply and sharply than anything positive, and these nasty characterizations tend to stay with us longer and to leave a deeper impression because of their more powerful impact. However, just because something unpleasant has been said does not mean they it reflects anything accurate about us. Indeed, many times criticism often more accurately reflects something about the critic than the one being criticized. One should avoid taking as personal what is best applied back to its source. Many people often beat on their own unknown, contemptible face but fling their accusations so as to strike innocent bystanders.

Avoiding Negative Criticism

It would seem to be normal to avoid people who constantly criticize us or who are abusive. Those who avoid negative criticism usually do so because they do not like it, because it does not meet any conscious or unconscious needs, and because they view it as harmful, destructive, unpleasant, and unfamiliar, although its familiarity may also drive us away, reminding us of individuals and experiences from the past that did not make us feel good about ourselves.

Similarly, those who tend to like themselves avoid it and the people who provide it owing to its aversiveness and unfamiliarity. They realize the abuse has nothing to do with them and they do not take it personally. Critical people are abrasive, and those who love or care about themselves do not subject themselves to this abuse. Avoiding critical abuse is not a flaw, it is a sign of psychological health. Those who take criticism or even seek it tend to believe that they deserve such shoddy treatment.

Some people fall into the trap of personalizing the critic's comments and try to change in order to feel good about themselves. Nevertheless, those who provide what is aversive and destructive should be viewed for what they are: aversive and destructive. Negative people offer negative advice, and the psychologically healthy individual avoids them altogether. Of course, there is constructive criticism, and many people avoid that as well.

Avoiding Constructive Criticism

Constructive criticism is not usually abusive or destructive. It requires us to stop behavior that is not growth-oriented or that is unhealthy, or to learn an easier, more effective way of accomplishing our ends. Constructive criticism places demands on us to achieve our full potential. Constructive criticism says, "Be successful." Negative criticism says, "Be a failure."

It is due to its generally helpful nature that some people may avoid constructive criticism and those who offer it. Helpful and genuinely concerned comments, like compliments, may feel unfamiliar. Some people just prefer those who criticize them in a destructive fashion, or those who truly care nothing about them at all. Similarly, if constructive criticism is offered it will be rejected because it offers the opportunity to break unhealthy patterns. In addition, helpful or truly constructive advice is threatening and is avoided as it is unfamiliar and requires one to cease to behave in a self-destructive manner. Hence, good or healthy people are often avoided.

Seeking Negative Criticism

Negative criticism is admittedly unpleasant. Some people find it familiar (since they have been treated this way before) and acceptable because they unconsciously feel that they deserve to be criticized. If their parents treated them badly, the criticism of others will feel "right," and they may even seek out people who treat them badly.

Some people need to be involved in punishing relationships, as lack of unpleasantness generates anxiety mixed with guilt. If they are treated well, they anxiously anticipate the bad that they know will happen sooner or later, so they seek punishment to ease this pervasive anxiety.

Some people will in fact provoke an abusive partner by making nasty, denigrating, or critical comments, or engaging in some behavior that they know will cause an outburst. They seek abuse because its absence is unusual and makes them feel anxious. The punishment, rejection, and criticism is thus a form of relief as well as the result of an attempt to maintain the familiar. When abuse is forthcoming, they no longer have to worry about it. Indeed, abuse, criticism, and fighting can be habit forming, and some people become anxious until they get their next "fix."

Abuse and Control

By attempting to criticize, control, or make a partner feel inadequate, the abusive spouse may resort to ultimatums, tantrums, denigrating comments, yelling, threatening, relentless criticism, or the cutting off or complete control over funds and finances. Essentially, such individuals are acting Parental as well as quite childish, because their Lesser Ego is activated and angry, and wants to get even.

Some verbally, emotionally, or physically abusive individuals will act on their threats or will break and hurl objects around the house to make their point or to punish their spouse for any behaviors which they view as rejecting or an attempt to resist being controlled. Its final form is actual physical assaults. Today the wall, tomorrow your face. However, it is often their own unknown self, the stranger within, that they are attempting to destroy.

They may be acting out their childish desire to control their mother, to retaliate against their father, or to treat someone else in the way they were treated as children. Some individuals, recognizing the hurt and angry little boy in their spouse, take on the role of a child themselves and try to rescue the spouse and thus win the love they themselves were always denied.

Jean never for a moment thought she was in an abusive relationship, although she was quite frightened of her live-in boyfriend, Mark, who sometimes slapped her and pushed her around. For several months, she had been dealing with his temper tantrums, which usually consisted of yelling and screaming and accusations about her morals and character.

Mark didn't like the way Jean dressed, the way she put on her makeup, her friends, or her family. When she got dressed up, he complained that she looked like a tramp and needed to "wipe that red shit off" her face or put on something that didn't make her look like a "prostitute." He called her friends low-life and trash, and he "couldn't understand why" she even talked to them. It seemed that no matter what she did, there was something wrong.

According to Jean, a day did not go by without some major blow-up. However, she remained convinced that these problems were actually her fault. Even when he slapped or shoved her, she believed the reason was her failure to be understanding and loving enough.

As the slaps and pushes were infrequent, not excessively violent, and more frightening than anything else, she was also convinced that she was not being abused. Besides, she knew what real abuse was, for her father was a drunk who had terrorized his family with extreme physical violence. Mark was an angel compared to "that bastard."

One evening, however, as he was preparing to shower, they agreed it would be great to have some ice cream. He told her the flavor he wanted and she left to buy some.

When she came home an hour later, he was livid: "You goddamned thoughtless bitch, where the hell have you been? Out screwing your old boyfriend. . . ."

As he continued his tirade, Jean tried to explain that she had needed gas, and that the oil light had then gone on, so she had had the attendant take care of that as well. And when she got to the store, she had run into an old girlfriend and they had started talking. And then there was only one lane open at the store, and there was a long, long line. . . .

Nevertheless, before she could finish, he slapped her hard in the face; yanked the packages from her hands, threw the ice cream against the wall, and then, taking a handful, grabbed her by the hair and, wrestling her to the floor, tried to grind it into her face and mouth, all the while yelling, cursing, and yanking her hair.

The next day, he apologized profusely and said that he had acted that way only because he had had a really bad day, was tense, and had been worried because he loved her so much. He began to cry, telling her that it would never happen again and that he was terribly ashamed of himself. However, although she wanted to forgive him, she was so confused and frightened that she did not know what to do. As he cried, she began to feel sorry for him. He was crying the tears that she couldn't, and she gave him the understanding and acceptance she craved.

Many women tolerate such abuse. It may be a resurrection of the familiar past. Owing to its familiarity it is tolerable, at least to the unconscious self-image. Many women stay in abusive relationships because they and their unconscious Child want to believe the apologies and the promises that there will be no more hitting or fighting. This same Child also believes that the abuse is deserved and is willing to tolerate it, despite the protestations of the conscious mind to the contrary. Sometimes even the conscious mind conspires against a person's best interests, for if the left brain and the language axis have heard something said enough times, they tend to accept it and any guilt that comes with it.

A person who had not been raised in an abusive, neglectful,

adulterous, or alcoholic environment, and who did not have enormous feelings of being "not OK," would find abusive behavior on the part of a spouse unacceptable at a conscious and unconscious level. It would not feel familiar and it would be avoided. That person would say "good-bye," leave, and congratulate himself for escaping before it was too late.

Manipulative Games: "It Didn't Happen" and "You're Crazy"

In abusive or dishonest relationships, one partner often attempts to control and manipulate the other by questioning the other partner's intelligence, the accuracy of her or his perceptions or memory, and sometimes even her or his sanity. Typically, this questioning takes the form of a series of games, which can be called, "You're Wrong," "It Wasn't Me," "It Didn't Happen That Way," "It Is All in Your Imagination," or "You're Crazy." These games are begun by the Parent, which demands that you respect its authority and denies your own perceptions, desires, or needs, as you are only a child and the Parent is the all-knowing deity that rules the family. Given the left brain's propensity to believe what it hears, it can often be persuaded that any feeling or awareness that contradicts what is being said should be ignored or dismissed.

When the event involving the ice cream was discussed with both Jean and Mark, he at first claimed to be unable to recall the incident. When Jean insisted and retold the story from start to finish, Mark began to argue that, when she had come home, she was acting strange and nervous and seemed upset. When he had begun to question her casually about her whereabouts she acted evasive, couldn't give him a straight answer, and started to become hysterical. He hadn't shoved or slapped her; he was trying to calm her down and get a straight answer out of her. When he had tried to help her with the bag, it ripped open and the ice cream fell out:

> Sure it splattered, and that made me mad. But I sure as hell didn't hit her with it. How can you rub hard ice cream into someone's face? I don't know why she got so scared. It makes me wonder what's she trying to hide. I mean, maybe

she's got a guilty conscience about something. But that rubbing ice cream in her face. That's crazy talk. Yeah, maybe she thought I was going to, but I didn't. That's all in her imagination. You know, it's weird how she always wants to blame me. All she ever can remember or talk about is the bad stuff. No wonder our relationship is on the rocks. What about all the good stuff? Why don't you ever remember that?

The person who begins this form of game manages not only to deny reality but to shift the blame onto the victim. "What's she trying to hide?" "All she can remember or talk about is the bad stuff." The Parent forces the partner to become the bad little child whom no one can love and the victim's own left brain begins to accept the words being thrown at it with such force. Indeed, both partners are self-deceived.

Insofar as Mark was concerned, Jean was to blame, or so he wanted me and Jean to believe. She was the source of all the problems in the relationship. With this attack, Mark tried what had previously been quite successful: projecting the blame onto his partner and forcing her consciously to acknowledge and accept it. He thus escaped all responsibility for his behavior. Indeed, her complaint was proof of her badness and inadequacy. She had acted her role as Child, and her spouse had acted the avenging Child/Parent, and the left half of each of their brains had accepted all his deceptions as valid.

Modeling and Mimicry

Many children consciously and unconsciously identify with and mimic the behavior of one or the other parent, or even both. The impressions made by parents or other authority figures constitute the Parent ego personality. Later in life, this complex of impressions, actions, and feeling states may exert tremendous influences on our behavior, emotional functioning, and the way we treat ourselves and loved ones:

Mary observed her father yell and scream at, occasionally hit and manhandle, and constantly belittle her mother. This behavior was always frightening and upsetting to her, although she herself was spared most of the abuse. It seemed that almost anything could set her father off, whereas her mother seemed always to be cowering and suffering attacks of anxiety, fearing what he might do or say, even when he was not home. Her mother's cowardice and anxiety and the mousy way in which she deferred to her husband bothered Mary as much as her father's outbursts. This behavior embarrassed her.

Whereas Mary feared her father, she felt utterly contemptuous of her mother for being weak and for taking the abuse. Thus, she learned to value the behavior of her father over that of her mother, and later in life, she began to mimic him in her own relations. In striving to not be like her mother, she sought to be like her father and learned that the way to not be trampled was to take the initiative by yelling, screaming, and browbeating one's mate.

Although she never made a conscious decision to follow this route, it was the one she took, and she ended up in several relationships in which she was "the boss" and would easily lose her temper over some trifle if things were not done her way. Indeed, even in her job in customer service at a local store, her argumentative Parent, which had always been right, would come to the fore and do its best to humiliate and thwart the desires of anyone with a complaint.

Mary's older sister learned exactly the opposite lesson. She learned that it was normal for a man to mistreat a woman, and for husbands to control their wives by fear, threats, and intimidation. Consequently, when she married she found a man just like the man that married dear old mom.

In contrast, the youngest child, their brother, Tim, although he, too, felt contempt for his mother, also felt sorry for her, particularly because she frequently turned to him for consolation. However, rather than act like his father, he maintained the role foisted on him by his mother and, as an adult, found himself attracted to helpless, inadequate, sometimes emotionally disturbed women in need of rescuing. Essentially he was still trying to rescue the mother whom he had failed as a child.

Nevertheless, rescue can be a form of control, and he was still playing one of the games he learned from his father. He learned that women need to be controlled and that the way for a man to deal with a woman is to control her by making her completely dependent on him. Indeed, by controlling others, whether his spouse or his children, he was saying, in effect, "You are inadequate. You are helpless. Without me you are nothing."

Abusive Relationships

The Hurting or Angry Child and the Rejecting Parent

Children who have been severely criticized, rejected, abused, or neglected sometimes harbor intense feelings of anger and resentment and a desire to strike back at the parent who hurt them. In many instances, such feelings are threatening to the child and are suppressed or never emerge from the unconscious into the conscious mind. The few children who dare to express these angry feelings are usually criticized or even beaten into submission. Others turn these bad feelings back on themselves and feel even worse.

Children who strongly feel this resentment and anger sometimes fully express them when they reach adulthood. These abused individuals sometimes become highly critical, rejecting, or abusive of others. One need not ever have expressed these angry impulses during childhood. Just the fact that they existed may give rise to their expression later in life. The Parent ego personality not only continues to reject, criticize, and abuse the internal Child, but it lies in wait for the opportunity to mistreat other victims as well. In addition, if the unconscious Child also harbors considerable anger and resentment, it may attempt to strike back at the unconscious Parent, represented in authority figures, loved ones, a future spouse, or anyone else who makes it feel inadequate and insecure. Thus, due to the combined angry influences of the unconscious Child and the unconscious Parent, people may behave in an abusive, neglecting, or controlling fashion. Their conscious and unconscious self-images tells them that this behavior is OK.

Unconscious Compatibility and Bad Needs

It is not at all uncommon for those who have incorporated the abusiveness of their parents in the form of a Parent ego personality to find a spouse who is similar not only to their unconscious Child, but to their unconscious Parent as well; that is, a parent who will abuse them back. Relationships between people who are similarly abusive are characterized by constant yelling, fighting, nitpicking, criticism, and attempts to control one another. At one moment, one spouse provokes the fight, and the next moment, the other spouse is getting even o.

voicing his or her own independent complaints and disappointment. In fact, they often trade off being victim and abuser.

Such a relationship may seem ideal, as both partners can simultaneously express their anger and experience the abuse that they unconsciously feel they and their partner deserve. Often, in fact, this is the only reason they are together, that is, to express and fulfill bad needs.

I have frequently had couples come to my office who are seemingly opposites and totally incompatible, at least as conscious needs are concerned. And yet, although the only thing they have in common is their fights and complaints, they stick together. For instance, he likes to travel and she likes to stay at home. He likes movies and she hates them. She likes to go for walks and he prefers to sit in front of the TV. She likes visiting with friends and he prefers to stay home and read. She is a teetotaler and her husband drinks himself into oblivion. The only thing such couples like to do together is fight and make each other miserable; in this regard, they are completely compatible, but for all the wrong reasons.

Most couples refuse to believe this when it is pointed out and often argue that this cannot be so. Of course, because they (unconsciously) enjoy arguing, this response is not completely surprising. Nevertheless, most respond incredulously:

Patient: Are you trying to say that I go home and have terrible fights because I have a need to be yelled at? Because I want to be yelled at?

Dr. J: Not exactly. However, at an unconscious level a bad need is being met. The need may be to fight, to yell, to act rejecting or to be rejected, or to strike out at those you care about because at some point in your life you were repeatedly criticized, rejected, made to feel inadequate, or badly hurt by people you loved and who supposedly loved you.

Patient: Well, I know what I feel, and I damn well know I don't need to be yelled at. The only needs I have are good needs. I don't even seen how there can be bad needs.

Dr. J: Consider a husband and wife who day after day, when they both come home from a hard day's work, hug, cuddle, and kiss and more often than not make love. Certainly, it seems reasonable to assume that they have a need to hug and kiss, and that they have a need to make love. Interacting in a loving manner meets certain needs. Otherwise, they would not treat each other, and themselves, this way. Right? Now consider a couple who day after day yell and scream and call each other horrible names; this goes on for years. Now, do they scream and fight because they don't want to? Why is it easy to believe that those who treat each other with love are meeting certain needs and hard to believe those who fight are also meeting certain needs, in this case, bad ones? The people who make love and those who fight do so because they want to.

The reason it is so hard to believe is that it sounds so bizarre to the left half of the brain, which is out of touch with its own feelings. It likes to believe its own words and explanations, which often have nothing to do with what is really going on. And of course, there are explanations for why people maintain a bad relationship: Brain tumors. Alien forces taking control over one's mind. Bad chemicals circulating through their brains. Genetics. Bad karma. The Devil. Drugs. And so forth.

And indeed, events and forces that having nothing to do with how one was raised are often responsible for fights and incompatibility. Traumas that occur during adulthood can also be the source of conflicts and complexes. People change. Some grow and some don't. People grow tired of one another, or they meet someone else who is more compatible. Twenty years of the same annoying habits may become intolerable.

Nevertheless, many couples who are unhappy, miserable, and consciously incompatible, who fight and incessantly argue, stay together because of similarities in the dynamics of their minds. Like belongs with like.

Because such couples are such a perfect fit—that is, insofar as unconscious needs are met—they are often unwilling or unable to separate and divorce. Indeed, they will invent innumerable reasons why they cannot separate—because they "love each other," they "need each other," they "do not want to be alone," or because of their religion or their children. However, if one asks them what they "love" or what they "need" or point out that their children might be better off after a divorce, they will be stunned or frightened.

Such people, like other victims and abusers, are reluctant to leave a bad situation because, unconsciously, they can reenact and relieve the turmoil and pain that characterized their young lives and, possibly, this time, finally achieve the love and acceptance they so desperately longed for then. This almost never happens. The love they want will never be forthcoming until these two abusers quit blaming each other. Once they accept responsibility for their own misery and stop their unconscious Parent from abusing themselves and each other, positive growth and a mutually satisfying relationship may be possible. However, as long as they continue to act "parental" or, conversely, "childish," the possibility of a happy relationship is essentially null—unless, of course, they are happy making each other miserable.

Child–Child Interactions: "Please Change and Make Me Happy"

Surprisingly, rather than simply tell one's spouse to act either like an adult or move on down the highway, some people attempt to appease this internal Child or acquiesce to the critical Parent. Paradoxically, rather than treat the Child for the child he or she is, by ignoring, restricting, and punishing it for its bad behavior, many of us allow the Child and the Parent (i.e., the Lesser Ego) to dictate how we should behave. The Lesser Ego says, "Make me happy and secure or I'll make you miserable." Individuals who fall into this trap begin to feel that the only way to appease this Lesser Ego is to act as it desires and dictates. As one young woman stated, "If I do what he tells me, I can usually make him treat me better."

There are several different fantasies at work here. One is that if "I" change and quit being such a bad "wife" or "husband," my partner will someday take me in his or her arms, hug and kiss me, tell me how much he or she loves me, and we will live happily ever after. However, this is not what the abusing spouse needs. He doesn't really want her to change, as his need is to regurgitate all his anger and hurt: If "I" can say things that hurt you, then I know that you care.

A related fantasy is that with love, tolerance, and understanding, the abused partner will somehow cause the abuser to change and will finally get the love he so desperately desires. That is, by proving her love and faithfulness through the acceptance of abuse, the abuser will come to realize that he is loved and can feel secure that the abused spouse will not hurt him (as he was presumably hurt by his last partner). Unfortunately, rescuers need someone to rescue, and they do not really want their abusive (or alcoholic, or drug-addicted, or criminal) spouse to change.

Some people respond to the unconscious Child or Parent residing in their mate by acting as a Child or Parent themselves. One Child/

Parent says to the other, "Is that all you can do? Just sit in front of that TV and rot? No wonder we never get along. Look at you, turning into a couch potato." The other Child/Parent responds, "Yeah, well at least I don't smoke like a chimney stinking up the whole house." And on and on.

More often than not, my patients say they want their mate to change, as if that would fix the problem. The notion that maybe they themselves should change or else quit the relationship usually draws a blank. If pressed further, these patients confabulate: "Well, I did try. I told her that, if she quit watching so much TV, we would have more time together, and I would quit smoking." Or "First you change, and then maybe I will too." In essence, each internal child is saying, "You first, you first," and the left half of her or his brain believes everything it is demanding and rationalizing. However, a spouse whose Lesser Ego demands that the partner change to suit his or childish or parental needs is really saying, "Allow me to say and do whatever I want to you, and if you take it, I will love you." Unfortunately, if the victimized spouse subjugates herself to his will, he will not love her; rather, he will despise her and hold her in contempt.

Sex and the Passion of a Good Fight

Love as a Reward for Abuse

Some people are reluctant to break off relationships with abusive partners because they view the little love they occasionally receive (and the occasional good sex, which seems, paradoxically, to go hand in hand with abusive interactions, at least during the initial stages) as a privilege that only this person can grant. Unconsciously, they may think: "Because I was never given love by the people I loved when I was little, this new person is not only important to me, but necessary. If I lose him, no one will love me again. I know this, because, no one loved me before."

In addition, many individuals perceive the "love" they receive from their abusive, critical partner as a reward for the abuse they have recently suffered. When abuse is eventually followed by "love," the love makes the abuse even more tolerable.

Take, for example, the principles of reinforcement as they are applied in gambling situations. You put your dollar in the slot machine and you lose. You do this again, and again lose. This is followed by a third and fourth try at which point you are about to give up and move on to the next machine. Losing is not fun. However, on the fifth try you win $10.00. You are now willing to put up with three or four losses because they pale in significance to the possible reward that you might receive. Indeed, the tension and excitement of knowing that a winner may be coming up drives you to continue putting money in this machine, until finally at some point you rationally realize that your losses now far exceed your winnings. Unless you are addicted to gambling and the pain of losing, you walk away.

The same thing happens in abusive relationships. The anticipation of being loved, that wonderful high that "only" one's partner can provide, drives some to try, try again, since they know that eventually their partner will provide them with the "fix" that they need.

Fortunately, logic sometimes prevails over delusions, and the abused spouse is able to end the relationship. Regrettably, sometimes the strength of the emotional attachment and the dependency win out, and the person is drawn back to the abusive relationship. Because they are addicted and have occasionally received a loving reward, they do not walk away.

When a Bad Fight Equals Good Sex

There are negative and positive arousal and excitement. The excitement of a fight, the perpetual uncertainty regarding an unstable relationship, the anticipation of the next blow-up, and the ever-present fear of and potential for abandonment and rejection—all give rise to considerable negative arousal, which is often misinterpreted as passion. This form of excitement is not only stimulating, but sometimes very attractive to those who seek the familiar negative drama of their dysfunctional childhood homes.

Fighting and Love

Children who are raised in homes where parents frequently fight and argue, or where there is a threat of possible physical violence, eventually begin to view fighting as a normal aspect of a loving relationship. One is supposed to yell at, and be yelled at by, people who "love" one. Such homes are also characterized by tremendous tension, which is usually released by fighting or lovemaking. In some families, arguments are, in fact, followed by equally passionate displays of love, affection, and sex, at least in the early phases of the relationship.

Children raised in such homes often are given the message that a normal home life is characterized by tremendous swings and extremes in emotion. Frequently, later in life, these children continue unconsciously to associate arguing and fighting with love: The more people fight, the more in love they must be.

Fighting and Sex

When one is experiencing high levels of negative excitement and the associated arousal, such as during and immediately following an argument, these tensions are sometimes released in sex. This association is not entirely abnormal. The nuclei of the limbic system control basic emotions such as anger, aggression, and sexual arousal, and when this region is activated, associated mood states may be triggered. That is, just as the right hand may be activated when one is speaking because hand and mouth control are located in immediately adjacent cortical areas, adjacent tissues connected with seemingly different emotions may also be aroused.

Sex is both psychological and biological and is mediated by the limbic system. Thus, it is based on an arousal of tension, and the release of that tension completes the sex act. The limbic system mediates not only sexual arousal, but feelings of love and affection, the need for intimate association, and aggression and the fighting response. Indeed, aggression and sexuality are linked not only by the primitive structures of the limbic system but behaviorally as well. Among animals, the male often must behave aggressively in order to fight off other males and win the female, who, in turn, must often be pursued. Similarly, human males often have to take the active, assertive, aggressive role or face the prospect of being alone and without a sexual partner. To engage in pursuit, one must feel a certain degree of arousal and tension.

Among couples who frequently fight, there is necessarily a buildup of tension. The tension increases in anticipation of a fight and during the fight as well. When they are not fighting, high levels of arousal and tension remain either in anticipation of the next fight, or as the residue of the previous aggressive encounter. These feelings must be dealt with, and many couples who fight interpret this tension as a desire for sex, and its release is often experienced as great sex. However, its source is not truly sexual, for what appears to be sexual arousal is often based on the conflicts and strains of the relationship. Indeed, if these couples did not fight, they would rapidly lose sexual interest in one another and would no doubt rapidly drift apart. If there were no battle, there would be no heat and no passion. The relationship would then be seen for what it is, empty and insipid.

Sometimes, it is important to one or both members to put a lot of energy into the lovemaking not because of the tension of a fight, but because each fears that he or she may have lost or irretrievably pushed away the other. Sex thus serves as a means of reestablishing the bond, as a form of reassurance that "you are mine" (or that "I am yours") because you gave yourself to me or accepted the love I offered. Sex thus serves to validate the relationship and to ease the fears of the shaky foundation loosened by the tremors of the latest battle. In effect, it is a statement: "Look how much we love each other, how close we are, how good it really is, how good we can make each other feel, how much we belong together." Yet just like any drug, the feelings of passion or the sexual high is short-lived, and the fighting must resume.

Moreover, having sex or passionately making up after having a fight inadvertently serves to reinforce by rewarding it, the fight, like getting a cookie for being bad. Indeed, the fighting may become a form of foreplay, and the couple may begin to look forward to the fight (at least unconsciously) as a prelude to some great sexual acrobatics or at least the chance to obtain temporary passionate reassurance.

Many of us have learned to equate sex with love, and good sex with real love. Hence, even if the relationship as a whole is quite unsatisfactory, the good sex more than convinces some of us that it is "right" and that the sex couldn't be so good and satisfying if it were not fueled by real love and affection. Indeed, sex may be the only way that two such individuals are able to successfully relate to each other intimately. Unfortunately, however, they have mistaken great pain for great passion.

Addictions to Bad Relationships

Sometimes, after a terrible argument or a series of fights, one or both partners are temporarily convinced that they never want to see the mate again and can happily live without her or him for the rest of their lives. They may decide that the relationship is over, begin to make plans to move out or on, and then, before a single day has passed, begin to feel anxious and lonely and want to forgive and forget. Such feelings are, again, under the yoke of the limbic system and in many ways are a replay of the infantile reactions of rage and anxiety that arose when Mom left the room.

It is not at all unusual for a person caught in an abusive love-hate relationship to want to leave and call it off but find that he or she cannot. No matter how bad the fights, no matter how many times the couple break it off and run away, they begin to feel even more terrible after they end it and are then drawn back for another "fix," and for another second chance to obtain the love they have sought since childhood. They may grumble that they"can't live with 'em and can't live without 'em"; they are just "too in love." This is a form of confabulation.

Although it may appear that they are addicted to "love" and seem to have an almost compulsive need to be with the "lover," they are, in fact, behaving at an infantile level and at the behest of their limbic system and the IDfant ego personality. It is also likely that they are addicted to reliving the painful familiarity of the past, experiencing the relief that comes from fighting and then making up. Indeed, after they break up or have a terrible fight, all the pain, rejection, neglect, and worthlessness from the past begins to overwhelm them. Consequently, they desperately seek the temporary reassurance that can be found in the lover's embrace.

Such a couple seems to be addicted to maintaining what to all appearances is (except for the fleeting passion, sex, and sporadic good times) a very destructive relationship. In truth, however, if not this abusive person it would be another, for such people seek out opportunities to relive the pain that has plagued them since childhood. In many cases, they are driven by an imbalance in the psychic relationship between the amygdala and the septal nucleus of the limbic system and are so desperate for contact comfort that they willingly establish relationships even with people who will abuse them. Frequently, the worse such relationships become, the harder it is for them to let go. Indeed, the more difficult it is to end a bad relationship, the more elements of one's early childhood it contains. The unconscious reasoning which sustains this behavior goes as follows: "I am terribly depressed and miserable without you. Although I am also miserable with you, at least I sometimes also feel really good; at least I feel loved." Unconsciously, there is also the fear "If I lose this person, no one else will love me, because no one has ever loved me before."

Unfortunately, this dependency is misinterpreted as a reflection of the importance of the spouse: "The more I need you (i.e., to fight and to give me abuse), the more wonderful you seem." However, it is not the partner that is important but the kind of relationship that is being maintained and the needs of the limbic system that are striving to be fulfilled.

Taking Responsibility and Gaining Self-Control

If one is no longer a child, no one has the right to act as that person's parent. If one is not interacting with a child, one doesn't have the right to treat that person as a child. However, to be treated as an adult, a person must be one. If one wants to be loved by an adult, one must be with an adult. If one wants to be treated with love, one must first love herself or himself and treat others with love as well.

If a person can begin to recognize his or her own Parent and Child and their critical, abusive, and hurting ways, and can learn to counteract these forces by rejecting abuse or punishment regardless of its source, he or she can assume self-control. However, one must also learn to question his or her own lies, confabulations, and deceptions, for often one's first explanation of his or her behavior is completely off the mark.

III

THE DEFENSE MECHANISMS

14

Reaction Formation and the Defense Mechanisms

We knowers are unknown to ourselves, and for good reason: How can we ever hope to find what we never look for? There is a sound adage which runs: "Where a man's treasure lies, there lies his heart." Our treasure lies in the beehives of our knowledge. We are perpetually on our way thither, being by nature winged insects and honey gatherers of the mind. The only thing that lies close to our heart is the desire to bring something home to the hive. The sad truth is that we remain necessarily strangers to ourselves; we don't understand our own substance. We must mistake ourselves; the axiom, "Each man is farthest from himself," will hold for us to all eternity. Of ourselves we are not knowers.

—NIETZSCHE

Since the invention of complex spoken language and the advent and eventual dominance of linguistic consciousness, many functions mediated by the right half of the cerebrum and the limbic system have often been viewed as dangerous, sinful, or irrelevant. Indeed, so autocratic and presumptuous is the left half of the brain that not only does it try not to be conscious of many of these natural abilities, impulses, and inclinations, but it often attempts to suppress or discard them as useless and unimportant.[1]

This attitude is very unfortunate, for the right brain has tremendous talent and innumerable capabilities, many of which were millions of years in the making and which enabled our very ancient ancestors to live in harmony with themselves, with others, and with the natural world surrounding them.

Unfortunately, because the left brain often refuses to fully use or is unable to acknowledge this supposedly hidden world and all the possibilities it represents, many (non-sports-related) natural non-linguistic abilities that prevailed for over 100,000 years have been allowed to wither. We have been increasingly taught to abandon the capacities that are associated with the right half of the brain, and instead, we have become slaves to the linguistic whims, confabulations, and rationales of the left brain and have increasingly lost touch with the original mind, which at one time dominated the mental system maintained by both the right and the left cerebral hemispheres.

Many of us have forgotten who and what we are and from whence we came and have been forced by prevailing logic, rationalizations, reasons, and preconceptions to deny the unconscious wealth which can be found within our own mind. This denial has given rise to unnecessary feelings of guilt and innumerable psychic and interpersonal conflicts and has increasingly disrupted the harmony and maintenance of what was over a million years in the making: the family and the capacity to see things as they are. All too often of ourselves we are not knowers.

Indeed, it is a curious thing about the conscious mind that if it is repeatedly told that something is unacceptable or to feel guilty for having or expressing certain thoughts, it and the whole brain may begin to feel guilty and will reject and condemn what in fact may be quite normal. In order to not feel guilty, the conscious mind begins to believe what it has been told, or what it tells itself, and then denies, represses, and explains away even the most natural of phenomena, even when they are an integral part of the self.

Many modern human beings of the Western world are so out of touch with who and what they are, and they have so willingly accepted others' judgments of what is acceptable and normal that they may view even the most natural of body functions such as breast feeding, sweat, body odor, urination, defecation, and even the sexual act, as abnormal, depraved, uncouth, and uncivilized, even when these actions or functions occur within the privacy of their own homes. At a minimum, these topics are a source of embarrassment for many, and there are thus numerous products on the market to make one appear and smell as if a person is other than who they are, even when what they look and smell like is perfectly OK (and by this I do not mean the unwashed body).

There is so much pressure, starting in childhood, to disavow certain tendencies, to wear certain clothes with the right label, to use the right perfumes or colognes, to drink certain alcoholic beverages, or to maintain a certain hair style or color, or a certain weight, waist size, or breast size, that the true self often becomes lost in the process. We are all bombarded with these messages—by advertisers, the media, our parents, our friends, and even the prevailing culture, which constantly tells us even what it is fashionable to think, feel, or believe. The overarching message is "You need help; you need to change," as what you are may not be acceptable.

But acceptable to whom? We must deny and be other than we are because otherwise we may offend someone else's sensibilities. Others may not like us or may reject us, and if others do not like or approve of us, how can we like ourselves? So, if we wish to be accepted, we must be other than what we are; at least that is what many are erroneously led to believe.

Sometimes, what is offensive to one person is another person's mental health, intellectual capability, independence, physical or athletic prowess, beauty, and sex appeal, or even their lack of concern as to what others might think. However, people who are offended, who demand that "you" change in order to appease their sensitivities, are often responding to their own insecurities, some of which have nothing to do with the person being attacked. As pertaining to advertisers, it should be obvious that they are not concerned about anyone's health or psychological well-being, but only in making money and will say or do whatever the law allows to get consumers to spend as much of it as possible.

Some people are easily offended not so much because they wish "you" to change so as to improve yourself as because what they see reminds them of their own hidden self, their own disowned and disguised impulses and desires, or perhaps, their own inadequacies.

The Defense Mechanisms

In the land of the blind,
those with eyes
are said not to see.

Humans are fraught with all kinds of desires and "needs." They have a need to eat, a need to make love, a need to go for a walk, a need to go to the bathroom, a need for knowledge, a need for companionship, a need for physical activity, and the list goes on. Although we may label our desires and needs in a variety of ways (e.g., physical needs, emotional needs), some might best be categorized as "good needs" whereas others would properly be described as "bad needs." For example, the need for drugs or the unconscious need to resurrect certain aspects of a painful childhood, as in the need to be hurt, rejected, or neglected, might best be described as bad needs. Although admittedly the notions of *good* and *bad* are relative and value-laden, what is meant here by *bad needs* are those which are harmful or self-destructive.

All desires, impulses, and needs, regardless of their being good or bad, are generated by the various regions of the brain and are responses to our internal and external environment and to how others treat us. Our needs and desires are thus shaped by both conscious and unconscious forces, and by the conscious and unconscious self-images.

We deny some of our needs and impulses, and we disguise or misinterpret others in accordance with how we see ourselves and how we think others view us. Some desires and unconscious impulses are not acceptable to our self-image. When we do this, conscious and unconscious conflicts sometimes result, and we may fail to recognize and meet our real needs. This failure may leave us in a state of deprivation and dissatisfaction, which may generate feelings of anger or depression as well as considerable tension.

Individuals employ a variety of defense mechanisms. A defense mechanism is a protective strategy most often used by the conscious mind and the left brain. Defense mechanisms serve to prevent conscious recognition of information that is in some manner threatening to the conscious self-image. However, the conscious mind has to have at least some notion of what is threatening in order to defend against it.

Some forms of information are simply too overwhelming, too threatening, and too difficult to deal with or to confront openly. The suspicion that one's spouse may be having an affair may be too threatening and painful to acknowledge. Consequently one may consciously decide to ignore any clues picked up by the right half of the brain which may force one to read the "writing on the wall." The person may engage in denial, in which case they simply deny the obvious. They may engage in suppression, in which they simply refuse to confront consciously even the slightest hint of betrayal. Or they may engage in "rationalization," such that they make up and confabulate "rational" explanations which explain away the obvious. The left brain is very good at fooling itself.

Reaction Formation

Mary had been raised in a small farming town by fairly strict parents who emphasized conservative moral and religious principles. Mary, in fact, saw herself as a fairly conservative, moral individual.

When she married, her husband was the first and only man she had ever made love to. Although she maintained her conservative viewpoint and lifestyle, Mary discovered that she loved sex and became completely uninhibited in the bedroom, to her embarrassment and the delight of her husband.

Two years into their marriage, Mary and her husband moved to a big city and bought a small house. On the day they moved in, Mary noticed that their next-door neighbor was in his yard working. His shirt was off, his chest was hairy, and she was attracted by his muscular arms and shoulders. He looked up at her and smiled, and she smiled back. That night, to her dismay, she found her thoughts drifting to him when she and her husband made love. These thoughts made her feel very guilty, so she tried to blot the neighbor out of her mind.

Two days later, he came over and introduced himself. Feeling extremely nervous and uncomfortable, she was not very nice to him (this behavior was a reaction to her original impulse and desire). That afternoon, she found herself thinking about him over and over, and that night, as she and her husband made love, she began to imagine that her husband was the man next door. She felt nervous and upset and tried to blot out the image. Then, midway in their lovemaking, much to her husband's chagrin, Mary suddenly said she was not feeling well.

The next day and the next, she refused to have sex with her husband, complaining of not feeling well and of having no sexual desire. She began dressing more conservatively than was her nature, and for the next three months she consented to have sex only once. During those three months, Mary and her husband had frequent arguments. She complained bitterly that he was just "using her as a sex object" and that she was "better than that," had not been "brought up that way," didn't have "any sexual desire," and had "never really wanted to make love anyway." She claimed she "just did it to please" him but was through with that because she was "not some wanton hussy" and he was no longer to treat her like a "whore." When he threatened a divorce, she sought counseling.

Essentially, Mary was using the defense mechanism called *reaction formation* as well as *denial* and *projection*. She denied having sexual feelings. She projected and blamed her bad feelings and guilt on her husband (thus enabling her to ignore the real source). And she protected her self-image and the image she presented to others by behaving and reacting completely opposite to what she was really (unconsciously) feeling; "I'm not a whore. I am a nun!" She reacted by forming feelings and ideas completely opposite to what she was truly desiring. By engaging in these defense mechanisms, she was able to defend herself from consciously confronting certain unconscious impulses and personal characteristics that she found abhorrent.

Although by using these defenses, she seemed to be countering a "bad need," she was actually confusing a good need (sexual desire) with a bad need (a desire for her neighbor), stifling both and thus damaging her relationship with her husband.

Essentially, Mary felt a desire which was completely contrary to her moral and conservative conscious self-image. Because she felt sexual desire for her neighbor, she in turn (i.e., her unconscious Parent and conscious Self) felt like a "wanton hussy" and a "whore" (feelings she projected and blamed on her husband instead of the fellow next door). She could have utilized a different defense mechanism such as projection. That is, she could have decided that it was her neighbor who had these desires so as to explain away these bad feelings by blaming him for making her uncomfortable.

Being a whore or screwing her neighbor were completely contrary to her conscious self-image. So to defend herself from feeling like a whore she had to react by denying her sexual desires. She reacted to these desires by claiming (and believing) that she had absolutely no sexual feelings. By denying her sexual desire she was able to deny her desire for the man next door and thus preserve her self-image. "I couldn't possible want to screw the man next door. I was not nice to him. I have no sexual desire. I am not a whore!"

Reaction formation allows people to maintain their conscious self-image by forming and engaging in behaviors or thinking certain thoughts which are in all ways opposite to their original unconscious impulse.

People who have an unconscious "inferiority complex" may also attempt to deal with their fears and feelings of inadequacy by reaction formation. They may develop a "superiority complex," the flip side of the same coin. Essentially, they are saying: "I am not inferior. I am superior." However, their exaggerated way of responding (i.e., by acting superior) is a pretty reliable indication that they are, in fact, fearful of being seen as inferior. Their conduct and attitude are designed to hide the truth from themselves and others by acting in a way that is the opposite of what they unconsciously believe and feel.

Suppressed Desires May Exert More Pressure and Become Exaggerated

For some people an unconscious need, desire, or tendency may be so contrary or upsetting to their conscious self-image that they go to great lengths to suppress, deny, compensate, or overcome their feared unconscious impulses. When this occurs, sometimes the continual effort to deny and suppress a desire results in its gaining strength, like steam in a pressure cooker.

Subsequently, the original impulse may burst forth in hugely exaggerated dimensions. This probably explains to some degree the cases of famous (and not so famous) evangelists who in the thick of the night seek out prostitutes or engage in other "scandalous" behaviors that during the day and among their flock they thunderously condemn; i.e., reaction formation: "I don't want to have sex. Sex is evil and bad."

Unfortunately, when an impulse is suppressed, it often gains strength and may actually increase its desirability by giving it a certain extra allure; i.e., forbidden fruit.

For example, the very reverend Thomas had a dutiful wife, a large congregation, his own radio ministry, and sometimes appeared as a guest on the show of one noted television evangelist. Where other preachers of his ilk were concerned with "Communism," "Democrats," and the need for more money, Thomas wanted to wash the world clean of the evils of sex.

"Sex and fornication," he would shout, his eyes bulging, "It is not the heat of passion, but the furnace of hell. Sex is a sty in the eye of God. Pluck out thine eye if it offends thee. The Devil walks among us, and his name is sex. Sex is the blight of humankind. We have become a cesspool of sinners. Pornography, abortions, disco dancing, adultery—sex is the key to the gate of hell."

Given his fervent shouting about the evils of nonreproductive sex, his congregation and ministry were aghast when they saw his picture in the local paper after his being arrested for soliciting sex from a police decoy. What was even more shocking was that the decoy was male.

"It was the devil," he sobbed. "The devil is trying to destroy God's work. Don't let him succeed. This is proof of the Devil's evil. Don't be fooled by the Devil. Send me your money and help me spread the word."

And he, like several other noted evangelists, actually seemed to believe his own denials.

It is indeed possible that those evangelists who have sought out prostitutes or members of their own congregation for adulterous sex (while vehemently condemning the very behavior they were engaged in) consciously believed their own lies. It is also possible they believed that "The Devil made them do it." In any case, these evangelists may have suppressed and denied normal sexual desire because they considered it sinful, but, denying the desire did not make it go away. The desire, like a hunger for the "forbidden fruit," only grew stronger and more intense, until it overcame them.

Compensation and Reaction Formation

It is not at all unusual for a person actually to employ a number of defense mechanisms simultaneously, such as reaction formation, compensation, self-deception, and denial.

For example, a man raised in an environment in which tenderness, nurturance, or femininity (that is, within a man) are treated with derision, may deny any "soft," "caring," or nurturing qualities within himself.

If "Rambo" has consciously accepted that a man must have no "feminine" characteristics and has fashioned his self-image accordingly, then any tendencies he possesses which could remotely be considered feminine would have to be reinterpreted, misinterpreted, denied, and suppressed. He will subsequently be forced to exert a tremendous amount of energy hiding these tendencies from himself. Consequently he will feel a considerable degree of tension as well as self-doubt.

What does he do when feeling such tension? He utilizes it in an adaptive fashion to eradicate his self doubt and any possible suggestion that he is not 100% prime beef MAN! He begins to run and lift weights on a daily basis, takes classes in martial arts, wears heavy gold chains around his neck (which, he would be surprised to discover, some people consider feminine), and joins the Air Force in order to become a jet fighter pilot.

Unfortunately, although he has succeeded in becoming an ace pilot and in developing muscles where no muscle has gone before, he still feels unhappy. Something is missing in his life. He continues to feel self-doubt and thus becomes depressed. His unexpressed needs continue to strive for expression and he feels tension and conflict. The pressure cooker is cooking.

How does he resolve this? He begins to take steroids and anyone who dares to even subtly question his masculinity is beaten to a pulp.

What has driven "Rambo" to become supermacho man? It could very likely be his unconscious femininity. Indeed, the more masculine or "macho" a man feels he must be, the more he may be trying to eradicate his "soft" inner feminine core. He thus overcompensates for a feared personal defect.

What this particular "Rambo" has done is misinterpret his desire and impulse to be "caring" and "nurturing" and reshape these tendencies in accordance with his conscious self-image. He has responded to these impulses by behaving in a manner completely opposite to his real nature. He has engaged in denial and reaction formation. He will not be caring, he will be brutal. He will be brutal and overly manly because he needs to prove to himself that he is not in any manner feminine.

If these impulses and the tensions driving them were not there initially, the energy to drive his masculine quest would have been missing. The tension of his original unmet needs thus continues to plague him, causing pangs of unhappiness, anger, and self-doubt. The tension mounts with each passing day.

Like the evangelist, "Rambo" feels increasingly threatened by these now exaggerated unconscious tensions as the pressure cooker boils. As they grow, he begins to fear that these impulses and tendencies represent something much more terrible than a desire to be "caring" and "nurturing." By being suppressed they have also begun to take on exaggerated importance and dimensions. Now his fear of what these impulses imply also increases. By feeling his own fear, the threatening nature of these impulses deepens. He has created a vicious circle and a downward spiral.

Though being a caring, nurturing human being is a natural and normal quality possessed not only by females, but by most members of the human race, he feels beside himself. Unfortunately for "Rambo" the exaggeration of these disguised impulses now threatens his entire self-image, and he begins to fear that he is a homosexual. If he had confronted his own unconscious tenderness, he would not now feel overwhelmed by impulses that he considers horrible. His response is to engage in "gay bashing."

He thus makes use of defense mechanisms such as compensation, denial, and reaction formation to protect his conscious self-image. He thunders to himself: "I am not a queer. I hate queers."

Isn't it curious that some men who are repelled by the notion of two men "touching" one another in an "intimate manner," feel compelled to make intimate physical contact with them, that is, in the form of physical blows. Is this yet another disguised impulse and need? As one homophobe told me: "I couldn't wait to get my hands on those queers!"

Rationalization

"Yes, there was lipstick on his collar and shirt, and he did have the aroma of perfume on his body, and he did come home rather late. But he rides a crowded elevator and lots of women work in his building. Besides he had a lot of work and meetings to go to, so that's why he was late, and that's why he couldn't have been with another woman; he was at a meeting. And I know he loves me. He brought home those nice roses, even though he did forget and leave them in the car until I found them . . ."

Projection

A person who engages in projection is usually blaming someone else for her or his own unacceptable desires, impulses, or bad conduct, as in, "Someone around here wants to have an affair. It must be you!"

John, a happily married, moral man, was flirting with his secretary, who a few days earlier had separated from her husband. She made a suggestion about his coming over for dinner to "go over these reports." Although he was tempted, he declined because he knew what might occur.

As he walked in the door of his home, still vaguely reminiscing about the possibilities with his secretary, he heard his wife, Jill, giggling on the phone. As he stepped into the kitchen, she looked at him with surprise and quickly told the other party good-bye. Immediately, John was suspicious: "Why was she giggling? She never giggles like that. Why did she get off the phone? Who was she talking to? Was that surprise or guilt on her face?"

John: Who were you talking to?
Jill: Oh, no one. Just Sally.
John: Sally? Why did you look so surprised and get off the phone so quickly?
Jill: Because I was just about to leave, and you're home early.
John: Just about to leave? Jill, are you having an affair?

Justification

A person who uses justification is attempting to dismiss or "justify" his or her own behavior by blaming someone else. When Jill left the house, John, having projected his desires onto her, began to feel very uneasy. He was convinced that she might be seeing someone else and, thinking back, realized that her behavior had been somewhat odd for the last week (beginning at about the same time his secretary had separated from her husband).

He became angry. If Jill was seeing someone else, why shouldn't he do the same? Why should he be a fool and hold back? Feeling rejected, upset, and now angry, he makes a conscious decision. He calls his secretary and asks if it is OK if he drops by after all. She says "yes."

Displacement

If one is treated badly, one may feel angry or hurt and yet be unable to express her or his feelings until later, at a safer time. These feelings may be placed on hold because someone has suddenly hung up and refuses to answer the phone when one calls them back, or the other individual is someone no one would dare retaliate against (e.g., an employer, a judge in court, or a policeman who has just issued a ticket). If this happens, one may take out his or her anger against the very next person or thing he or she happens upon in order to safely vent this rage. He or she may kick the family dog, yell at some kids playing in the street, or honk and swear at the driver in the next car.

When Jill came home laden with the ingredients for the special dinner she was planning for John that night, she found a note in the kitchen. It stated matter-of-factly that he had gone to a business meeting and told her not to wait up. At that moment, her son walked in and threw his books on the kitchen table. Jill screamed at him , "How many times have I told you not to throw your books on my table? Now get the hell out of here!"

Her son slunk away feeling guilty and wondering what in the world was wrong. Seeing the hurt look on her son's face, Jill resolved to apologize for being so irritable. However, when she looked again at the note, a series of possibilities occurred to her, and she began to wonder about John's odd comment about an affair. Her anger flared again, and she tossed all the groceries into the garbage, forgetting that she had yelled at her son.

Not all defense mechanisms require that information be completely submerged in the unconscious. However, all require that feelings, needs, desires, and other impulses be misinterpreted, denied, or suppressed to varying degrees. Defense mechanisms are not always bad; sometimes, they serve an adaptive purpose. For example, they may prevent a person from being overwhelmed. However, they may also serve to prevent a person from getting a closer look at her or his own unknown face.

15

The Misinterpretation of Needs
Limbic Needs

Many needs, but by no means all, are mediated by the limbic system, particularly biological needs, such as hunger, thirst, and the need for social, emotional, and physical contact with others. Even the need for love or, conversely, the need to get even by striking back has limbic roots. Thus, many needs originate in the most primitive unconscious regions of our mind.

Some of these needs, originating in the most inaccessible realms of the psyche, must go through a laborious process of transformation and interpretation before they can be recognized by the conscious mind. Because the conscious and unconscious minds do not speak the same language, misinterpretation is a natural by-product of the translation process. Some impulses cannot even be consciously recognized by the language-dependent left half of the brain.

For this reason, some needs remain unconscious and never achieve conscious scrutiny, whereas others in fact become wholly misinterpreted in the process of linguistic translations and categorization. In either case, these needs go unfulfilled. Fortunately, many needs that remain unconscious do not require conscious assistance to be taken care of and many more do not even have to be acted on; they just fade away. For example, an individual may be feeling extremely angry or sexually aroused, but then continues with whatever she or he is engaged in, and the desire for sex or violence slowly dissipates.

Most people attempt to satisfy their needs without understanding them. When thirsty, they drink. If hungry, they eat. They do not find it necessary to understand body chemistry or to engage in philosophical arguments about why one feels hunger or thirst. The person feels thirst, drinks some water, and continues with whatever has been occupying his mind at that point. Indeed, such needs are often fulfilled without any conscious thought; most people respond according to habit, or the right half of the brain takes over and directs their actions.

Limbic Limitations

The limbic regions of the brain and mind are relatively primitive and have no capacity to devise alternatives or to engage in creative or detailed problem solving. The limbic system does not think; it just acts. If hungry, it seeks food; if sexually aroused, it seeks sex.

If a need cannot be easily satisfied by traditional, habitual means or in a direct, simple, and straightforward manner, then the limbic system is stymied and can respond only by generating physical and psychological tension through its control over the capacity to feel pleasure or pain. When this occurs, we may feel tension without knowing why. If the need has reached conscious awareness, new means of satisfying the need may be devised and acted upon, should habitual modes prove inadequate. If hungry and without food or money, a person may consciously decide to go to the bank and either steal money, borrow it, or if there is money in his or her account, take it out. If none of these alternatives is available or acceptable, the person may get a job.

Sometimes, the left brain and the conscious mind are also stymied if an often-used means of satisfaction is suddenly not available. Habits are hard to break, and when people have been responding according to habit, their ability to improvise diminishes from lack of use. They lock themselves into a particular mode of responding and may even be proud of their lack of ingenuity, as in, "This is the way I have always done it, and this is the way I will do it now," or "This is the way it is supposed to be done, so don't argue with me," or "It's company policy" (as if God handed down these rules).

Consciously, the left half of the brain believes its explanations and is satisfied. However, unconsciously, people who behave autocratically and inflexibly are usually responding in the mode of the Parent ego personality, which likes to make rules and exert authority. If they appear closed-minded, it is because they have, in fact, closed their minds to possibility and are responding to needs and desires that are not apparent to them. Even if apparent, the conscious mind and left half of the brain often prefers to ignore the mysteries of the unknown and relies instead on its own authority and well-ingrained way of doing things.

Take as an extreme example some city folks who become lost in the desert and have depleted their water supply. Although there may be abundant cacti or other desert plants around, it may never occur to our city slickers to partake of the water that all these plaints contain by sucking or chewing on them: "Water comes from a faucet (or maybe from a stream or creek), not from plants." Many a stranded sailor has died of thirst in the ocean because it never occurred to him to catch a fish and suck its watery juices.

Tension and Pleasure

Because many needs are influenced, monitored, and controlled by the limbic system, we usually feel unpleasant sensations when they are unmet, or conversely, we experience pleasure (or at least satisfaction) when they are fulfilled. The reason is that the limbic system controls the most basic aspects of emotional functioning, such as pleasure, displeasure, and even rage. By rewarding a person with a feeling of pleasure when needs are met, or conversely by generating sensations of displeasure and discomfort when they are unmet, the limbic system motivates the person to act on limbic impulses and desires.

Tension Reduction

At its most basic level, whenever a need remains unfulfilled, there is initially a corresponding sensation of deprivation and arousal, which (for lack of a better word) is felt as tension. This may be experienced as an unpleasant sensation of hunger, a pleasurable feeling of sexual arousal, or even nervousness and anxiety. In some instances, we may have "no idea" what is bothering us and try either to figure it out, to suppress the feeling entirely, or to reduce the discomfort. If the tension is unpleasant and we are conscious of what we need, we try to reduce the tension by fulfilling the need. If the sensation is pleasurable (e.g., sexual tension), we often act to increase the tension level to the point of optimum pleasurable arousal, where tension release is obtained.

Consider hunger. The hypothalamus monitors the nutritional needs of the organism through various receptor cells that sample, for example, the glucose levels in the blood. If these levels fall, the hypothalamus becomes aroused and generates a discomfort that motivates the person to eat. Because the increased arousal and tension in this instance are unpleasant, the person is motivated to reduce the tension by eating, and the hypothalamus triggers feelings of pleasure to maintain this activity.

However, once the brain is signaled that the nutritional requirements have been met, the sensation of pleasure decreases rapidly so that, even if the sandwich one just finished tasted fantastic, one does not simply keep eating; the pleasure is gone. If the person does keep on eating, it is because the primitive hypothalamus is dysfunctional, has been overruled by other brain structures, or it has been taught to eat in response to some other need (like Pavlov's dog that would salivate to the sound of a bell which had formerly been associated with the presentation of food). Of course, there is more to it than that. Some of us have too many fat cells that constantly cry out: "Feed me."

Pleasurable or uncomfortable, identifiable or unknown, the optimal aim of the limbic system is thus tension reduction or release and the fulfillment of one's desires. It should be emphasized that I have simplified a very complex principle and that I use the term *tension* for convenience. This use of the term is not universal; some people use such terms as *drive* and *arousal* instead. Sigmund Freud referred to this process of tension reduction as being based on the pleasure principle; that is, reducing tension feels good and generates pleasure, so that we are motivated to perform the same action again and again.

The experience of tension is only part of a very complex process of homeostatic monitoring, neural-chemical interaction, and limbic system functioning. And the term *tension*, as used here, refers very loosely to biological interactions that motivate certain actions. Similarly, need satisfaction cannot be fully explained by drive theory or the pleasure principle. Humans are more complex than that. Because of this complexity and the development of our neocortex, we are not concerned solely with maximizing pleasure and minimizing tension. The need for power, influence, knowledge, money, prestige, new cars, new clothes, new toys, and so on transcends unconscious limbic motivations.

The Misinterpretation of Tension

Often, the need for power, influence, money, and so on, although seemingly based on objective conscious desires, is, in fact, motivated by need deprivation or flaws in the conscious or unconscious self-image. For instance, a man who never felt he received enough love from his mother may try to fill his sense of emptiness with money. Instead of marshaling his resources to obtain the love he desires, he attempts to build a sense of security through the accumulation of wealth. If he obtains enough funds, of course, he may be able to "buy" the love and affection he so desperately desires, as there are certainly plenty of women (and men) who also seek money as a form of security and as a substitute for love, intimacy, and self-respect. Love is not always for sale, however.

If he still does not find love, or if his need for love is never consciously satisfied, he will "think" at a conscious level that what he really needs is more money. The quest will continue forever because he does not consciously know what he really desires (love) and instead focuses on fulfilling an indirectly related desire (accumulated wealth). One desire is confused for something else.

The feeling of emptiness when one has been rejected by a loved one may be misinterpreted as a need to go shopping, or as a need to eat. That is, the need for companionship and love and the sense of loss that is now associated with this need may be too painful to scrutinize consciously. Hence, the person engages in some unrelated behavior that results in a temporary reduction of the feelings associated with loneliness and also results in a feeling of pleasure. Feeling good about one thing masks the pain being felt elsewhere.

The Confusion of Needs

A need that is largely unconscious may come to be confused with an entirely different need simply because both were repeatedly satisfied at the same time. This occurs through the process of association. Two different needs become linked and blended together.

If an infant that craves its mother's touch is held only during feeding (if he or she is so unfortunate as to have a noncuddling mom), the need to be touched and to be nourished will become closely associated. Indeed, touch and feeding are in fact closely linked, as the infant seeks food (i.e., the breast) through tactile sensations near the mouth and lips and is usually (or should be) held when fed. Later, the need for companionship may be confused with a need to eat.

As another example, a young boy who is continually teased or tormented by other children may run home and seek the security of his mother. Later in life, as an adult, whenever his feelings are hurt or he is feeling insecure, he also feels a need to run to his mother. However, because this action is not acceptable to his self-image, he instead has a drink (seeking the breast) or gobbles up some food because his mother not only used to comfort him but usually gave him cookies or other goodies when he came home crying and upset.

Many people eat and drink when they are not actually hungry or thirsty. They may think they're hungry or thirsty, but what they are hungry for is love and affection. From early learning experiences, they have associated eating (or drinking, or the receiving of gifts and presents) with love and caring. Such people respond to a disguised, unrecognized, or misinterpreted need out of habit. They feel a tension that is associated with and often satisfied, at least temporarily, by eating or drinking, so that is the behavior they engage in.

Need misinterpretation is not always the result of associational processes originating in the limbic system. It may also involve the mental system of the right half of the brain and the interactions of the unconscious Parent and Child:

Trudy had very wealthy parents, both of whom worked and had many responsibilities with clubs and other groups to which they belonged. Her father also traveled frequently and was sometimes gone for weeks at a time. Thus she saw very little of her parents and instead had a succession of nannies and maids who served as surrogate mothers. She felt sad and neglected, that is, at an unconscious level.

Trudy was an only child, and her parents, feeling guilty about their prolonged absences, often brought her lavish gifts. Her father in particular always made it a point when he had been out of town to bring her a wide assortment of toys and clothes. However, what she really wanted and really needed was their love and attention. These were not forthcoming, as her parents were just "too busy." As her starvation for love and her feelings of being "not OK" and rejected were closely intertwined, she suppressed these feelings in order to feel nothing. They remained confined to the right brain and the limbic system.

As she grew into a teenager and young adult, Trudy was often viewed by her few friends as spoiled, stuck-up, and distant, but worst of all, as selfish and greedy. She had severe difficulty maintaining friendships, and when on a date, she expected to be taken only to the most expensive places in town. Trudy also had much difficulty keeping boyfriends because of this attitude, as she not only expected but demanded that money be spent on her lavishly. Even when they treated her in the manner to which she was consciously accustomed, she would grow bored after a couple of dates and would end the affair before it began. Although she had long ago consciously learned to equate affection, caring, and even love with having lavish sums spent on her (a gift became a substitute for love), even when a male met her monetary demands, she could not feel truly close to him. Something always seemed to be missing. What she was missing was the love that her parents had never given her. However she could not recognize this lack because it was too painful. Gifts and lavish spending were not a substitute for love; in her unconscious, they were associated with being unloved.

Why didn't she try to find the love she desired elsewhere? Because love had become associated unconsciously with absence; that is, being neglected and rejected. Hence, she dealt with these feelings via reaction formation; "It is not I who am being rejected. I am rejecting you." Her unconscious Parent thus treated her suitors as she had been treated in childhood, while her love-starved Child continued to demand the gifts that were associated with the love she unconsciously craved. The perfect man for Trudy would have been someone who was wealthy and generous, but absent and neglectful most of the time. Unfortunately, if she had formed such a relationship, she would have been responding to a bad need, i.e., to be rejected.

Later, after a failed marriage to a very attentive husband, she sought counseling and complained that she felt too obsessed with money and was also terribly lonely:

> I always feel alone. I can be at the grandest party and be surrounded by the most eligible of men and the most beautiful of people, and I feel nothing. I feel totally alone. I may consent to an evening with a man whom everyone describes as rich and charming, and although his attentions flatter me,

I feel as if something is missing, as if we are not truly making contact.

When I tried to point out that what was missing was the love she had needed as a child, she became quite upset and adamantly insisted that she had been treated like a princess by her parents and that no little girl could have been more loved. As far as she was concerned, the reason she could not find real love or even maintain a close friendship was that no one could compare to her parents or give her the love they had lavished on her. In truth, however, she had never experienced real love, and thus, when it was offered by her attentive husband, she was unable to recognize it.

Rather, when Trudy felt the first stirrings of the need for love, they were immediately misinterpreted as a need to go shopping, to have men buy her things, or to have others wait on her. Her need for love had long ago become associated with getting gifts. Of course, when these misinterpreted needs were fulfilled, she continued to feel miserable; the original needs had only become more intense because they were still being unfulfilled. It was not more gifts but love that she desired, the love denied by her parents.

Similarly, misinterpretations of tensions and needs plague many people. They do not consciously recognize what their real needs are. Sometimes, old hurts and unfilled hopes and desires are just too painful to confront consciously.

Jenny was the youngest of a large, seemingly close and loving family and had many older brothers and sisters. Her mother's last pregnancy was unexpected (having occurred after she had reached her forties), and Jenny had not truly been wanted.

Because her parents were employed and her brothers and sisters went to school or worked, she was often neglected and did not receive much attention. Indeed, her mother returned to work when Jenny was just a few months old, and from that point on, she was passed from one family member to another in shifts, as they went about their various schedules. Although no one was mean to her, she was ignored. No one really played or interacted with her much, other than caring for her basic needs and making sure she stayed out of harm's way. Consequently, her social skills suffered, as well as her self-confidence and feelings of self-worth.

Nevertheless, breakfast and dinner were times of great joy, as the whole family came together, laughed, talked, and teased one another good-naturedly. Jenny looked forward to these special times, as they always made her feel close to her family. In her unconscious mind, socializing, feelings of closeness, and a sense of belonging became associated with eating.

Many years later, when her parents died and her brothers and sisters moved away, Jenny remained single and felt very depressed. She could never understand why she was depressed because she had "a good job" and "a nice apartment." At one point, she decided she must be bored and tried to take up a hobby. The depression lingered. Her only solace was to eat. Although she realized that she wasn't hungry, she tried to fill the gap in her life with food. The consumption of food always made her feel better, at least while she was eating.

When questioned about any possible links between her overeating and her family life, she adamantly insisted that her childhood had been wonderful and filled with togetherness and love. As far as she (i.e., her left brain) was concerned, her eating problem had nothing to do with how she had been brought up. Loneliness, lack of self-confidence, and feelings of low self-worth were not problems, she claimed, as she had "a good job and a nice apartment": "My goodness, how could loneliness be in any way associated with eating? Besides, I might be alone, but I am not lonely."

During our third session, she began to admit that her self-esteem might be a little low. Nevertheless, she was convinced that her low self-worth was a result of her overeating, not vice versa. She insisted that she felt bad because she ate too much. It was silly even to consider the possibility that she ate too much because she felt bad: "If I weren't so fat I would feel better about myself and would have more self-confidence. Then I know I could go out and make lots of friends."

In truth, however, she was fat because she had substituted one need for another. She ate when she was lonely and her self-esteem was so low that she could not bring herself to venture out and make friends. Her only solace was to eat. She ate too much because she had a low self-concept, and not vice versa. She had long ago learned to associate food with socializing and love.

Sometimes, in even the most seemingly close and loving of families, particularly when there are many siblings, there is just not enough love to go around. Food, drugs, alcohol, violence, or sex may become an irresistible substitute. In one very large and famous political family, following the assassination of their father, several of the boys became heroin addicts, and one (who frequently complained of feeling lonely and isolated) died of a drug overdose (or committed suicide).

Confusing Means with Ends and Secondary Needs with Primary Needs

It sometimes happens that over time, we may begin to lose sight of a particular need and instead focus on the means by which it was satisfied, so that the means becomes confused with the ends; and the means of satisfying a particular need are equated with the need itself. Soon, the original need is forgotten or no longer recognized, and the person begins to believe that the manner in which he has satisfied his needs is the need. The original need may then go unfulfilled as what is sought is the opportunity to indulge in the behavior meant to satisfy the original need.

Peter had a drink after a bad day at work, and it made him feel better. Two days later, he had an argument with a co-worker, came home upset, and had another drink, which made him feel better. A few days later, his boss complained about the quality of Peter's work during that past week. Peter consciously decided that his problems with his co-worker were interfering with his job performance and resolved to concentrate on his work. He went home feeling upset and worried, poured himself a "tall one," and then another, and felt much better.

The next day, someone from another department called and yelled at him for something that was not Peter's fault. He became upset and immediately felt a need for a drink. Although his actual need was for things to go well at work, he had now confused feeling better about his job with feeling better after having a drink. Besides, alcohol is an excellent anesthetic and source of oblivion and self-destruction. It can deaden the pain and the ability to feel, at least while one is under its influence.

Everyone feels a need to associate, socialize, and make contact with others. This need is mediated by the limbic system and the right hemisphere, and it is present from birth. When we do not socialize, we not only are lonely but experience an increase in tension. The tension is the drive to go out and interact with others.

Jessica and Stephanie were best friends and spent all their time together laughing, talking on the phone, going to movies, and hanging out at the mall. When Stephanie moved away, Jessica ceased to go out, became very depressed, and just moped around the house. When her parents encouraged her to go out, she responded that she just didn't feel like it because she was too depressed. In her mind, she had ceased to go out because she was depressed, when in fact she was depressed because she had ceased to go out and socialize with others.

In effect, she had misinterpreted her need to socialize and associate with others with her enjoyment of Stephanie's company because her friend had satisfied that need. Moreover, she confused Stephanie's satisfying her need to socialize with her enjoyment of Stephanie's company. That is, originally, Stephanie had fulfilled her need to be with others, and over time, Jessica came to believe that her need was to socialize with Stephanie. Thus, this need ceased to be fulfilled after her friend moved away, and the resulting increased tension was misinterpreted as depression. She had not only confused and lost sight of her original need but lost her ability to satisfy it in other ways. Of course, she also misses her friend.

Seeking to Fulfill Misinterpreted Needs

Sometimes, a behavior that has fulfilled a need becomes a need in itself. The original need is forgotten, although it forms the basis of the secondary need. If it is forgotten, and the secondary need is no longer being satisfied, the ability to attend to the original need is almost completely lost. It is difficult to satisfy a need if one does not know what it is or even realize that it exists.

This is what had happened to Jessica. Her one friend, Stephanie, had satisfied her need for association and friendship. Over time, Jessica developed a need to be only with Stephanie. However, if she did not have a need to socialize in the first place, Stephanie would never have become her friend.

When primary and secondary needs become confused, people become locked into a particular mode of interacting, behaving, and thinking, and lose the capacity to satisfy their original need by some alternative action. Their ability to improvise or devise alterative problem-solving strategies is diminished:

Tim and Donna had been married for over twenty years. For the past fifteen years, they had been arguing, yelling, screaming, or, conversely, ignoring one another. Each admitted not having touched or said anything nice or affectionate to the other in years.

Nevertheless, just as a child who is often neglected or ignored will behave badly so that Mom or Dad will finally pay attention to her or him (that is, being yelled at is better than being ignored), Tim and Donna treated each other badly to get each other's attention. When being ignored became too upsetting, they met their needs for human contact by yelling at one another. However, their need wasn't to fight; it was to interact. They had confused their original need for association with fighting merely because it satisfied the original desire.

Because fighting met their desire for contact, the need for interaction was forgotten, and they began to act on this secondary need (which satisfied the original desire) as if mutual hate were actually the basis of their relationship. If they had been prevented from fighting they would have had no idea how to interact and would have felt doubly miserable.

People who become locked into erroneous modes of need satisfaction so that their original needs go unsatisfied, experience a buildup in tension that is often misinterpreted as depression, anger, hunger, or a desire to go shopping, take drugs or a drink, fight, argue, and so on. They will spend their time futilely attempting to obtain satisfaction by engaging in behavior that is either no longer available or that no longer satisfies the original need, and the need will go unfulfilled because it is no longer consciously recognized.

Misinterpretation of Needs and the Inappropriate Discharge of Tension

As we've noted, when needs go unmet there is a buildup of tension, and the individual's level of arousal increases. Energy stores are released and mobilized for the purposes of need satisfaction. Sometimes, the resulting tension is experienced as separate from the original need. A person who is hungry and goes for a long time without eating begins to feel an increased level of tension. As the person becomes more tense, he or she may draw from that energy store and become irritable and then angry (e.g., aggressive for the purposes of obtaining food; hunting). Hence the need to eat triggers a feeling of arousal and aggression which may then be interpreted or felt as anger, all of which are mediated by the hypothalamus. In consequence, tensions that are only indirectly related to the original need are experienced and expressed. When this occurs, people sometimes lose sight of their original needs and begin to act on the misinterpreted or secondarily induced tension. The hungry man, instead of searching for food, may begin to yell at and threaten those around him. Instead of eating, he gets into an argument. This behavior does not reduce the original tension levels, however, and is ultimately self-destructive.

Projection, Displacement, and Rationalization: Blaming Others

Many people believe that others are the cause of their bad feelings and the associated discomfort. Sometimes, they act on their projections rather than on their real desires. When this occurs, their misdirected behavior acts only to increase their tension levels as their original needs continue to go unmet:

John had an important early-morning appointment that, if it went well, could result in a substantial increase in his salary. A higher salary was a secondary need that would meet his primary need of security. He got up the next morning and hurried through his routine of exercising and getting dressed. Although he did not show it, he was also feeling nervous and tense. The tension represented the motivational force (or drive) associated with his need for all to go well and to obtain the security required by his psyche.

As the time got closer for him to leave for the appointment, he began to hurry to get all the necessary paperwork together. Stepping into his private study, he did not see the needed papers, which he was sure he had left on his desk next to the typewriter. His tension level increased. The IDfant proclaimed, "I want it now." His drive to find these papers and the tension and arousal level associated with his need to do well increased.

John misinterpreted the tension as anger and did not recognize it as an increase in energy that would support his need to find his papers. Moreover, like many people, he began to look for an external cause of his anger. He projected his responsibility for his own predicament onto his family and engaged in the defense mechanism displacement and projection: "Someone has misplaced my papers. It must be you!"

"Who the hell has been in my study?" he yelled and then began feverishly to rummage through the papers on his desk and, like a frustrated child or infant, knocked several piles onto the floor. He then stomped into the kitchen to confront his wife, whom he began to blame for "not watching the kids," "for letting them get into his stuff."

John had not only misinterpreted his tension, which he experienced as anger, but had acted on his misinterpretation, which, in turn, interfered with his ability to function effectively (i.e., to find his papers). Instead, he was distracted, and his energy was misdirected and wasted on yelling and blaming his wife. He now believed that his wife and kids were the obstacle preventing him from getting to his appointment, when, in fact, the only obstacle was himself and his failure to continue searching for his papers.

Unfortunately for his wife and kids, John typically satisfied his need for tension release by abusing them (through displacement). This was a function of his limbic system's becoming increasingly aroused, as well as of his unconscious Parent, which liked to blame and criticize. This behavior also took care of his sense of inadequacy (through projection by blaming them) and removed the tension associated with his failure to achieve (his unconscious Parent yelled at them instead of at him). Indeed, he was so consumed by his sense of inadequacy that he did, in fact, sabotage himself by misplacing his papers (through activation of his unconscious Child, which saw itself as a failure and inadequate). He was thus also acting out a self-fulfilling prophecy.

By yelling and screaming at his wife, John temporarily reduced his feelings of tension. But his need to find his papers remained. Unfortunately, he had wasted his time and energy by engaging in unrelated behaviors, such as yelling, which temporarily made him feel better because his tension levels had been reduced. The energy stores of the hypothalamus and other brain structures had temporarily been depleted. Nevertheless, his feelings of tension would again increase as he had not satisfied his original need.

The ridiculousness of such behavior is exemplified in a parable presented by Snell and Gail Putney in their excellent book *The Adjusted American*: It is a very hot day and a great stag with huge antlers slowly makes his way down a hillside toward a pond surrounded by a dense thicket of brambles. As he attempts to push through the little deer trail leading to the pond, his antlers become caught in the brambles. He attempts to push forward and then backward with little success. His entanglements, coupled with his thirst, lead to an increase in arousal and energy mobilization; adrenalin is released, his pulse quickens, and blood is shunted to the large muscles. With a sudden surge forward, he breaks through the brambles, steps up to the pond, and begins to drink.

If the stag were to have behaved like John instead, he would have become angry and would have thrashed and torn at the brambles, kicking and stomping them. Indeed, once he broke free, instead of going on to the pond, he would probably have continued to kick, stomp on, and destroy the brambles, all the while thinking about getting the guy who planted them in the first place. That is, the stag would have misinterpreted the tension associated with his need to drink as anger at the temporary obstacle.

Some people, fearing to behave in an outwardly destructive, aggressive, and angry manner, instead turn the anger back on themselves as depression. If caught in the brambles, instead of surging forward and breaking through or attacking and destroying the brambles in a rage, they sit down in the thicket, give up, and become depressed; hence their unconscious Child takes over.

In either case, a particular need creates a mobilization of energy, the person becomes aroused, and the arousal is felt as tension. Faulty interpretation of the tension or faulty interpretation of the need leads to projection, displacement, misdirected behavior, and increased tension. The increased levels of arousal then drive the misdirected behavior. In John's case, not only had he acted shabbily by yelling at his family, but he had used yet another defense mechanism, justification, to justify his mistreatment of them: If they hadn't fooled with his papers, he wouldn't be yelling at them.

Misinterpretation within the Brain

The misinterpretation of needs is often due not only to a tendency to protect the conscious self-image, but to learning, faulty association, and the fact that some brain regions may misinterpret or fail to perceive impulses arising from other regions. Some forms of information cannot be recognized by the conscious mind of the left brain, and the left brain must attempt to guess at or fill in the gaps in the information it receives. As a result, some types of information may be misinterpreted. Consequently, the person's behavior may be misdirected, or her or his original needs or desires may fail to be met or even recognized.

In other instances, the neocortex of the right brain may misinterpret the information it receives. For example, although it is closely aligned with the limbic system in terms of functional specializations, many limbic impulses are transmitted only in the form of raw emotional energy, and the right neocortex must guess at and interpret the subcortical impulses received. If the right brain correctly perceives a particular emotion, it still may not be able to determine what is causing it. Both the right and the left brains may then begin to search for a cause as well as attempt to discharge the tension that the whole brain is experiencing.

16

Self-Deception and Denial

He knows it, because he believes it.
Believes it, because he knows.
But alas, he knows only that he believes.

Confabulation

When the language axis (Wernicke's and Broca's area, the angular gyrus) and thus the linguistic conscious mind become isolated or are prevented from receiving information from other regions of the psyche, an individual may begin to engage in denial, projections, rationalization, and displacement. However, when a brain injury causes a disconnection of the language axis from yet other areas of the brain and mind, the person may engage in denial, including elaborate confabulations, denial of the existence of the left half of the body, or claims that she or he can see when she or he cannot.

As previously noted, patients who have sustained injuries involving the right brain and who are subsequently paralyzed on the left side may claim that their paralyzed extremities belong to the doctor or a person in the next room. In less extreme cases, they may claim that their paralyzed limbs are normal even when they are unable to comply with requests to move them. For example, one woman, when asked why she could not move her arm, replied, "Somebody has hold of it." Another patient, when asked if anything was wrong with her arm, said, "I think it's the weather. It's just a little cold and stiff. I could warm it up, and it would be all right." One women whose left leg was paralyzed continued to claim she could walk, even when she obviously could not. "I can walk at home," she stated, "but not here. It's slippery here." Another patient claimed that he could not raise his arm because of the stiff shirt he was wearing.

With severe right-cerebral injuries (which, in turn, result in a massive reduction in information exchange between the two halves of the brain), patients may make up extremely bizarre explanations regarding their injuries, which they seem to believe. For example, one patient who had suffered burns as well as a significant brain injury in an auto accident claimed that his rocket ship had crashed and the result had been an atomic explosion. He repeatedly insisted that he was filled with radioactive fluid and thought that his various scars were the result of this fluid's being removed from his body. However, these ideas were not due to some form of psychosis; they were the product of a loss of inhibitory restraint and of confabulation, an attempt to make sense of what he was experiencing.

Yet another patient, who had received a gunshot wound to the right brain, repeatedly and vehemently complained about being kept in a hospital, believing his confinement to be due to a plot by the government to steal his inventions and ideas; he had been a grocery store clerk. When it was pointed out that he had undergone surgery for the removal of bone fragments and the bullet, he pointed to his head and replied, "That's how they're stealing my ideas."

Moreover, these confabulators truly believed their own stories and sometimes adamantly argued if one tried to persuade them otherwise. Rather than relinquish an incorrect belief when confronted with contradictory information, such individuals sometimes simply adjust their story slightly.

In fact, the same thing often occurs when information stored in memory (whether in the right or the left brain) is not available to the language axis and the conscious mind. Take, for example, a patient who was suffering from partial amnesia and who, although having been a patient for over two weeks, had no recall of why he was in the hospital. He could accurately recall much of what had occurred before the onset of his condition. When the nursing staff attempted to give him medication, he would respond, "What is this for? I've got to get out of here," or "I just stopped by for a checkup. I've got to get back to work." When I informed him that he had been a patient for over two weeks, he became very hostile and upset and angrily replied, "I just came here for lunch. I do not belong here. I'm supposed to work in the garage. Now would you please tell me how to get to the garage?"

Essentially, the left half of the brain, being denied access to various sources of information, responds to the gap in the information received with the ideas and associations that are available, even when they are irrelevant or ridiculous. The left brain fails to correct these erroneous statements because disconfirming evidence is not available to it. The language axis and the conscious mind have been disconnected without knowing it and act to fill the gaps in the incomplete messages received with ideas that are related in some way to the fragments available.

When the conscious mind is disconnected from another brain area due to injury or a functional, psychic resistance or repression, then the language axis, with which it is closely associated, will be unable to describe what is going on in that cerebral region. Instead, it will describe only what is available to it, which it then believes. Moreover, it will not know that it does not know, or that its explanations are erroneous, because the area of the brain that does know has been either destroyed or psychologically disconnected and cannot alert the language axis to any discrepancies.

In extremely severe and rare cases of brain injury, the language axis and the conscious mind may be completely disconnected from the rest of the cerebrum, so that almost no information is available to them. When this occurs, the associations available are so limited that it cannot think, cannot act, and can only respond to the simplest of questions in a stereotyped and very limited way. However, the right brain may still engage in singing, praying, and its normal routine, that is, if it is not too severely damaged.

In an interesting case described by Geschwind, Quadfasel, and Segarra,[1] a twenty-two-year-old woman with massive destruction of neocortical tissue due to gas asphyxiation was found to have a preserved language axis when all the surrounding brain tissue was destroyed; thus, the language axis was completely disconnected from the rest of the cerebrum. Once the patient regained "consciousness," "she sang songs and repeated statements made by the physicians. However, she would follow no commands, resisted passive movements of her extremities, and would become markedly agitated and sometimes injured hospital personnel."[2] In all other regards, however, she was completely without comprehension or the ability to communicate:

The patient's spontaneous speech was limited to a few stereotyped phrases, such as "Hi daddy," "So can daddy,"
"Mother," or "Dirty bastard." She never uttered a sentence
and she never replied to questions and showed no evidence
of having comprehended anything said to her. Occasionally,
however, when the examiner said, "Ask me no questions,"
she would reply, "I'll tell you no lies," or when told, "close
your eyes" she might say, "go to sleep." When asked, "Is this
a rose?" she might say, "roses are red, violets are blue, sugar
is sweet and so are you." To the word "Coffee" she sometimes
said, "I love coffee, I love tea, I love the girls and the girls
love me."

An even more striking phenomenon was observed early
in the patient's illness. She would sing along with songs or
musical commercials sung over the radio or would recite
prayers along with the priest during religious broadcasts. If a
record of a familiar song was played the patient would sing
along with it. If the record was stopped she would continue
singing correctly both words and music for another few lines
and then stop. If the examiner kept humming the tune the
patient would continue singing the words to the end. New
songs were played to her and it was found she could learn
these as evidenced by her ability to sing a few lines correctly
after the record had been stopped. Furthermore, she could
sing two different sets of words to the same melody.[3]

In other words, although the language axis was disconnected, the
right brain (although also damaged) was still able to engage in some of
its normal expressive routines, singing and praying. The left brain,
however, being totally isolated, could produce almost nothing except
well-learned and stereotyped phrases. The left brain could repeat only
what was available to it.

Such conditions are quite rare. Nevertheless, even when the brain
has not been damaged, the language axis is not always able to gain
access to other brain areas. When this occurs, it will make up explanations as the need arises. One need not suffer a brain injury to confabulate; it is a normal part of left-brain psychic functioning.

Quite frequently, because of one's conscious self-concept, or other
concerns, information that is available but undesirable is suppressed,

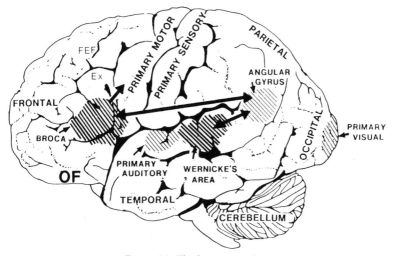

Figure 36. The language axis.

ignored, or denied. When this occurs, the language axis and the conscious mind may generate whatever story they wish to explain away their actions, feelings, or behavior or the behavior and feelings of others. Having no access to the correct information, perceptions, memories, and emotions, the language axis and the conscious mind cannot be alerted, except indirectly (by the generation of tension) to any discrepancies, unless, of course, the right brain or the limbic system completely overwhelms them and takes over. However, even then, the left brain may be denied or may deny itself access to whatever motivated its behavior and, like the brain-injured individuals discussed above, will generate whatever explanation is acceptable, even if it is implausible or ridiculous.

Hiding Information from the Conscious Self

Desires, impulses, and a variety of emotions arising from the unconscious may be reinterpreted or defended against in many ways. Some desires just fade away or are replaced by more important concerns. Others simply cannot be perceived or even recognized by another region of the brain or the mind. Others are held in check by the frontal lobes, for example, until the appropriate time, or until a means of fulfilling them emerges.

However, some tendencies and personal characteristics are pervasive and deeply ingrained. They are part of who we are. Sometimes, these personal traits are largely unconscious and associated with the unconscious self-image, such as the ego personality of the Child. This constellation of traits may include feelings of hurt, jealousy, anger, guilt, cowardice, weakness, incompetence, clumsiness, fearfulness, and childishness, and of selfishness and greed such as often occur when children are spoiled.

Although accepted by the unconscious self-image, these unconscious tendencies may be completely unacceptable to the conscious self-image, that is, what we have described as the Adult self or the Adult ego personality. Being unacceptable, they are ignored, denied, projected onto others, or inhibited. At a conscious level, a person may not even know he or she possesses these traits.

Indeed, many of us have had experiences in which a good friend or a lover has behaved selfishly or inconsiderately. However, when we have complained, he or she may have steadfastly denied ever having behaved this way and vehemently denied even having such disagreeable tendencies. Incredibly, this person will deny what is obvious to us and will believe his or her own denials. Essentially, the language axis and the conscious mind have become disconnected from the sources of this undesirable input. When the left brain tells these "lies," it is telling the truth as it perceives it.

When unconscious needs, desires, or personality traits are contrary or upsetting to the conscious self-image, people may go to great lengths to suppress, deny, compensate for, or overcome them. Even when people have engaged in some action that is contrary to their conscious self-image, they may deceive themselves into believing that it did not occur, or that it was a minor aberration that meant nothing. If they are shocked by what they have done or said, they may shrug it off with a simple "I didn't mean it," "I was tired," "I was drunk," or "I don't know what came over me," as if some alien force has taken over the brain for a few minutes. These are defense mechanisms used by the conscious mind, and they are referred to as *rationalization, denial, justification*, and *self-deception*.

Self-Deception

Denial and Believing Your Own Lies

A person has lied when she or he knowingly possesses the truth and consciously attempts to deceive another person. Curiously, people sometimes half-persuade themselves that a lie they have told is true and then behave accordingly. When this occurs, we may complain that they "believe their own lies." They have engaged in the production of confabulations.

Consciously, we often attempt to hide disagreeable information from ourselves, including unconscious personal traits and tendencies that we consciously find disagreeable. Sometimes, these characteristics are too difficult to confront consciously.

Denial, however, does not make a desire, impulse, or personal shortcoming vanish. Often, we remain unconsciously aware of its presence. Moreover, through the generation of various associated tensions and feelings, the conscious mind continues to be influenced, if not controlled, by these unconscious forces. We may even act on these impulses and not know why or "what came over us." Feeling these tensions or observing our actions, the left half of the brain may confabulate explanations for why we did what we did.

It is precisely because of these "mysterious behaviors," emotional tensions, and influences, including the dual nature of our brain and mind, that the left half of the brain must engage in denial in the first place. We are *aware* of what we are hiding. Denying something means simultaneously knowing that it exists. We cannot deny something if we have not been confronted by it in the first place. Hence, whenever we attempt to deny or hide information from ourselves, we are simultaneously acknowledging that we are aware of how and why the information is threatening. We would not deny and hide information from ourselves if we did not know that it would bother us. However, it is the right brain that knows; the left brain is in denial. How is this possible? Through the frontal lobes, which are the senior executives of the personality and the ego, and which exert significant inhibitory restraint on information processing within the brain.

The dual organization of the human brain and mind enables the development of countless conflicts, some of which are resolved by denial, projection, or self-deception. Because we have two brains, we can know, yet not know, and can deceive ourselves about our own unconscious tendencies and impulses.

Alan was pampered and spoiled by his two fairly well off, but not quite rich parents, particularly his mother, who took a special delight in indulging her son. Although the second son in a family with four children, he was the apple of their eye and remained so even after he had been placed on trial for rape. Paradoxically (or so it seems), he always felt angry at his parents, whom he believed never did enough for him. Unconsciously, he always felt cheated, even though he was the only one of their four children who never had to hold a job and had his way paid through college. He held his mother in particular contempt, as he could recall several instances in which he had had to beg or plead for some special toy or favor before she would yield. In other words, he was the typical spoiled child who took love and affection for granted but was reluctant to return it, in part because of feeling cheated, but mostly because it had never been required for him to win favor. His unconscious Child, therefore, was quite selfish and greedy, and his unconscious Parent told him he deserved to have his own way.

His unconscious resentment, however, soon came to include most others, for although he was highly intelligent, he was very short, thin, and frail and was often left out or was the last to be chosen for sports and games at school. He had never had a girlfriend or a date until he became a graduate student in college. He claimed, however, that this lack of girlfriends did not bother him because most people were idiots, women in particular. He stated that all women were selfish, mean-spirited, untrustworthy bitches and whores, who would do anything to "clip a guy" and "screw men over." The only thing they were good for was sex, and insofar as he was concerned, he could take them or leave them. Moreover, he thought quite highly of himself. He said he was a strong believer in fair play and honesty, that is, as long as he was treated in kind. His conscious self-image was that of a "great guy." Indeed, this was the conscious message he had always received from his parents, his mother in particular.

Much to his pleasant surprise, when he was a graduate student and teaching assistant, undergraduate females began to show him some interest, which he exploited miserably up to the time he was arrested for rape. This arrest came as a crushing surprise and confirmed his worst fears about women, as he said he had acted no differently with this woman than he had with several other undergraduate students that year.

"The prick-teasing bitch wanted it," he told me, and he could not see what he had done that was wrong. "She knew what was going on. She hardly struggled. She didn't scream. If she didn't want sex, why did she come to my house? Like all women, she just wants it both ways. Get what she can and then screw everybody else. I'm the one who is getting raped. You just can't trust women. They try to trick you and deceive you, and then take everything you've got without giving you anything in return except grief. I just can't believe this is happening to me."

According to Allen, it was one of his typical first dates. He had taken the woman to a drive-in movie, where he plied her with malt liquor, followed by "whiskey chasers." Once he was sure she was fairly intoxicated, they left the movie and returned to his apartment, where the drinking continued. When he was sure she was too intoxicated to resist, he began removing her clothes. Although she still had enough presence of mind to try to resist and beg him to stop, he "knew" that she really meant "yes" and simply ignored her protests and, overpowering her, took his pleasure.

When I tried to point out that all the cruel things he ascribed to women in general actually fit his own behavior (e.g., untrustworthy, taking advantage, screwing someone over), he became angry and stated that this was what she had done to him by pressing the rape charges. He had only acted in a normal manly manner and had seduced her.

In actuality, however, he had seduced himself. He believed his own lies and had projected his own horrible tendencies onto women in general. This self-deception allowed him to justify and rationalize his mistreatment of them, while simultaneously meeting the needs of his Child and enhancing his poor self-image and conscious feelings of masculine superiority in the process. Nevertheless, he was so cut off from his true self-image and selfish tendencies that he was unable to recognize his actions for what they were, and he believed his rationalizations and explanations.

Even when I asked him how he would feel if he went out with some men, became very intoxicated, and was then sodomized or forced to engage in oral copulation, he could not see the connection. "Men don't do that to each other," he stated, "unless they're both queers, in which case it is OK."

Dr. J: It's OK when someone is too intoxicated to resist or to know what's going on?

Allan: Hey, she could talk. She could walk. So what, she was drunk? I didn't rape her. I seduced her. Nobody forced her to keep drinking.

Dr. J: You're twenty-five. She's seventeen. Who supplied her with the drinks? Who got her so drunk she could barely resist?

Allan: Hey, man. That's the way it's done. You have a sex problem or something?

Dr. J: No. I have a problem with rape.

Allan: Rape! I didn't use force. I didn't beat her up.

Dr. J: Sure you did. You used the force of alcohol. You numbed her brain so that she hardly knew what was going on. It sure wasn't your sparkling personality that got her into that bed. You might as well have hit her with the bottle.

Knowing, Yet Not Knowing

Melody has a conservative background and well-meaning, upper-middle-class, semireligous parents, who have impressed on her the need to maintain certain values and morals. Melody is in her first semester at college and is living in a dorm. One day, a handsome, muscular, sexy, upperclassman whom she has seen about campus, invites her out. She consents. They have dinner, stop at a local nightclub, dance close together, and have a drink. So far, it has been an exciting, romantic evening.

On the way back to the dorm, he invites her to stop by his place to "check out the new CD player and stereo equipment" he has just purchased. She knows, or rather, is *aware* of the possibility of his intentions and what he may really be up to. Indeed, the look in his eyes, his tone of voice, and his body language have made his real interests completely clear to the right half of her brain long before they even began dinner. However, insofar as her left brain is concerned, this guy just wants to show her his stereo. Because of her *awareness* of his intentions, including the manner in which she may respond, her left brain begins to invent reasons why this is OK. He seems real nice. they had a real good time. He has not acted inappropriately. Why not?

Once at his home he makes certain advances; sitting close to her, looking in her eyes, he tells her she is "really beautiful." Her right brain, responding to his tone, body language, etc., is now almost certain of his desires and what he means by his words. Moreover, she is not only aware of what might occur next, but also aware that she may have to make a decision regarding it. However, she does not want to "think about that" (she is a nice girl).

Because of her conscious self-image, this information is suppressed and ignored. The left brain does not gain access to what the right brain is fully aware of. The left brain thinks: "This guy is only being nice."

Slowly, he places his hand on her knee. The act risks changing the situation by calling for an acknowledgment and an immediate decision about its meaning. To leave his hand there is to consent, to acquiesce to his desires, and to consciously acknowledge them willingly. Yet, to withdraw his hand is not only a recognition of what it implies, but a refusal. Her aim is to postpone her decision, to stall conscious recognition. She has "no idea" what he is up to. She leaves his hand there because she "doesn't notice it."

And yet, as they talk, he has moved closer to her, his hand inching its way past her knee. But she is concentrating on what he is saying, the curve of his lips, the white of his teeth, her reply to his question; this is what she is conscious of and to what she addresses herself. She reacts as a personality that does not know that a hand lies there upon her leg. She is being seduced.

She is also being self-deceived. She has restricted her consciousness and her thoughts to only selective aspects of what is occurring. She is failing to acknowledge consciously what she is fully aware of. Nevertheless, she is aware that his actions are sexual and has been aware of everything that might occur since the moment he asked her out.

However, if someone were to have told her that this guy would take her home and make possible sexual advances, she would have turned down the date. After all, her conscious self-image is that of a "good girl," a "nice girl." Although she is not a "prude," as she has always told her friends, she will have sex only with a man she loves. Indeed, the possibility of "sex" was so far from her (conscious) mind that it never even occurred to her that this guy would make certain advances.

So why did she go out with him? Why is she not taking his hand and indicating that his behavior is not acceptable: "Please stop." One would have to assume that her limbic system was "turned on," that she was sexually attracted to him and that the possibility of his making sexual advances was not only acceptable but welcome; at least on an unconscious level. However, what is acceptable on an unconscious level is not acceptable to her conscious mind and conscious self-image. Unconsciously, she is in fact yielding to his and her desires. Consciously, she still does not know what is going on. His behavior seems harmless.

If we accept Jean-Paul Sartre's position,[2] there is a possible unity in a situation such as this, for her actions necessitate a recognition of the intentions of her male friend as well as a disavowal. Otherwise she would have asked, "What is your hand doing on my thigh? Please remove it." Indeed, even if unable to verbalize her conscious disapproval (as some young women presumably become too intimidated to speak up), she would have made some attempt to back off or disengage herself if that is not what she wanted. However, consciously, she is disconnected from what is really occurring and from what she is fully aware of, so she does nothing.

This is the beauty of self-deception; it allows a person to engage in a certain behavior while simultaneously preserving his conscious self-image. It preserves the possibility of an event's occurring or a desire's being acted on while the person acts as if the event or action is not occurring.

Unconsciously, a person may be fully aware of what he is doing or intends to do, while consciously he may behave as if nothing of the sort was going to occur. This allows a person to deceive himself into thinking he is not responsible for his own behavior or for what happens.

He takes her hand, leans close, and kisses Melody on the lips. The kiss takes her by "surprise." He snakes his other arm around her and kisses her again. She does not know what to do. She is confused. And yet, her limbic system is proclaiming, "I want it, now!" Her right brain is saying, "Go for it," and the left brain, in all innocence, is saying, "What's going on?" He begins kissing and touching her passionately, and she feels herself yielding as her limbic system begins to overwhelm her conscious inhibitions. Her limbic system, in effect, has taken over, and her left brain, essentially, has gone "off-line," and is now a passive observer of behavior over which it has no control.

Indeed, it often happens that people behave in a certain manner, or say certain things that are shocking even to themselves. They don't know "what came over" them. They may see themselves as moral, just, kind, and loving, and for some unknown reason act cruelly, selfishly, or spontaneously. Yet they retain the conviction that what they have done or said is not a true representation of how they really feel or who they really are. Often, however, once the limbic system has been satisfied and the left brain regains dominance, it may begin to impose feelings of guilt on itself. The left brain may then begin to confabulate explanations to defend the conscious self from the truth or to minimize these self-induced feelings of unnecessary guilt. Such behavior, they reason, has an explanation outside themselves, being only a rare and momentary lapse; is justified by certain mitigating circumstances; or is due to the shocking behavior of someone else: "He pushed me too far that time!"

And, if Melody was "taken advantage of" by "that rat," "that scoundrel," "that womanizer," then she is of course not responsible for what happened. She "had no idea as to what he was up to," or that "he was that kind of guy." And besides, she "had too much to drink," (although she, in fact, had only one drink) and "was too tired to fight him off" (although she never tried), and before she "knew it, it was too late," and so on. Her conscious self-image remains untarnished. She is free of conscious guilt. As pointed out by Sartre, she has deceived herself. And yet, she knows and is aware of the truth. It is precisely because she knows that she invents innumerable excuses for her behavior.

Fortunately, for all concerned, however, "that rat" called her the very next day and brought over a dozen roses. She was enthralled, they began dating, and he turned out to be not such a bad guy after all.

Projection

Consciously, we often attempt to hide unpleasant information from ourselves. This includes personal traits and tendencies that we consciously find disagreeable. We also sometimes fail to recognize consciously certain disagreeable traits possessed by others, such as loved ones and potential mates. Sometimes, these characteristics are too difficult to face. This is why it is said that "love is blind." However, within the domain of the right half of the brain, we are often very *aware* of the presence of these personality features, whether they are possessed by others or ourselves.

When we meet someone attractive, who is nevertheless flawed in some way, this *awareness* may be realized in the form of a "gut feeling," "alarm bells," and other warning signs that may be ignored (by the conscious mind) until it is too late. It is only later, after we have been repeatedly hurt, used, or abused, that we become conscious of what we were aware of. It is at this point that our good friends also begin to remind us cheerfully, "I told you so!" And we know they are right. We were aware of it the whole time. The knowledge, however, was unconscious, part of the domain of what I have referred to as the *unconscious awareness* of the right half of the brain.

Denial and self-deception often imply a *knowing* coupled with a *not knowing*. Although the right brain may know, the left brain is in the dark about what is actually occurring. However, ignorance may not be bliss, as the left brain is still subject to tensions associated with these needs and desires, and may correspondingly experience some degree of discomfort.

Although information exchange between the two halves of the brain or the limbic system and the rest of the cerebrum may be incomplete and prevented to various degrees, there is often a seepage of information from one mental system to another. Even though, as we've seen, information may be incomplete and possibly misinterpreted and mistranslated, that seepage, coupled with the experience of tension, arousal, and any other associated feelings, often alerts the left brain to the possibility that certain information may be unpleasant and disagreeable. This initiates the activation of various defense mechanisms. A person is thus able to protect himself from gaining complete conscious knowledge by engaging in denial or self-deception, or through the confabulation of various explanations and justifications.

Indeed, even when we act on these impulses and desires, make the cruel comment, do and say the wrong thing, the left brain and conscious mind is still able to disavow what is obvious. It makes up reasons, rationalizes, justifies, denies, or even projects blame onto someone else. When others are blamed or when traits which we unconsciously possess are thought to be possessed by someone other than ourselves (such as in the case of Allan the rapist), we have engaged in projection. "Projections change the world into a replica of one's own unknown face."

17

Projection and the Modeling of Abuse

There are quantities of human beings, but there are many more faces, for each person has several. There are people who wear the same face for years; naturally it wears out, it gets dirty, it splits at the folds, it stretches, like gloves one has worn on a journey. These are thrifty, simple people; they do not change their face, they never even have it cleaned. It is good enough, they say, and who can prove to them the contrary? The question of course arises, since they have several faces, what do they do with the others? They store them up. Their children will wear them. But sometimes, too, it happens that their dogs go out with them on. And why not? A face is a face.

Other people put their faces on, one after the other, with uncanny rapidity and wear them out. At first it seems to them they are provided for always; but they scarcely reach forty and they have come to their last. This naturally is something quite tragic, as their last is worn through in a week, has holes, and in many places is thin as paper; and then little by little the under layer, the no face, comes through, and they go about with that.

—R. M. RILKE[1]

Knowing, Yet Not Knowing

Just as we might avert our eyes if something particularly gruesome were to come into view, sometimes we refuse to consciously scrutinize unpleasant information if it is especially disagreeable. This includes information that we already possess and may be fully aware of. Sometimes, we just don't want to know, particularly if we have a desire or impulse that is not acceptable to our conscious self-image. Thus, the left brain may actively deny harboring certain personality features or feelings (which are, in fact, possessed by the right brain). It accomplishes this partly through the inhibitory action of the frontal lobes.

In fact, because of the two halves of the brain and the limbic system, we may simultaneously possess three different and sometimes opposing attitudes. However, regardless of the left brain's denials, within the right half of the brain we are often fully aware of the presence of these personality features and impulses.

When impulses are inhibited but continue to strive for expression, whether they are limbic or right brain in origin, we may feel discomfort. Although these forces may be suppressed, misinterpreted, or denied, they continue to exist and can exert tremendous influences on the conscious mind.

If these influences, memories, or feelings remain unconscious, the left brain may not recognize that these forces originate in the other half of the brain. Hence, the left brain knows they exist, but not where. The question for the left brain is: "Who do these feelings belong to?"

The left brain may inappropriately look elsewhere for the source and cause of these feelings. It will look for them in other people. That is, the left brain will project feelings, attitudes, intentions, impulses, and personality traits and characteristics onto others: "Someone is harboring unpleasant tendencies! It must be you!" "Someone is thinking sexist thoughts. It must be him!" "Someone is contemplating adultery. It must be her!" This is a defense mechanism, and it is referred to as *projection*.

Claiming That the Left Half of the Body Belongs to Someone Else

The projection of an unrecognized characteristic is not confined to impulses and personality traits. As noted earlier, in the case of a stroke, when the left brain is prevented from receiving information normally processed by the right brain, it may go so far as to deny not just emotions, personality traits, and characteristics, but the ownership of arms and legs, claiming that they belong to the doctor or a patient in the next room.

One such patient engaged in peculiar erotic behavior with his "absent" left limbs, which he believed belonged to a woman. He claimed that a woman was lying on his left side; he would utter witty remarks about this and sometimes caress his left arm. One woman who was confronted with her paralyzed left arm said it belonged to another person, whom she thought was in bed with her and whose arm had slipped into her sleeve. Another declared (speaking of her left limbs), "That's an old man. He stays in bed all the time."

Some patients (i.e., their left brains) develop a dislike for their left limbs, try to throw them away, become agitated when they are referred to, entertain persecutory delusions regarding them, and even complain of strange people sleeping in their beds when they bump into their left limbs during the night. One patient complained that this person on her left tried to push her out of the bed; she insisted that, if it happened again, she would sue the hospital. Another patient, after bumping into her left arm and leg all night, complained about "a hospital that makes people sleep together." She expressed not only anger but concern lest her husband should find out; she was convinced there was a man in her bed.

If a person's left brain (which, unlike the right brain, is not damaged in these cases) can deny ownership of a leg or arm and claim it belongs to another person, it is not at all surprising that a left brain can claim that emotions, feelings, and personal characteristics also belong to someone else.

The Scapegoat

Long ago, in parts of the Middle East, Africa, and even Europe, it was the custom to hold periodic ceremonies in which sacrifices were made to the lord god, so the local village would be cleansed of evil forces they believed to be causing some pestilence. Usually, the members of the entire village were gathered together around an altar or a great fire by their shaman, priest, or witch doctor. After a few solemn rites, there might be dancing, the beating of drums, singing, screaming, cursing, and writhing, until finally a goat or some other creature was brought forth to be sacrificed.

Following more curses and incantations, the witch doctor or village priest sometimes chanted and performed spells. Sometimes, the villagers went up one by one and touched the goat. A highly religious atmosphere prevailed, and the purpose of the ceremony was to transfer and cast the collective evil of the village onto this innocent creature. In this way, the gods could be appeased, or the sins of the villagers could be removed, and any misfortune or other ill plaguing the village could be alleviated or prevented from occurring.

As the witch doctor, priest, or village shaman cast his spell, each villager was cleansed of his evil as his sins were cast into the goat. Now it was not the villagers or their village but the goat that was evil and filled with sin. Once the creature was assumed to possess all of the collective evil of the village, the people set upon it with sticks and stones, or the priest took a knife and sacrificed it.

Today, we would call such a creature a *scapegoat*. However, it is not a goat but other people on whom we willingly cast our disowned "evil spirits."

Anyone can be a recipient of projected evil and unsavory characteristics: a loved one, a hated enemy, an ethnic minority, or race of people. The end result is the same: They are all potential recipients of hated stereotypical or individual characteristics that are being disowned by the projecting perpetrator. That is, we project and cast onto others traits, thoughts, motives, attitudes, and impulses that we find too repulsive, despicable, or uncomfortable to face rationally within ourselves. Indeed, many people intensely and irrationally hate qualities in others which they despise in themselves, but, rather than hate themselves for possessing these characteristics, they instead hate someone else. Of course, in some instances these unwitting recipients in fact possess these traits, which makes them all the more suitable as an object of anger or revulsion.

Projection need not just encompass feelings of hate, someone can project any qualities that are viewed as unacceptable to the self-image, be it weakness, ugliness, selfishness, or sinfulness. Some of these projected qualities are even good. It is just that we don't want to possess them.

Nevertheless, as simply disowning these personal qualities still leaves a feeling of tension, as well as an awareness that the quality continues to exist, the person begins to look for the cause of the tension outside himself. He may then engage in projection and find a scapegoat. Once he finds a potential victim and casts his sins accordingly, he is then able to decide that it is this other person and not himself who harbors these despicable and unwanted qualities.

Instant Dislike

Many of us have had the experience where, for example, we see a stranger walking into a room and instantly dislike him. It may be something in his walk, the look on his face, the tone of his voice, the manner in which he has combed his hair, his style of clothes. Although we cannot readily identify what it is about the person which is irritating, we find ourselves turned off to him—even though we truly know nothing about him.

Although it is possible that the right brain has detected something that is truly undesirable, more often than not, particularly if the feelings of dislike are very strong (perhaps even bordering on the irrational), this other person has either reminded us of something we have experienced in the past or reminded us of ourselves, or has simply become a convenient target upon which we have flung our own unknown face. When this occurs, we feel indignation, irritation, and contempt when he walks into the room.

However, if we were actually to approach this seemingly unsavory person and talk with him, we might discover he is actually quite a decent person. The key here is the intensity of one's feelings and the sometimes irrational nature of the projection. Of course, the right brain may actually pick up bad "vibes" from a stranger and then alert us to his presence. In these instances, however, our feelings of discomfort do not border on hate nor are they irrational or very intense, and there is seldom a desire to retaliate or for harm to befall the stranger.

Sexual Projection

George was raised in the Midwest by God-fearing parents who taught him about the evils of dancing, drinking, loose women, and rock and roll. His father was also the local minister. George followed in his father's footsteps and also became a minister. He married a very attractive young woman, and being bright and articulate, he moved up in the church's hierarchy. Eventually, he was transferred to a big city, where he thundered every Sunday about the evils of television, liquor, and rock and roll.

One Sunday afternoon after services, a new member of his church stopped on the steps and introduced herself. She was in her early twenties, blond, sexy, and recently divorced. George was immediately attracted. His limbic system became activated, and his conscious mind was suddenly overwhelmed by strange and unwelcome impulses. He interpreted these feelings as nervousness and decided that it was something in her manner that made him feel uncomfortable. He not only felt uncomfortable but, noting her free and easy manner and the shortness of her hem, decided that he did not really like her.

The next day, as he was shopping in the local supermarket, he saw her coming down the aisle toward him. She was wearing tight shorts and a snug blouse with a revealing neckline. Again, he was overwhelmed by the strange and unwelcome impulses from the day before.

Recognizing him, she walked up smiling. He could smell her perfume and could see the faint outline of her nipples pressing against her tight blouse. The strange sensation increased. As she began to chat in a very friendly tone, he began to feel so uncomfortable that he decided there was something wrong, perhaps even evil about this woman. He decided he did not like her. And yet, he lingered, smiled when she smiled, laughed when she laughed.

He became confused. "What am I doing? Why am I talking to this wanton hussy?" his left brain asked. His right brain and limbic system, however, were fully aware of what was going on and what he was really feeling. He wanted to have an affair with her. He wanted to throw her down right there on the supermarket floor.

His limbic system was proclaiming, "I want her right now," whereas his unconscious Parent was criticizing him for having these feelings, thus causing great unconscious confusion. His left brain and conscious self-image, however, found such notions completely unacceptable. He could never acknowledge to himself that he had such desires. He began thinking that this woman might be up to no good, and it occurred to him that he should suggest she join a different parish.

Glancing around the store, he spied one of the church elders, Joshua, smiling and walking toward them. George decided, however, that Joshua was smiling in a very odd manner and was looking not at George, but at the young divorcée. Joshua introduced himself, and he and the young woman began to chat.

As the two talked, George, still feeling very uneasy but being unsure why, suddenly decided that Joshua was being overly friendly and was looking at the young woman in a way that was not quite acceptable. Indeed, he thought he saw the older man's eyes linger on her breasts and decided he was standing a bit closer than was necessary. It occurred to George, "The old geezer is thinking impure thoughts!"

George was shocked. Here was Joshua, a respected elder of the church and a married man with children, flirting and carrying on with this young hussy. Again, George decided that the woman was evil, and he was additionally grievously disappointed in Joshua. It occurred to him that maybe he should bring this up with some of the other church elders.

As Joshua and the woman continued to talk, now ignoring him, George thought that maybe these two were already having an affair. Feeling disgusted and indignant, he excused himself, walked off, and made a firm decision to take some kind of action. After all, he had the welfare of his congregation to protect.

George had engaged in projection. Consciously, George could not help but detect his own obvious sexual impulses, which were being transmitted and experienced unconsciously. However, his conscious self-image almost completely prevented him from consciously realizing that these sexual messages were originating within himself.

As the sexual arousal was so pervasive, he could not completely deny its existence. Instead, he opted to protect his conscious self-image: It is not "I" who has this undesirable impulse; it is Joshua. Because George's conscious self-image also considered such impulses evil, he again looked for a source outside himself and projected these feelings (which he has interpreted as "evil") on this "evil" woman.

Jealousy and the fear that one's mate may be having or at least contemplating a sexual affair is often based on the same principle, that is, projected desire. Indeed, it is always easier to condemn others for our own faults than to deal with them in ourselves.

Projection and Love

We all require association with others. Contact and physical intimacy maximize personal growth and physical health and serve as an outlet for sexual tension. Even the limbic structures require contact in order to continue functioning appropriately.

We also require association as a means of self-discovery. We learn who and what we are by comparing ourselves to others, and by recognizing how others treat us. Our siblings, parents, friends, and lovers serve as mirrors. However, we sometimes do not recognize the reflected image as belonging to us. The face that stares back may seem evil, selfish, critical, abusive, rejecting, unloving, hateful, passive, or dependent, and these qualities do not correspond to our conscious self-image. Who is it that we see? It is our own unknown face.

Many of us require association not only as a means of self-discovery, but as a means via which disowned and latent qualities that we do not find personally acceptable can be projected onto others, such as a member of the opposite sex, and then either loved or hated. Again, however, the traits projected do not necessarily have to be bad.

That is, we may consciously believe that certain traits and feelings are inappropriate for a man but not a woman, or a construction worker but not an accountant. A woman may consider it inappropriate for a woman to be "tough," or "hard," or a "heavy drinker." Similar qualities are acceptable in the guy next door, who is a construction worker, but not in her husband, who is an accountant. We find many qualities and traits acceptable in others, but not in ourselves, and although we may readily recognize these tendencies in others, we may be blind to them even when they are part of our own hidden nature. In this regard, sometimes people are able to project not only their own unknown qualities, but characteristics that their spouse or children possess. A parent can then convince herself that it is not her children who are bad, but those neighbor kids. Her husband is just relaxing with a drink or two after dinner; the guy next door is a drunk.

Sometimes, however, these projected qualities are quite good, sweet, sexy, and loving (depending on the person's perspective), but are still unacceptable to a particular individual. He may even find these traits as quite endearing and attractive but not in himself. Being unable to accept or recognize that these characteristics exist within us (if they are unacceptable to the conscious self-image), we instead project or discover them in someone else, such as a friend or a lover. We are then able to enjoy these projected characteristics when in the company of the recipient. That is, by projecting these traits onto another person, we can enjoy those traits which are in fact our own.

Suddenly, we find that by being with this woman or that man, we are able to experience aspects of ourselves that we did not know existed. However, rather than recognize that it is our own projected characteristics which have given us this enjoyment, we instead believe it is the person we have projected them onto, and then need to be with him or her. We project these qualities onto our partner and then decide they are exciting and wonderful to be around; and they are, but mostly because we are now allowed to express and enjoy our own unknown potential in their company.

By projecting these aspects of themselves, some people are able to enjoy or love themselves at their leisure without feeling discomfort. Indeed, some people see their partners as an extension of themselves, and projecting these traits onto their partner, they then fall in love, not with their partner, but with their own projected fantasies.

For instance, a man who unconsciously needs and desires to be pampered, protected, and spoiled may feel quite unhappy if these needs are contrary to his conscious self-image. As the awareness of these needs may still plague him, he may relieve himself of the pressures he feels by projecting these desires onto some willing recipient. Thus, he may satisfy these needs by finding a woman either who has similar needs, or who can serve as the recipient of these projected desires. As a result, he may feel an enormous degree of love, as he is able to meet her needs and meet his own at the same: "It is not me who needs to be held and kissed it is her. I am just doing it to please her."

Until the 1970s, many women (particularly when they were children) were told that it was wrong or unfeminine to be assertive, aggressive, bold, domineering, or sexually or physically active. Once an adult, a woman who had been brought up in this way, and who had such traits, may have been unable to acknowledge them consciously because they were contrary to her self-image. Because these tendencies existed, she felt some discomfort when they tried to emerge. She may have attempted to deal with these feelings through compensation (that is, becoming overly "feminine" or helpless), or she may have projected these assertive qualities onto other women and then despised them.

As these needs continued to strive to be met, she was likely to project these qualities onto a male recipient. She would then have found herself admiring and being attracted to these same qualities when they were seemingly possessed by the man and would have been able to love her own rejected characteristics.

Loving in Others Qualities That We Reject in Ourselves

In the Middle Ages, long before it was demonstrated that physiological and chemical agents of both sexes exist within us all, it was said, "Every man carries a woman within himself." Indeed, many men have nurturing, caring, sensitive qualities often erroneously described as "feminine." The great analytical psychologist Carl Jung argued long ago that the more "tough," "rough," and "masculine" a man may appear to be, the more "feminine" are his hidden qualities. As being super "macho" is the converse of ultrafeminine, one might assume that such a man is trying to compensate for these feminine qualities. He is engaging in reaction formation.

Compensation and reaction formation are defense mechanisms. However, even a successful defense does not eradicate one's unknown face. There is always some degree of awareness, although it may be meager. Depending on the degree of awareness, the supermacho male may look for a source outside himself for his despised feminine elements, which are struggling to be expressed. He may see a weak, nerdy-looking man and, projecting his own despised femininity onto this unfortunate fellow, feel like kicking sand in his face, and yet it is in reality his own unknown face that he so ardently wishes to destroy.

Fortunately, not all men feel a need to kill and destroy people who are unwilling recipients of their projections. Sometimes certain qualities that they do not find personally acceptable can be projected into a friend or member of the opposite sex and then loved or treated with loving concern as they attempt to help their friend or mate overcome these deficiencies.

Indeed, although the unconscious Child and Parent attempt to exert their influences when we meet members of the opposite sex, sometimes men fall in love with women who in some manner correspond to their own unconscious femininity. Failing to find such a person, the second choice is often a women who can unhesitatingly receive the projections emanating from his soul; his own unconscious. Unfortunately, over time he may become increasingly disenchanted if it becomes apparent that the projections do not fit; or worse, if his mate purposefully or inadvertently discovers his hated self and points it out to him.

In either case, finding a woman who resembles one's own unknown face, or who can act as a recipient for these projections, may seem rather ideal. Unfortunately, as pointed out by Jung, it often turns out that the man has fallen in love with his own worst weaknesses. Often the man has in fact married the projected image of his mother, or worse, a composite of his unconscious Parent and then goes through the rest of his life being treated accordingly.

Similar misfortunes, of course, befall many women. In some instances they fall in love with a "masculine" ideal, and then find a man upon whom they can project these qualities; only to become sorely disappointed when the real man emerges from behind this projected mask. Or they may seek their father or a father ideal, and find a man who can best mirror "daddy's" worst and best qualities. A composite of the unconscious Parent is also a major attracting force.

In any case, women who marry men who correspond to their unconscious Parent or idealized male image, or men who are really nothing more than a recipient of projections, more often than not find that they have bonded to potentially very destructive forces. Like their male counterparts, it often turns out that these women have married their own worst fears and weaknesses.

Men and Women Who Marry Their Projections

Love and Projected Emotion

Many individuals are uncomfortable with emotional displays in others or in themselves. They may believe that emotions may overwhelm them, and represent a loss of control. One goal of socialization, of civilizing children, is to teach them self-control. How parents deal with emotional displays in their children will greatly affect how these children will behave emotionally when they reach adulthood.

By example and through punishment children learn what, when, where, and to what degree to express their emotions. Some emotional displays are encouraged, and others are discouraged, depending on what parents see as appropriate for boys or for girls. Characteristically, boys are allowed if not encouraged to express overt aggression, particularly on the playground, in team sports (football), and in make-believe (cowboys and Indians, soldiers at war, and so on). Girls, however, are often discouraged from engaging in this type of overt aggressiveness and must use indirect methods to express their feelings of aggression or anger, such as gossiping, sarcasm, crying, tattletaling, name calling, and pouting. Yes, even in our enlightened society these sexist attitudes still prevail, if not overtly, then subtly.

When we are restricted in our expression of emotions, the emotions don't just go away, they will be expressed in other, less appropriate ways. Sometimes, aggressive and angry feelings are turned inward, and the child engages in self-destructive activities and feels considerable anxiety, guilt, and depression because of having forbidden feelings. In self-destructive activities, the child is seeking punishment for having these qualities.

Sometimes, children are recipients of projections from their parents, and if these projections are contrary to their own characteristics, they must assume a false persona and find other outlets for their real feelings.

Some children who are not allowed to express their feelings (be they tenderness, dependency, hate, anger, resentment, jealousy, selfishness, aggressiveness, assertiveness, or assaultiveness) will seek out partners later in life who can and will express these same types of feelings and emotions. A woman might link up with a man who will vent emotions which she possesses unconsciously but could never express as a child or adult.

Sharon's parents were very strict, religious, and overly concerned about "what the neighbors might think." They had very definite Old World views about what was appropriate for a lady and discouraged any form of emotional expression, physical activity, or intellectual curiosity that did not fit in with this idealized image. Politeness, respect for authority and tradition, good manners, and good taste were always being pressed on Sharon. Because all these qualities were actually contrary to her nature, she often felt disapproval in the way her parents interacted with her, and she felt somewhat rejected. This rejection made her feel quite angry and resentful, although these feelings remained locked away in her unconscious. Wanting to be loved and accepted, she continually attempted to assume the role of the good girl and to suppress her bitter feelings of anger, resentment, hate, and destructiveness for fear of being further rejected. However, her true self often made its presence known, much to the mortification of her parents, who would then withdraw even further from her.

As she discovered that her quick wit, sarcastic tongue, and sometimes domineering manner resulted in looks of pain, embarrassment, and discomfort from her parents, she attempted to disavow these qualities as well. She not only tried to be "good" but to behave in a passive, nonintrusive manner so that no one would find fault with her. She sincerely attempted to become the perfect little lady. But to no avail.

Although admired by her friends and teachers, she never felt that she had truly obtained the love she hungered for from her parents. Unconsciously, she knew there was something wrong and unacceptable about her. This knowledge only increased her unconscious feelings of resentment and anger. She countered these feelings by suppressing them even further and becoming more and more sedate in her actions and speech.

Seemingly a passive person with wonderful manners and social etiquette, Sharon began dating and then married Jeffrey, who was highly domineering, aggressive, angry, and bitter. Jeffrey did not hesitate to give someone a verbal lambasting if he felt slighted, thwarted, or in any way frustrated. He always seemed to be at war, ready for battle, and ready to take on all comers, even if they had the power to damage his career or fire him. He tried to contain himself when interacting with these latter individuals, as he was not completely self-destructive.

Everyone seemed surprised that Sharon would marry such a man, as he seemed to be in all ways everything she was not. Some of her friends even began to feel sorry for her (which unconsciously gave her great satisfaction because of the hurt she had felt as a child).

Although Sharon always acted embarrassed by Jeffrey's volatile nature and, in fact, tried to help him be less obnoxious, she actually felt enormously and secretly (unconsciously) pleased whenever he lit into some offending person. She found his volatile, hostile nature quite endearing, although not consciously.

If someone asked her, "Why him?" she would reply that she loved "the little boy in him" and the "sweet, caring way he treats me." "He makes me feel safe," she would sometimes say. Unconsciously, what she was really saying was "He can safely express my unconscious feelings, and I love him for it."

Jeffrey was a perfect match for Sharon's unconscious projections. He was able to express her feelings of anger and resentment, as well as her sarcastic and aggressive wit, and she was able to enjoy their expression without ever having to suffer the consequences. Indeed, she was able to continue playing the role of the good girl, with the added bonus of finally being recognized as the victim (for having to put up with this obnoxious bastard) that she had always felt herself to be. However, it wasn't Jeffrey who had victimized her, it was her parents.

One might wonder: What was the payoff for Jeffrey? What did this dynamic, aggressive, verbally assaultive person find attractive in the meek and passive Sharon? Was he with her because "opposites attract?"

There is an old adage (once uttered by Rhett Butler to Scarlett O'Hara) that "Like belongs with like." Seeming to be opposites insofar as the conscious aspects of their personalities were concerned, unconsciously Jeffrey and Sharon were a perfect fit. Jeffrey was with Sharon because she was a willing recipient of his own unconscious projections. Projection was not the only defense mechanism he used, however, as he was also compensating for flaws he perceived in his character, flaws that although unacceptable to him were acceptable as part of Sharon's persona.

The assertive, aggressive warrior that Jeffrey presented himself as was in part a reaction to his own sense of insecurity, low self-esteem, and feelings of possible weakness and worthlessness. It was a reaction formation. Although even as a child he had seldom behaved passively or weakly, he had often been beaten into submission by his father. The father beat him until he was forced to cower in order to survive. Jeffrey hated himself for submitting almost as much as he hated his father for beating him. He also feared that he might actually be the passive, weak person that his father had attempted to make him into. Despising this feared aspect of himself, he went out of his way to prove himself powerful and aggressive.

He also sought out a woman who could safely mirror and accept the aspects of himself that he despised. That is, Sharon was weak and passive, not he, and he could safely look at her and feel assured that in no way was this he. In other words, knowing that someone was weak, passive, and inferior, he could project these feelings about himself onto a ready recipient, who was willing to act out his own unconscious fears.

Thus, although they seemed like opposites, Jeffrey's conscious personality was a perfect match for Sharon's unconscious personality. Conversely, her consciously maintained self-image meshed with his own unconscious fears.

Projected Hate and Love

People frequently condemn their mates (and even their children) for having traits that they do not, in fact, possess. What they are attacking is their own despised projections. Indeed, I have sat through session after session with married couples who repeatedly hurled unfounded accusations at each other that were little more than projections. Although the "shoe fit," they refused to wear it and instead tried to force it on the feet of their spouse.

Often, however, I hear each partner accusing the other of the same thing as if each is projecting the same own unknown face onto the spouse. They call each other the same names and accuse each other of the same "sins." It quickly becomes apparent that although each is engaging in projection and is simultaneously disavowing what the unconscious Parent is saying, they are together because they are a perfect fit. They have the same bad needs and the same unknown face. Instead of loving this face, however, they hate it. Because they do not want to hate themselves (they prefer to love themselves), they hate someone whom they claim to love. What they love is the ability to safely express the self-hate that they have long felt. Instead of a "mutual admiration society," they have created a "mutual hate society" and fling identical projections at one another. They are flinging self-hate:

Gary: She is always trying to control me.
Mary: He is always trying to tell me what to do.
Both: You're the one who's controlling!

Michelle: He doesn't care about anybody.
Roy: All she cares about is herself.

Bob: She is always nagging and complaining.
Linda: He is always picking on me and complaining.
Both: You're the one who can't stop complaining!

Sarah: He never listens.
Rick: She always interrupts me.

Ginger: I am not allowed to have an opinion.
Thomas: She disagrees with everything I say.

George: She is always putting me down.
Carol: He is always tearing me down.
Both: You're the one who does the tearing down.

Nannette: He never wants to do what I want.
Troy: She never wants to do what I want.

Charlie: She's a selfish bitch.
Carla: He's a selfish bastard.
Both: You're the one who is selfish!

It is surprising that neither partner recognizes the circular nature of these accusations and their sameness. However, as both are engaging in projection, this realization is almost impossible. The conscious self-image must be protected at all costs, and the left brain and the language axis believe everything that it conjures up.

Sometimes during the course of an argument, one half of the couple may become aware that he (or she) is behaving in an unreasonable fashion. This recognition, however, does not mesh with his self-image, and it also engenders additional tension. This tension might be labeled as "guilt" or even "anger." Because he cannot be angry at himself and cannot accept the possibility that it is he who is behaving in an unreasonable fashion, he decides he is angry at his partner because it must be his partner who is behaving unreasonably. This decision enables him to discharge his tensions through projection. It also allows him to justify the bad way in which he is treating his partner. Consciously, he proclaims: "I am arguing with you because you are unreasonable and because you make me angry. I hate people who are unreasonable, and such people are not only wrong but despicable. I am therefore completely justified in treating you badly and should treat you even worse because that is what you deserve." In truth, however, he is yelling at himself.

Projection of Misinterpreted Tension

In many relationships, if one partner harbors desires or feelings that are contrary to her or his conscious self-image or that cause feelings of guilt, she or he may try to alleviate these tensions by projecting the guilt onto the mate and by confabulating some form of justification: "I called her a bitch because she insulted me in front of my friends"; "I refused to make love to him because he's been ignoring my needs." Sometimes, in fact, fights are provoked so that guilt and projections can be safely hurled from one partner to the other. The fight justifies the projection and relieves the sense of guilt:

After Sally's husband, John, had gone off to work, she decided to sleep until the afternoon. When she finally got up, she still felt too tired to clean up the house or even to wash the clothes that should have been washed days before. Instead, she decided to lie on the couch and watch TV, as she had done for two weeks since she had been laid off. As the afternoon wore on, and as soap opera after soap opera passed before her eyes, she began to feel more and more tension. She knew she should clean up. Looking at the clock, she realized that her husband would be home soon, and this realization made her even more uncomfortable. She should have started dinner long ago.

Feeling tense, she imagined what he would think if he saw her lying here, the clothes unwashed, the house a mess, and no food on the table. She got up feeling irritated and began to pick things up quickly. Glancing again at the clock, she realized that he would be home any minute.

She rushed to the kitchen and again imagined what he might say and how he might criticize her for letting another day go by without accomplishing anything. She began to feel extremely tense, a tension only made worse by the depression she unconsciously felt regarding her loss of work and the effect it had had on her self-esteem. However, she misinterpreted her tension as anger. Consciously and unconsciously, she realized she had no reason to be angry. So she projected her anger and instead saw her husband as angry at her, which only made her feel more tense. Now, she began to feel angry because he would be angry at her. She was not angry at herself; she was angry at him. How dare he criticize her?

Frequently (but not always), when we *imagine* we know what another person might be thinking about us, we are probably engaging in projection. If Sally was positive that John would criticize her if he knew how she had spent her day, she was probably projecting her own self-criticism onto him (even if he would indeed criticize her). As a result, she might attack him for thinking such thoughts and create an argument where none might have occurred. Had she realized that the source of her criticism, tension, and anger was within her own mind, she might have used the tension to get her work done and to go out and find another job.

However, if he later dared actually voice any of the criticism she had been projecting, the argument would be likely to be doubly intense and quite terrible:

Still working feverishly in the kitchen, nursing her anger, Sally heard the door open and her husband's footsteps as he headed toward the kitchen. She turned and looked at him. He smiled, looked at the kitchen table, back at her, and then said:

John: Where's dinner?
Sally (feeling tense and defensive): Where do you think it is? I haven't made it yet!
John: So I see. So . . . What have you been up to today?
Sally: What do you mean by that comment?
John (feeling defensive): Just asking.
Sally: You know what I've been doing.
John: No I don't. So, when's dinner going to be ready?
Sally: What do I look like? A slave? You think all I've go to do is wait on you hand and foot?
John: Hey. What the hell is the matter with you? You don't have to get hysterical!
Sally: You're the one who's hysterical. You act like a spoiled child: "Feed me, feed me. When's dinner?" All you think about is you. You go sit in your air-conditioned office, and you don't have to get off your lazy ass all day and then you come home and act like a spoiled brat.
John: You're the one acting like a brat.
Sally: And you're acting like an asshole!

"Child," "lazy," "spoiled brat," "hysterical"—the accusations Sally hurled at John were nothing more than embarrassing qualities that Sally harbored in the secrecy of her soul. She alleviated her sense of guilt by projecting her doubts and ill feelings onto John, where she could then attack them. It is easier to attack someone else than to attack the real source of one's own shortcomings. Indeed, many accusations (when hurled in a fit of anger) are best applied to their source.

Projection and Children

Often, we inadvertently not only misinterpret tension and feelings of arousal but use denial and projection to mistreat loved ones, including our own children:

Mary asked her son and daughter to please be quiet and play in another room while she worked on a proposal for her job that was due the next morning. She was already feeling anxious because she had procrastinated and was irritated at herself. However, she believed that she did best under pressure.

As she was typing furiously, the door suddenly swung open, and her son barged in, asking: "Where is my baseball glove?" As he spoke, she made several typing errors. She turned on her son, yelling: "See what you made me do! I told you to leave me alone. Now get out of here." He slunk away feeling very guilty.

However, it was not his barging in that caused her to make a mistake, but her own feelings of tension and irritation and the need to hurry. She also felt tension because she had put off her report and was neglecting her children. Being a neglectful mother was not part of her self-image, so her object was to displace this tension, discharge it, and alleviate her feelings of guilt. She used the old standby, projection, and the misinterpretation of tension.

The solution: "I am not the one who is guilty. I have nothing to feel guilty about. It is my son who is guilty. He has barged in here asking silly questions and has caused me to make these mistakes, bad, wicked boy!" She was thus able to justify her behavior and alleviate her guilt by projecting her guilt onto her son. It was not she who had done something bad by ignoring him and not giving him the love and attention he needed; it was he who had done something wrong by bothering her and making her goof up on her job. He deserved to be neglected and banished.

Unfortunately, because he accepted her projections and responded as if "the shoe fit," the possibility that his mother (or he) might come to realize that she behaved badly is greatly reduced. He inadvertently reinforced her bad treatment of him.

Many married couples do the same thing. They project their own guilt, their own sense of inadequacy, their own shortcomings, and their own selfishness and then use these projections to justify mistreating the ones they love. In some instances, even the love is a projection, and the spouse serves as little more than a blank screen on which they can project their own unknown face and then either love it or hate it.

Modeling Abuse

Children who are abused, who have mothers who were frequently beaten, or who grew up where there is a great amount of tension, rage, anger, and fighting in the household, sometimes find it quite difficult to cope with their own feelings and fears regarding what they observed. Consequently, their own emotions are often expressed in inappropriate ways. This expression may take the form of bed-wetting, difficulties at school, depression, fighting with peers, psychosomatic complaints, or other forms of inappropriate acting out, all of which may represent a cry for help. Sometimes, however, children just model what they see at home, whether feelings or behavior, and then act out what they have learned, sometimes in a self-destructive manner or in highly inappropriate settings. Often they engage in behaviors that are self-punishing. Feeling angry hurt, frightened, and upset, they may turn these emotions on themselves in the form of depression. Or they may place themselves in punishing situations or punish others by acting out cruelly and callously, becoming bad kids. Later, as teenagers, they may continue to express these emotions inappropriately via substance abuse, promiscuity, or criminal activity. Essentially they engage in behaviors that are self-abusive or abusive to others or that result in some form of punishment.

However, by continually engaging in behaviors which are self-destructive, which may result in punishment, or by becoming involved in situations where there is a considerable degree of turmoil, they are in effect recreating the familiarity of their former family atmosphere, but outside the home. They thus reexperience the familiar hurt, anxiety, rejection, and other fears experienced as a child, and which the unconscious Child within the depths of their mind continues to feel.

When children observe their parents fighting, attacking, accusing, belittling, and blaming one another for life's various woes, and when they are infrequently exposed to compassion, love, respect, and mutual caring, they are likely to see such behavior as normal and to reenact it when they grow up and have their own families.

Frequently, such children, when they become adults, provoke others to abuse them or, conversely, find someone they can verbally and physically assault on a regular basis. They also tend to misinterpret the innocent behavior of others as provocative so that the comfortable familiarity of childhood can be maintained, and so that a fight will ensue. Indeed, responding to real or imagined slights they may activate the unconscious Parent, respond in a highly abusive manner, and end up arguing, abusing, or being abused by a "loved one."

Physical Abuse

Abusers deal with their insecurity and rage by attempting to control and by criticizing or abusing others. However, they are also engaging in projection and are, in fact, attacking their own unknown face. That is, these individuals often have an unconscious self-image that is embarrassing and unacceptable to their conscious mind. The conscious mind, being affected by this image, attempts to deal with it by projecting it onto a not always willing recipient. Usually, what they project is the face of their own unconscious Child. In addition, as they are in possession of an unconscious Parent that urges them to beat up on others, once they have found their unknown face hanging on another victim, their Parent can begin abusing that person as well.

Victims

A few, but certainly not all (or even most), victims almost willingly become recipients of the projections hurled by their abusive spouse, simply because they have a familiar ring and they seem to fit. Additionally, some victims are trying to win that elusive love that was denied them during their childhood. Thus, the projections that the abuser hurls match the victim's own unknown face, and corresponds to the unconscious self-image maintained by his or her Child.

Similarly, some victims, as we've noted, have an unconscious Parent that tells them they deserve to be abused, as well as an unconscious Child that believes it is bad and should be treated poorly. Hence, some victims accept the abuse of a spouse because their own unconscious mind is conspiring against them.

Abusing Children

Abusers who project their own inadequacies on a spouse and then mistreat him or her sometimes project onto their children as well. An abuser's unconscious Child usually feels bad and knows it deserves to be hit. The unconscious Parent tells the person it is OK to hit children if he or she was hit when young. Indeed, the unconscious Parent often beats the unconscious Child in the privacy of the mind, and often in dreams. Abusers also project the face of their own unconscious Child onto their children, and then the activated Parent punishes these innocent victims for the abuser's own secret fears, inadequacies, and frustrations.

Not only are these innocent children being abused, they are learning that this is the way fathers and mothers are supposed to treat their children and each other. Thus, abuse, like feelings of inadequacy, alcoholism, drug abuse, and self-hate, is passed on from generation to generation until someone seeks help to put an end to this repetitive pattern of self-destruction.

IV

APPLICATIONS

18

Choice and Responsibility

To those who say:
"You can't," "You won't."
I say:
"I can," and "I will."

In considering the multitude of forces that act on human beings—selective pressures over the course of evolution, heredity, genetics, biochemical fluctuations, how our parents raised us, and the cultural biases that we are all subject to—it may well seem that all our behavior is determined by causal factors set in motion so long ago that we are helpless to alter their course. Certainly order exists in the universe, we are affected by our environment, and forces such as fate, God, or even our astrological sign (if we wish to believe in such things) may exert some influences on our behavior. Nevertheless, we are not completely at the mercy of forces over which we have absolutely no control. We are still capable of making choices.

Admittedly a person is capable of choosing only what he is able to conceive. One cannot explore a road he does not know exists. If a person is brought up in an environment where mom and dad are alcoholics, drug addicts, or thieves who are on welfare or frequently in jail; in an environment where people are shot to death for merely being on the street; in a society that worships sex, violence, and easy money and that mocks those who strive for something better, such an individual may be incapable of choosing to live his life any differently. It simply would not occur to him. It would be as alien as living on the planet Mars and is thus not an option and is almost impossible for him to consider.

Often, the only way out for individuals raised in these circumstances is to become educated about the alternatives available. However, they must also be able to scrutinize their own lives so as to recognize the maladaptive, self-destructive patterns they have been engaging in. It will be very difficult for them to change their behavior if they cannot recognize it so as to avoid it.

Although it can be a most difficult task, men and women are capable of searching their souls and exploring their own unknown face so as to discover the stranger within. Throughout our lives most of us remain capable of learning new strategies for living and can alter and extinguish repetitive patterns that are self-destructive and defeating. The tape recording can be turned off, and a new one can be put in its place; the brain can learn new patterns of responding and thinking and can reorganize itself in the process. Change is possible. For some it is just not very easy.

The capacity to make choices and to take control over one's life is not unlimited. We cannot choose to soar with eagles by merely flapping our arms, and we cannot will away our problems. However, in order to act on our choices, as limited as they may seem, one must be aware of what we are choosing and why we are choosing it. We must also be aware of whether we are making choices and not just responding to a habit or to the expectations of others. Even choosing not to choose is a choice.

Choices may be made by different aspects of the psyche. The conscious mind may have one desire, and the unconscious mind another. This difficulty may seem insurmountable given the power of certain regions of the brain. However, even when beset by powerful limbic forces that may urge us to eat, or to drink, or to have sex, human beings are capable of and have been capable of controlling those impulses so as to eat when they wished, or to not eat at all and starve to death.

Choice and Social-Emotional Intelligence

Some people who consider themselves highly intelligent have tremendous difficulty establishing or maintaining successful relationships; others repetitively experience failure at work; and still others are simply unhappy. In defining intelligence, it is useful to include the ability to be happy, to lead a contented life, or to maintain a good relationship. If we are failing in our relationships, if we are unhappy, then perhaps we are not as intelligent as we would like to believe. Certainly, one may be intelligent and not use that intelligence effectively, or use it to make the wrong choices. However, the wise use of choices is an indicator of intelligence. In this regard, sometimes it is not *why* we choose that is important but *what* we choose. Some people choose to be depressed, some choose to be angry, some choose to be happy, and others choose not to care. Choice is something we can freely engage in, and we can even choose what we want to believe about the ability to make choices.

There is no denying that our genetic inheritance, how we were raised, and our neurochemical internal environment play a large role in how we behave and think. Brains become damaged, hormonal and neurochemical fluctuations and abnormalities occur, and physical limitations sometimes result from genetic endowment. However, what we think can actually change our neurochemical environment as well as the actual structure of our brain. If we think new thoughts or think old thoughts in a different way and apply different meanings and interpretations to them, we can expand and stretch the limits that our biological makeup has imposed. This process is called learning. Even a person who is mentally retarded is capable of making choices and being happy.

Admittedly, of course, some individuals are so damaged and biologically impaired that discussions regarding choice and responsibility are not applicable. This does not apply to anyone reading this book, however.

Some of us choose to believe that our friends, parents, children, or spouse are responsible for our choices and even our emotions. However, choosing to believe that others control us is an abdication of our personal responsibility. The person who says, "She made me do it," "I can't help it," "It's a habit," or "That's the way I was raised" is choosing to remain in the same rut. He or she is just making up excuses or is confabulating.

Frequently, I ask my patients to imagine there is a little person in a little room somewhere (perhaps inside their head) who has been controlling their lives. This person has been pushing and pulling little levers or buttons and making all their choices, controlling all their behavior, putting them in every bad or happy situation they have ever experienced. I then ask:

Dr. J: If we found out there was such a person and that he or she is responsible for all your choices, all that has ever happened to you, both the good and the bad, can you tell me, does this person like you? Does this person have your best interests at heart, or has he or she been purposely trying to mess you up, to ruin your life and hurt you?

Most of my patients reply, "He's been trying to screw me over." Indeed, one recent patient told me he would kill that person if he ever got his hands on him.

Dr. J: This may come as a surprise, but there is no little person who has been controlling you. That man in the room is you. It is you who has been making every choice. It is you who is in control, and it is you who is purposely trying to ruin your life. If you want to hurt or get even with that person, it is time for you to realize you have been doing that all along.

Many of my patients become surprised at this revelation. Some will also argue against this conclusion or plead for more information, and this is what I say: "Since we know it is you and that there is no little man in a room controlling your life, perhaps it is time for you to begin making choices which are good instead of bad for you. You are in charge, you make the choices. It is your decision how you want to live your life."

Yet, many people are afraid to be in control of their own lives, to take charge of themselves, for the responsibility is too frightening. Although some people may wish to believe otherwise and may be convinced that they have relinquished or have never had this responsibility, ultimately, they are in charge; they make the choices. After all, the right, left, and limbic brains and the unconscious mind belong to the person who possesses it.

Certainly, events sometimes overwhelm us and affect people in a manner that is almost impossible to control. If someone's house burns down and he experiences profound misery, it would be quite difficult, at least initially, for him to choose to feel differently. Many things can happen that are not our fault, and in these instances maybe it is best to believe that our lives intersected with fate leaving us as a victim or beneficiary of random or preordained events. We cannot accept responsibility for everything that occurs, and sometimes, our choices have no

Figure 37. The stranger within.

bearing on a particular outcome. Again, there is always an interaction between what some people may call fate (or their genes, environment, etc.) and their decisions.

Many people feel they are affected by the behavior of others, and this, too, is a correct assumption. Indeed, others can influence our feelings and our actions such that we sometimes lose control. Complete self-control is a rarity.

If Jake's girlfriend has an affair and he gets depressed, it does him little good if his psychologist tells him that he has made a choice to be depressed and could just as easily be happy.

In situations such as these, it is a normal reaction to become depressed, it was not his choice. What was his choice was to choose a girl who would do such a thing to him and develop a relationship with her. This is what he is responsible for. Moreover, should he choose to dwell on her behavior, berate himself, and maybe even beg her to come back to him or plague himself with thoughts of her actions for months to come, then he has chosen to continue in a destructive pattern of behavior, to think certain thoughts, which will result in his continuing to feel depressed.

On the other hand, he may choose to *see* things differently, to apply a different *meaning* to what has happened, to reinterpret his entire relationship with her—what he got out of it, what kind of person she was, how lucky he is not to have married her—and thus to lessen his depression if not make it go away altogether. He may not be able to change the past, but he can certainly alter what it means. This choice is what he can control, and what he is responsible for. But, to simply change emotions as easily as changing a light bulb or flipping on a light switch is a feat that few of us are able to accomplish on the spur of the moment. Nevertheless, ultimately we are responsible for our choices.

Those who prefer to blame others, God, or fate for whatever wrong happens in their lives are simply shirking their own responsibility. Although they may feel and perhaps even believe that they are not responsible for their feelings or their behavior, ultimately they are simply self-deceived. By blaming others, they have accomplished nothing except preventing themselves from focusing on the true source of their difficulties: their repetitive self-destructive thoughts and maladaptive behavior. Their choosing to blame others prevents them from taking and exercising healthy control over their own lives and keeps them from focusing on the forces within their own mind which have contributed to their self-destruction. (However, it is still their choice.)

Indeed, those who blame others for misfortune often attribute whatever success or happiness they experience to forces outside their control as well. "It was dumb luck." Unfortunately, that attitude simply prevents them from recognizing their own responsibility and prevents them from marshaling their own resources to achieve happiness and success in the future.

By contrast, those who believe they are responsible for and control their own lives and emotions are able to feel good about themselves without attributing their happiness to "luck" or other mystical forces and can thus choose to behave in a way that will enable them to feel good again in the future. If things turn out well, if they change their lives for the better, it is because they have handled things well and have made things turn out as they wanted them to.

Similarly, those who accept responsibility for their misfortunes can also change their lives because they know that if they feel bad, angry, or depressed about some misfortune, it is usually because they made the mistake themselves. By taking responsibility for their behavior, and by examining the unconscious forces affecting their lives, they are also assuming control of their future so that they will not have to make the same mistakes again. They learn from their errors because they know that, ultimately, they control their own feelings and behavior, within the limits imposed by fate, hormones, genes, and childhood experiences.

Gaining Control over Emotions

Individuals can gain control over themselves by becoming aware of their true needs, their true motives, the influences exerted by the unconscious Parent and Child, and their defense mechanisms. They then can change the way they think and behave, what they believe and even what their own past means to them. Indeed, what they believe to be true, the way they think, and the meanings they apply to their experiences can ultimately lead to happiness, anger, or depression.

Although language is in some ways very limiting, we can use language and verbal thoughts to reprogram our brains and to give new meaning to our lives and expectations for the future. Unfortunately, it is not easy to think in new ways or to unlearn unhealthy habits. Well-learned belief systems are entrenched and not easy to abandon. Nevertheless, with effort it can be done. As we've seen, the very circuitry of the brain can be changed by how we think, and by thinking in new ways. Again, this process is called learning.

Those whom I've seen gain control over their lives and their feelings have first made themselves aware of their choices. Many of them have learned to ask themselves more probing questions through therapy, or through the wise and objective counseling of friends or teachers. Nor do they believe their very first explanation when examining their behavior. This is very important, for it is necessary to be aware that the left brain likes to confabulate and believe its own explanations, contradictory information not withstanding.

Confronting Our Parents

Part of changing thought and behavioral patterns requires that we confront the internal Parent and provide some sympathy and understanding to the internal Child. Most people take personally the way in which they were treated when young, as if it reflects their true value and worth as human beings. Many feel responsible for being treated badly and thus internalize this badness as part of their unconscious self-image. Unconsciously, many believe that they caused their parents to criticize, torment, or beat them, as if for some reason they deserved to be treated poorly.

However, the child does not cause the bad behavior of a parent. Indeed, if a different sperm and a different egg had been fertilized, in all likelihood the different child who was born would have been treated identically. Children who are mistreated, neglected, and abandoned by their parents are really nothing more than innocent bystanders; they are victims and not perpetrators.

Fortunately, once people stop taking responsibility for their parents' treatment of them, they can begin to accept responsibility for themselves and can minimize any hurt they may feel. They can then change the meaning they have applied to their childhood experiences and can change their thoughts, change their brains, and thus change their minds. Once they have taken responsibility for themselves, and only themselves, they can begin to free themselves from the labels, deceptions, and projections of their parents and others. This alteration in feeling and attitude brings about a surge of relief. When they refuse to take responsibility for someone else's bad behavior, they have taken a giant step toward freedom.

Looking at Language

As scientific understanding has grown, so our world has become dehumanized. Man feels a stranger to himself, because he is no longer involved in nature and has lost his emotional "unconscious identity" with natural phenomena which have slowly lost their symbolic implications. Thunder is no longer the voice of an angry god, nor is lightning his avenging missile. No river contains a spirit, no tree is the life principle of man, no snake the embodiment of wisdom, no mountain cave the home of a great demon. No voices now speak to man from stones, plants, and animals, nor does he speak to them believing they can hear. His contact with nature has gone, and with it has gone the profound emotional energy that this symbolic connection supplied.

JUNG

In writing this book, I have tried to tell a story about the unconscious and the right half of the brain. We have more than one mind, and much is hidden from us because of the constraints of language and the difficulty that the two brains have in communicating and sharing their secrets.

We read about the Tower of Babel and how God presumably confounded the minds of humans through the invention of tongues and dialects, that is, language. In consequence, people were no longer able to communicate and lost sight of what they were striving for. They wandered about in confusion.

There is still so much confusion because people continue to mistake language for reality and labels for facts, and they have lost sight of the original way in which communication took place. However, most people cling to their confusion as they fear that if they explore beyond the spoken word, the reality they find may be too unnerving to bear. However, this too is a consequence of the spell cast by language.

The language axis of the left brain has a strongly developed propensity to believe whatever lies and confabulations it produces, and it is very hard for many people, in their language-dependent conscious minds, to see through the fog created by the labels we call words. Again, I always tell my patients not to believe the first explanation they produce when exploring the motives and reasons for their behavior. Sometimes, it is necessary to ponder what seems to be the least plausible and most undesirable explanation in order to get at the truth.

Language can be a weapon that we turn back on ourselves to hide our real selves, or to deceive others. However, it is possible to remove the blinder called language so that we can see things for what they are, so that we can see ourselves for what we are, and so that we can recognize lies, self-deceptions, and other warning signs when interacting with friends, lovers, co-workers, and strangers.

To remove the blinders requires that we learn to doubt, to consider alternatives, and to recognize things for what they are rather than for what they should be or what we hope they might be. This requires that we rely less on language and more on those skills so well hewn by the right brain over the course of the last few hundred thousand years. This requires that one start listening to the way things are said, the body language and facial expressions that accompany a person's actions and speech, and particularly past behavior. Indeed, sometimes those who claim to be "reborn" have in fact only engaged in a relabeling.

Alpha and Omega

In gaining greater control over one's life, it certainly can be useful to understand the roles of the right and left brain and limbic system, the interactions between the unconscious Child and Parent, the deceptive uses of language, and the defense mechanisms. Once a person begins to gain an understanding of these forces, negative and self-destructive tendencies can be diminished and brought under control.

By practicing a willingness to explore and examine one's limitations, deficiencies, and inadequacies, and to work on strengthening one's true strengths, one will come to recognize, understand and accept their own unknown face and the stranger within. Consequently, the capacity to trust right-brain judgments, inferences, intuitions, and gut feelings will be maximized. This is called practicing self-acceptance and exercising one's right brain and ability to make healthy choices. In this manner, one may be better able to navigate through life and will be able to better utilize their right and left brains for success.

Facing the past, facing the manner in which we were raised, and facing one's own unconscious Child and Parent are only half the battle. To begin to achieve inner harmony and to begin to live a more healthy life, one must begin not only to tap into the amazing capabilities of the right half of the brain, but to learn to use the whole brain to its fullest. Once one begins down that road, they may perchance begin to learn how to see, to cease to take personally so much of what has nothing to do with them, and to accept both the bad and the good about themselves. When this occurs, maybe all will become what it is: an infinity of possibility.

Notes

Introduction

1. J. B. Watson, *Behaviorism* (New York: Norton, 1930); B. F. Skinner, *Science and Human Behavior* (New York: Macmillan, 1953).
2. J.-P. Sartre, *Being and Nothingness* (New York: Philosophical Library, 1956).
3. R. E. Nisbett, and T. D. Wilson, "Telling More Than We Know," *Psychological Review* 84 (1977), 231–259.

Chapter 1

1. S. Kramer, *History Begins at Sumer* (Philadelphia: University of Pennsylvania Press, 1956); C. L. Wooley, *Ur of the Chaldees* (New York: Norton, 1965).
2. It is noteworthy that the tablets unearthed from the ruins of ancient Sumer refer to several great cities and civilizations preceding their own. They claim that these ancient population centers were destroyed in a great flood that had occurred long before their own time. Based on studies of erosion and weathering, it has been estimated that the Great Sphinx in Egypt was built at least seven thousand years ago, and perhaps more than ten thousand years ago. Moreover, the Egyptian pharaoh Khufu, who supposedly built the Great Pyramid of Giza, apparently left an inscription on a limestone

stela (discovered in the mid-1800s) that suggests that the Sphinx and the Great Pyramid were there before he became king, thus pushing their origins much further back in time. It has also been claimed by Zecharia Sitchin, in his book *The Stairway to Heaven* (New York: Avon, 1980), that the red paint markings (found behind some stones blasted away in the Great Pyramid of Giza) that indicated this pyramid was the tomb of Khufu, were forged by the man who discovered them, as the writings were a mixture of scripts from various time periods after Khufu.

3. R. S. Solecki, *Shanidar: The First Flower People* (New York: Knopf, 1971).

4. E. Trinkaus, *The Shanidar Neanderthals* (New York: Academic Press, 1983).

5. Abraham, who became the father of Judaism, Christianity, and Islam, was born and raised in Babylon and was apparently greatly influenced by its myths and religious dogma. However, a thousand years before, ancient Babylon and the city in which he was born had been known as Sumer, and many of the beliefs and myths of this great civilization were still fashionable during Abraham's time.

6. W. James, *The Varieties of Religious Experience* (New York: New American Library, 1958).

7. R. Joseph, *Neuropsychology, Neuropsychiatry, and Behavioral Neurology* (New York: Plenum Press, 1990).

8. D. Bear, "Temporal Lobe Epilepsy: A Syndrome of Sensory Limbic Hyperconnexion," *Cortex* 15 (1979), 357–384.

9. D. M. Bear, "Hemispheric Specialization and the Neurology of Emotion," *Archives of Neurology*, 40 (1983), 195–202.

10. C. G. Jung, *Man and His Symbols* (New York: Doubleday, 1964); C. G. Jung, *Psyche and Symbol: A Selection from the Writings of C. G. Jung* (V. S. de Laszlo, Ed.) (New York: Doubleday, 1958); C. G. Jung, *The Portable Jung* (J. Campbell, Ed.) (New York: Pantheon, 1956); C. G. Jung, *Symbols of Transformation* (New York: Pantheon, 1956); C. G. Jung, *The Structure and Dynamics of the Psyche* (New York: Pantheon, 1960); C. G. Jung, *The Archetypes of the Collective Unconscious* (New York: Pantheon, 1963).

11. The Aztecs viewed the cross as having celestial significance. Moreover, it even appeared in pictures of Quetzalcoatl, the plumed serpent god, as an emblem on the shield he carried in his left hand. Interestingly, the Aztecs believed that Quetzalcoatl had been driven out of their land and into the east by the god of war, and that he would return on the day of his birth in the year of "1 Reed," that is, in 1519, the exact year in which Cortés appeared in his ships from the East. This coincidence must have been very confusing and troubling to the Aztecs and probably contributed to their downfall.

12. S. Freud, *The Interpretation of Dreams*, Standard Edition, Vol. 5, 1900, 339–622.

13. S. Freud, *Five Lectures on Psychoanalysis*, Standard Edition, Vol. 11, 1910, 1–56.

14. S. Freud, *Formulations Regarding the Two Principles in Mental Functioning*, Standard Edition, Vol. 11, 1911, 409–417.

15. S. Freud, *Repression*, Standard Edition, Vol. 14, 1915, 141–158.

16. S. Freud, *The Unconscious*, Standard Edition, Vol. 14, 1915, 159–204.

17. S. Freud, "The Ego and the Id," Standard Edition, Vol. 19, 1923, 1–59.

18. S. Freud, *Project for a Scientific Psychology*, Standard Edition, Vol. 1, 1895, 283–391.

19. D. Galin, "Implications for Psychiatry of Left and Right Cerebral Specialization," *Archives of General Psychiatry* 31 (1974), 572–583.

20. K. D. Hoppe, "Split-Brains and Psychoanalysis," *Psychoanalytic Quarterly* 46 (1977), 220–244.

21. L. Miller, *Inner Natures* (New York: St. Martin's Press, 1990).

22. L. Miller, *Freud's Brain* (New York: Guilford Press, 1991).

23. R. Ornstein, *The Psychology of Consciousness* (San Francisco: Freeman, 1972).

24. R. Joseph, "The Neuropsychology of Development," *Journal of Clinical Psychology* 38 (1982), 4–33.

25. R. Joseph, "The Right Cerebral Hemisphere: Emotion, Music, Visual-Spatial Skills. Body Image, Dreams, and Awareness," *Journal of Clinical Psychology* 44 (1988), 630–673.

26. R. Joseph, "The Limbic System: Emotion, Id and Unconscious Mind," *The Psychoanalytic Review* 79 (1992).

27. Joseph, 1990.

28. E. DeRenzi, *Disorder of Space Exploration and Cognition* (New York: Wiley, 1982); H. Hecaen, & M. L. Albert, *Human Neuropsychology* (New York: Wiley, 1978); Y. Kim, L. Morrow, D. Passafiume, and F. Boller, "Visuoperceptual and Visuomotor Abilities and Locus of Lesion," *Neuropsychologia* 2 (1984), 177–185.

29. D. Breitling, W. Guenther, and P. Rondot, "Auditory Perception of Music Measured by Brain Electrical Activity Mapping," *Neuropsychologia* 25 (1987), 765–774; M. P. Bryden, R. G. Ley, and J. H. Sugarman, "A Left-Ear Advantage for Identifying the Emotional Quality of Tonal Sequence," *Neuropsychologia* 20 (1982), 83–87; C. Knox and D. Kimura, "Cerebral Processing of Nonverbal Sounds in Boys and Girls," *Neuropsychologia* 8 (1970), 227–237; S. J. Segalowitz and P. Plantery, "Music Draws Attention to the Left and Speech Draws Attention to the Right," *Brain and Cognition* 4 (1985), 1–6.

30. Joseph, 1990; A. Smith, "Speech and Other Functions after Left Hemispherectomy," *Journal of Neurology, Neurosurgery, and Psychiatry* 29 (1966), 467–471; A. Yamadori, et al., "Preservation of Singing in Broca's Aphasia," *Journal of Neurology, Neurosurgery, and Psychiatry* 40 (1977), 221–224.

31. G. Constable, *Neanderthal Man* (New York: Time-Life Books, 1973); F. H. Smith, "Upper Pleistocene Hominid Evolution in South Central Europe," *Current Anthropology* 23 (1982), 667–703; E. Trinkaus, *The Shanidar Neanderthals* (New York: Academic Press, 1983); I. Davidson & W. Noble, "The Archaeology of Perception. Traces of Depiction and Language," *Current Anthropology* 30 (1989), 125–155; P. Lieberman, *The Biology and Evolution of Language* (Cambridge: Harvard University Press, 1984); L. R. Binsford, *In Pursuit of the Past: Decoding the Archaeological Records* (London: Thames & Hudson, 1983); L. R. Binsford, "Faunal Remains from Klaisis River Mouth," *Journal of Anthropology* 4 (1984), 292–327; L. R. Binsford, "Comments, Rethinking the Middle/Upper Paleolithic Transition," *Current Anthropology* 23 (1982), 177–179.

32. Constable, 1973.

33. Trinkaus, 1983; J. Hawkes and L. Wooley, *History of Mankind* (New York: Harper & Row, 1963).

34. W. H. Calvin, "Comments: Traces of Depiction and Language," *Current Anthropology* 30 (1989), 138–139; G. W. Hewes, "Comments: Traces of Depiction and Language," *Current Anthropology* 30 (1989), 145–146.

35. R. L. Holloway, "Human Paleontological Evidence to Language Behavior," *Human Neurobiology* 2 (1983a), 105–114; R. L. Holloway, "Human Brain Evolution: A Search for Units, Models, and Synthesis," *Canadian Journal of Anthropology* 58 (1983b), 1101–1110.

36. Calvin, 1989; Hewes, 1989.

37. Constable, 1973.

38. T. Prideaux, *Cro-Magnon Man* (New York: Time-Life Books, 1973); H.-G. Bandi, *Art of the Stone Age* (New York: Crown, 1961); L. R. Binsford, 1982, 1983, 1984; Davidson and Noble, 1989.

39. D. Falk, "Brain Lateralization in Primates and Its Evolution in Hominids," *Yearbook of Physical Anthropology* 30 (1987a), 107–125; D. Falk, "Hominid Paleoneurology," *Annual Review of Anthropology* 30 (1987b), 107–125.

40. Prideaux, 1973; Bandi, 1961; L. Leroi-Gourhan, *Treasures of Prehistoric Art* (New York: Abrams, 1960); J. Maringer and H.-G. Bandi, *Art in the Ice Age* (New York: Praeger, 1952); M. H. Woolpoff, *Paleoanthropology* (New York: Knopf, 1980); Binsford, 1982, 1983.

41. Holloway, 1983a, b.

42. Binsford, 1982, 1983.

43. R. Joseph, "Reversal of Cerebral Dominance for Language and Emotion in a Corpus Callosotomy Patient," *Journal of Neurology, Neurosurgery and Psychiatry* 49 (1986), 628–634.

44. R. Joseph and V. A. Casagrande, "Visual Field Defects and Recovery Following Monocular Lid Closure in a Prosimian Primate," *Behavioral Brain Research* 1 (1980), 150–178; V. A. Casagrande, & R. Joseph, "Morphological

Effects of Monocular Deprivation and Recovery on the Dorsal Lateral Geniculate Nucleus in Galago," *Journal of Comparative Neurology* 194 (1980), 413–426.
45. Dr. Vivian Casagrande and I have shown that, if neurons from one sensory system are given a competitive advantage, they will grow into, make connections with, and take over and suppress neurons from a rival system. For example, if one eye is prevented from receiving patterned visual input early in life, the sensory fibers from the normal eye will functionally suppress the cells from the disadvantaged system, which are located in the thalamus. Later the disadvantaged eye will fail to respond to normal visual input and will seem blind. It is for these reasons that doctors are reluctant to place patches over one eye in young children for extended periods of time. However, if the normal eye is for some reason destroyed or removed, vision will return in the formerly deprived and blind eye. That is, once the cells in the thalamus that receive visual information are no longer being used by one eye, the remaining eye, via its nerve fiber projections into the brain, will take these cells over so that vision is restored. The nerve cells supporting vision in the formerly deprived eye will consequently also grow much larger. Hence, there is much competition within the brain for functional representation, and once cells are taken over, later appearing abilities can be displaced and prevented from functioning.

Chapter 2

1. H. Goodglass and E. Kaplan, *Boston Diagnostic Aphasia Examination* (Philadelphia: Lea & Febiger, 1982).
2. D. N. Levine and E. Sweet, "Localization of Lesions in Broca's Motor Aphasia." In *Localization in Neuropsychology*, A. Kertesz, Ed. (pp. 185–207) (New York: Academic Press, 1983).
3. R. Joseph, *Neuropsychology, Neuropsychiatry, and Behavioral Neurology* (New York: Plenum Press, 1990); R. Joseph, "The Right Cerebral Hemisphere: Emotion, Music, Visual-Spatial Skills, Body Image, Dreams and Awareness," *Journal of Clinical Psychology* 44 (1988a), 630–673.
4. A. Yamadori, U. Osumi, S. Mashuara, and M. Okuto, "Preservation of Singing in Broca's Aphasia," *Journal of Neurology Neurosurgery, and Psychiatry* 40 (1977), 221–224.
5. One famous example was the composer and pianist Maurice Ravel, who suffered an injury to the left half of his brain in an auto accident. This injury resulted in apraxia, agraphia, and Wernicke's aphasia. Nevertheless, he had no difficulty recognizing various musical compositions, was able to detect

even minor errors when compositions were played, and was able to correct those errors by playing them correctly on the piano.

6. Joseph, 1988a, 1990; R. Joseph, "The Neuropsychology of Development: Hemispheric Laterality, Limbic Language and the Origin of Thought," *Journal of Clinical Psychology* 33 (1988), 3–33.

7. Joseph, 1982, 1990.

8. H. M. Ehrlichman, J. S. Antrobus, and M. Wiener, "EEG Asymmetry and Sleep Mentation during REM and NREM," *Brain and Cognition* 4 (1985), 477–485.

9. D. Hodoba, "Paradoxic Sleep Facilitation by Interictal Epileptic Activity of Right Temporal Origin," *Biological Psychiatry* 21 (1986), 1267–1278.

10. N. H. Kerr and D. Foulkes, "Right Hemisphere Mediation of Dream Visualization: A Case Study," *Cortex* 17 (1981), 603–611.

11. L. Goldstein, N. W. Stolzfus, and J. F. Gardocki, "Changes in Interhemispheric Amplitude Relationships in the EEG during Sleep," *Physiology and Behavior* 8 (1972), 811–815.

12. R. Broughton, "Human Consciousness and Sleep/Waking Rhythms: A Review and Some Neuropsychological Considerations," *Journal of Clinical Neuropsychology* 4 (1982), 193–218.

13. J. P. Banquet, "Inter- and Intrahemispheric Relationships of the EEG Activity during Sleep in Man," *Electroencephalography and Clinical Neurophysiology* 55 (1983), 51–59.

14. Joseph, 1990.

15. S. Blumstein and W. E. Cooper, "Hemispheric Processing of Intonational Contours," *Cortex* 10 (1974), 146–158; F. Boller, M. Cole, P. B. Vrtunski, M. Patterson, and V. Kim, "Paralinguistic Aspects of Auditory Comprehension in Aphasia," *Brain and Language* 9 (1979), 164–174; D. Breitling, W. Guenther, and P. Rondot, "Auditory Perception of Music Measured by Brain Electrical Activity Mapping," *Neuropsychologia* 25 (1987), 765–774; M. P. Bryden, R. G. Ley, and J. H. Sugarman, "A Left-Ear Advantage for Identifying the Emotional Quality of Tonal Sequences," *Neuropsychologia* 20 (1982), 83–87; A. Carmon and I. Nachshon, "Ear Asymmetry in Perception of Emotional Non-Verbal Stimuli," *Acta Psychologica* 37 (1973) 351–357; A. Gates and J. L. Bradshaw, "The Role of the Cerebral Hemispheres in Music," *Brain and Language* 3 (1977), 451–460; H. W. Gardan, "Hemispheric Asymmetries in the Perception of Musical Chords," *Cortex* 6 (1970), 387–398; H. R. McFarland and D. Fortin, "Amusia due to Right Temporal-Parietal Infarct," *Archives of Neurology* 39 (1982), 725–727; K. M. Heilman, D. Bowers, L. Speedie, and H. B. Coslett, "Comprehension of Affective and Noneffective Prosody," *Neurology* 34 (1984), 917–921; K. Heilman, R. Scholes, and R. T. Watson, "Auditory Affective Agnosia," *Journal of Neurology, Neurosurgery and Psychiatry* 38 (1975), 69–72; F. L. King and D. Kimura, "Left Ear

Superiority in Dichotic Perception of Vocal Nonverbal Sounds," *Canadian Journal of Psychology* 26 (1972), 111–116; C. Knox, and D. Kimura, "Cerebral Processing of Nonverbal Sounds in Boys and Girls," *Neuropsychologia* 8 (1970), 227–237.

16. About the relations between music, math, and geometric space, Pythagoras, the great Greek mathematician, argued almost two thousand years ago that music is numerical, the expression of number in sound. In fact, long before the advent of digital recordings, the Hindus and Babylonians and then Pythagoras and his followers translated music into numbers and geometric proportions. For example, by dividing a vibrating string into various ratios, they discovered that several very pleasing musical intervals could be produced. Hence, the ratio 1:2 yielded an octave, 2:3 a fifth, 3:4 a fourth, 4:5 a major third, and 5:6 a minor third. The harmonic system used in the nineteenth century by various composers was based on these same ratios. Indeed, Bartok used these ratios in his musical compositions. These same musical ratios, the Pythagoreans discovered, were also able to reproduce themselves. That is, the ratio can reproduce itself within itself and form a unique geometrical configuration, which Pythagoras and the ancient Greeks referred to as the *golden ratio* or *golden rectangle*. It was postulated to have divine inspirational origins. Indeed, music itself was thought by early humans to be magical, and musicians were believed by the ancient Greeks to be "prophets favored by the Gods."

This same "golden rectangle" is found in nature, for example, in the chambered nautilus shell, in the shell of a snail, and in the cochlea of the ear. The geometric proportions of the golden rectangle were also used in designing the Parthenon in Athens, and by Ptolemy in developing the "tonal calendar" and the "tonal Zodiac," the scale of ratios being "bent round in a circle." In fact, the first cosmologies, such as those developed by the ancient Egyptians, Hindus, Babylonians, and Greeks, were based on musical ratios. Pythagoras and, later, Plato applied these same "musical proportions" to their theory of numbers, planetary motion, and the science of stereometry, the gauging of solids. Indeed, Pythagoras attempted to deduce the size, speed, distance, and orbit of the planets based on musical ratios as well as estimates of the sounds (e.g., the pitch and harmony) generated by their movement through space, that is, "the music of the spheres." Interestingly, the famous mathematician and physicist Johannes Kepler, in describing his laws of planetary motion, also referred to them as being based on the "music of the spheres."

Thus, music seems to have certain geometric properties, such as are expressed via ratios, and Pythagoras, the "father" of arithmetic, geometry, and trigonometry, believed music to be geometric. As we know, geometry is used in the measurement of land and the demarcation of boundaries, and, thus, in the analysis of space, shape, points, lines, angles, surfaces, and configuration—capacities mediated by the right brain. In nature one form of musical expression—that is, the song of most birds—is also produced for geometric purposes: A bird sings not "for joy" but to signal others of impending threat, to attract mates, to indicate direction and location, to stake out territory, and to warn away others that attempt to intrude on the bird's space. If we assume that long before humans sang their first song, the first songs and musical compositions were created by our fine feathered friends (sounds that inspired mimicry by humans), it appears that musical production was first and foremost emotional and motivational, and directly related to the geometry of space, that is, the demarcation of one's territory. Emotion and geometry are characteristics that music still retains, and all are linked to the right half of the brain.

17. D. Benson and M. Barton, "Disturbances in Constructional Ability," *Cortex* 6 (1970), 19–46; A. Benton, "Visuoperceptive, visuospatial and visuoconstructive disorders." *Clinical Neuropsychology*, K. M. Heilman and E. Valenstein, Eds. (pp. 186–232) (Oxford: Oxford University Press, 1979); F. W. Black and B. A. Bernard, "Constructional Apraxia as a Function of Lesion Locus and Size in Patients with Focal Brain Damage," *Cortex* 20 (1984), 111–120; R. Calvanio, P. N. Petrone, and D. N. Levine, "Left Visual Spatial Neglect Is Both Environment-Centered and Body-Centered," *Neurology* 37 (1987), 1179–1183; E. DeRenzi, *Disorder of Space Exploration and Cognition* (New York: Wiley, 1982); H. Hecaen and M. L. Albert, *Human Neuropsychology* (New York: Wiley, 1978); Y. Kim, L. Morrow, D. Passafiume, and F. Boller, "Visuoperceptual and Visuomotor Abilities and Locus of Lesion," *Neuropsychologia* 2 (1984), 177–185.

18. Joseph, 1988, 1990.

19. H. Gardner, H. H. Brownell, W. Wapner, and D. Michelow, "Missing the Point: The Role of the Right Hemisphere in the Processing of Complex Linguistic Materials." In *Cognitive Processing in the Right Hemisphere*, E. Perceman, Ed. (New York: Academic Press, 1983); H. H. Brownell, H. H. Potter, and A. M. Bihrle, "Inference Deficits in Right Brain-Damaged Patients," *Brain and Language* 27 (1986), 310–321; M. Cicone, W. Wapner, and H. Gardner, "Sensitivity to Emotional Expressions and Situations in Organic Patients," *Cortex* 16 (1980), 145–158.

20. Joseph, 1988, 1990; E. DeRenzi, "Prosopagnosia in Two Patients with CT-Scan Evidence of Damage Confined to the Right Hemisphere," *Neuropsychologia* 24 (1986), 385–389; T. Landis, J. L. Cummings, L. Christen, J. E. Bogen, and H.-G. Imhof, "Are Unilateral Right Posterior Cerebral Lesions Sufficient to Cause Prosopagnosia? Clinical and Radiological Findings in Six Additional Patients," *Cortex* 22 (1986), 243–252; D. N. Levine, "Prosopagnosia and Visual Object Agnosia: A Behavioral Study," *Brain and Language* 5

(1978), 341–365; R. G. Ley and M. P. Bryden, "Hemispheric Differences in Processing Emotions and Faces," *Brain and Language* 7 (1979), 127–138.

21. J. L. Bradshaw, N. C. Nettleton, and K. Spher, "Braille Reading and Left and Right Hemispace," *Neuropsychologia* 20 (1982), 493–500; A. Carmon and H. P. Bechtoldt, "Dominance of the Right Cerebral Hemisphere for Stereopsis," *Neuropsychologia* 7 (1969), 29–39; A. Carmon and A. L. Benton, "Tactile Perception of Direction and Number in Patients with Unilateral Cerebral Disease," *Neurology* 19 (1969), 525–532; H. D. Cohen, R. C. Rosen, and I. Goldstein, "Electroencephalographic Laterality Changes during Human Sexual Orgasm," *Archives of Sexual Behavior* 5 (1976), 189–200; S. Corkin, B. Milner, and T. Rasmussen, "Somatosensory Thresholds: Contrasting Effects of Post-Central Gyrus and Posterior Parietal-Lobe Excisions," *Archives of Neurology* 23 (1970), 41–58; E. DeRenzi and G. Scotti, "The Influence of Spatial Disorders in Impairing Tactual Discrimination of Shapes," *Cortex* 5 (1969), 53–62; J. E. Desmedt, "Active Touch Exploration of Extrapersonal Space Elicits Specific Electrogenesis in the Right Cerebral Hemisphere of Intact, Right-Handed Man," *Proceedings of the National Academy of Sciences* 74 (1977), 4037–4040; A. G. Dodds, "Hemispheric Differences in Tactuo-Spatial Processing," *Neuropsychologia* 16 (1978), 247–254; J. Hom and R. Reitan, "Effects of Lateralized Cerebral Damage on Contralateral and Ipsilateral Sensorimotor Performance," *Journal of Clinical Neuropsychology* 3 (1982), 47–53.

22. K. Heilman, "Neglect and Related Disorders." In *Clinical Neuropsychology*, K. Heilman and E. Valenstein, Eds. (pp. 300–320) (New York: Oxford University Press. 1979); R. Joseph, "Confabulation and Delusional Denial: Frontal Lobe and Lateralized Influences," *Journal of Clinical Psychology* 42 (1986), 507–518; Joseph, 1982, 1988, 1990.

23. H. H. Brownell *et al.*, 1986; Cicone *et al.*, 1980; J. W. Dwyer and W. E. Rinn, "The Role of the Right Hemisphere in Contextual Inference," *Neuropsychologia* 19 (1981), 479–482; Gardner *et al.*, 1983; H. Gardner, P. K. Ling, L. Flamm, and J. Silverman, "Comprehension and Appreciation of Humorous Material Following Brain Damage," *Brain* 98 (1975), 399–412.

24. Joseph, 1988, 1990; P. B. Gorelick and E. D. Ross, "Aprosodia," *Journal of Neurology, Neurosurgery, and Psychiatry* 37 (1987), 727–737; Heilman *et al.*, 1975; A. M. Mahoney and R. S. Sainsbury, "Hemispheric Asymmetry in the Perception of Emotional Sounds," *Brain and Cognition* 6 (1987), 216–233; E. Ross, "The Aprosodias: Functional-Anatomic Organization of the Affective Components of Language in the Right Hemisphere," *Archives of Neurology* 38 (1981), 561–589; Bryden *et al.*, 1982; Carmon and Nachshon, 1973; D. Tucker, "Affective Discrimination and Evaluation of Affectively Toned Speech," *Neurology* 27 (1977), 947–950; S. Weintraub, M.-M. Mesulam, and L. Kramer, "Disturbances in Prosody: A Right Hemisphere Contribution to

Language," *Archives of Neurology* 38 (1981), 742–744; D. B. Hier, J. Mondlock, and L. R. Caplan, "Behavioral Abnormalities after Right Hemisphere Stroke," *Neurology* 33 (1983), 337–344; E. Hillbom, "After-Effects of Brain Injuries," *Acta Psychiatrica Scandinavica* 142 (Supplement) (1960); R. A. Jack, N. T. Rivers-Bulkeley, and P. L. Rabin, "Secondary Mania as a Presentation of Progressive Dialysis Encephalopathy," *Journal of Nervous and Mental Disease* 171 (1983), 193–195; R. C. Jamieson and C. E. Wells, "Manic Psychosis in a Patient with Multiple Metastic Brain Tumours," *Journal of Clinical Psychiatry* 40 (1979), 280–282; V. C. Jampala and R. Abrams, "Mania Secondary to Right and Left Hemisphere Damage," *American Journal of Psychiatry* 140 (1983), 1197–1199; W. A. Lishman, "Brain Damage in Relation to Psychiatry Disability after Head Injury," *British Journal of Psychiatry* 114 (1968), 373–410; M. W. Otto, R. A. Yeo, and M. J. Dougher, "Right Hemisphere Involvement in Depression: Toward a Neuropsychological Theory of Negative Affective Experiences," *Biological Psychiatry* 22 (1987), 1201–1215; H. A. Sackheim, R. C. Gur, and M. C. Saucy, "Emotions Are Expressed More Intensely on the Left Side of the Face," *Science* 202 (1979), 424–435.

25. R. Joseph, "Dual Mental Functioning in a Split Brain Patient," *Journal of Clinical Psychology* 44 (1988), 770–779; J. Levy, "Psychological Implications of Bilateral Asymmetry." In *Hemisphere Function in the Human Brain*, S. Dimond and J. G. Beaumont, Eds. (pp. 121–183) (London: Paul Elek, 1974); J. Levy and C. Trevarthen, "Metacontrol of Hemispheric Functions in Human Split-Brain Patients," *Journal of Experimental Psychology* 2 (1976), 299–312.

26. D. Falk, "Brain Lateralization in Primates and Its Evolution in Hominids," *Yearbook of Physical Anthropology* 30 (1987), 107–125; D. Falk, "Hominid Paleoneurology," *Annual Review of Anthropology* 30 (1987b), 107–125.

27. Joseph, 1982.

28. P. F. MacNeilage, M. G. Studdert-Kennedy, and B. Lindblom, "Functional Precursors to Language and Its Lateralization," *American Journal of Physiology* 246 (1984), 187–201.

29. L. Leroi-Gourhan, *Treasures of Prehistoric Art* (New York: Abrams, 1960); J. Maringer and H.-G. Bandi, *Art in the Ice Age* (New York: Praeger, 1952); M. H. Woolpoff, *Paleoanthropology* (New York: Knopf, 1980); T. Prideaux, *Cro-Magnon Man* (New York: Time-Life, 1973).

30. Falk, 1987a, b.

31. R. L. Holloway, "Human Paleontological Evidence to Language Behavior," *Human Neurobiology* 2 (1983a), 105–114; R. L. Holloway, "Human Brain Evolution: A Search for Units, Models, and Synthesis," *Canadian Journal of Anthropology* 58 (1983b), 1101–1110.

32. Ibid.

33. L. R. Binsford, *In Pursuit of the Past: Decoding the Archaeological Records* (London: Thames & Hudson, 1983).

34. I. Davidson and Noble, "The Archaeology of Perception: Traces of Depiction and Language," *Current Anthropology* 30 (1989), 125–155.
35. Prideaux, 1973; Bandi, 1961.
36. Ibid.; J. Halverson, "Art for Art's Sake in the Paleolithic," *Current Anthropology* 28 (1987), 63–71; P. Mellars, "The Character of the Middle-Upper Paleolithic Transition in Southwest France." In *The Explanation of Culture Change*, C. Renfrew, Ed. (pp. 255–276) (London: Duckworth, 1973); R. White, "Rethinking the Middle/Upper Paleolithic Transition," *Current Anthropology* 23 (1982), 169–192; Binsford, 1982, 1983.
37. Ibid.

Chapter 3

1. R. Joseph, "The Neuropsychology of Development: Hemispheric Laterality, Limbic Language and the Origin of Thought," *Journal of Clinical Psychology* 38 (1982), 4–33.
2. R. Joseph, *Neuropsychology, Neuropsychiatry, and Behavioral Neurology* (New York: Plenum Press, 1990).
3. Joseph, 1990.
4. E. Bisiach, "Line Bisection and Cognitive Plasticity of Unilateral Neglect of Space," *Brain and Cognition* 2 (1983), 32–38.
5. D. Kimura, "Right Temporal Lobe Damage," *Archives of Neurology* 18 (1963), 264–271.
6. J. Levy, "Language Cognition and the Right Hemisphere," *American Psychologist* 38 (1983), 538–541.
7. K. J. Meador, D. W. Loring, D. Bowers, and K. M. Heilman, "Remote Memory and Neglect Syndrome," *Neurology* 37 (1987), 522–526.
8. B. Milner, "Memory and the Medial Temporal Regions of the Brain." In *Biology of Memory*, K. Pribraum and D. E. Broadbent, Eds. (pp. 122–137) (New York: Academic Press, 1970).
9. B. Milner, "Visual Recognition and Recall after Right Temporal Lobe Excision in Man," *Neuropsychologia* 6 (1968), 191–209.
10. A. F. Wechsler, "The Effect of Organic Disease on Recall of Emotionally Charged versus Neutral Narrative Texts," *Neurology* 23 (1973), 130–135.
11. P. J. Whitehouse, "Imagery and Verbal Encoding in Left and Right Hemisphere Damage Patients," *Brain and Language* 14 (1981), 315–332.
12. R. Joseph, "The Right Cerebral Hemisphere: Language, Music, Emotion, Visual-Spatial Skills, Body Image, Dreams, and Awareness," *Journal of Clinical Psychology* 44 (1988), 630–637.
13. Joseph, 1982.
14. Joseph, 1990.

15. J. Levy, "Psychological Implications of Bilateral Asymmetry." In *Hemisphere Function in the Human Brain*, S. Dimond and J. G. Beaumont, Eds. (pp. 121–183) (London: Paul Elek, 1974).

16. J. Levy and C. Trevarthen, "Metacontrol of Hemispheric Functions in Human Split-Brain Patients," *Journal of Experimental Psychology* 2 (1976), 299–312.

17. H. Gardner, "Comprehension and Appreciation of Humorous Material Following Brain Damage," *Brain* 98 (1975), 399–412.

18. Joseph, 1990.

19. K. M. Heilman and T. Van Den Abell, "Right Hemispheric Dominance for Mediating Cerebral Activation," *Neuropsychologia* 17 (1979), 315–321.

20. K. M. Heilman and T. Van Den Abell, "Right Hemisphere Dominance for Attention," *Neurology* 30 (1980), 327–330.

21. D. M. Tucker, "Lateral Brain, Function, Mood, and Conceptualization," *Psychological Bulletin* 89 (1981), 19–46.

22. E. DeRenzi and P. Faglioni, "The Comparative Efficiency of Intelligence and Vigilance Tests Detecting Hemispheric Damage," *Cortex* 1 (1965), 410–433.

23. Joseph, 1988, 1990.

24. R. C. Jamieson and C. E. Wells, "Manic Psychosis in a Patient with Multiple Metastatic Brain Tumours," *Journal of Clinical Psychology* 40 (1979), 280–282.

25. V. C. Jampala and R. Abrams, "Mania Secondary to Right and Left Hemisphere Damage," *American Journal of Psychiatry* 140 (1983), 1197–1199.

26. W. A. Lishman, "Brain Damage in Relation to Psychiatry Disability after Head Injury," *British Journal of Psychiatry* 114 (1968), 373–410.

27. M. W. Otto, R. A. Yeo, and M. J. Dougher, "Right Hemisphere Involvement in Depression: Toward a Neuropsychological Theory of Negative Affective Experiences," *Biological Psychiatry* 22 (1987), 1201–1215.

28. R. G. Ley, "An Archival Examination of an Asymmetry of Hysterical Conversion Symptoms," *Journal of Clinical Neuropsychology* 2 (1980), 1–9.

29. Ibid.

30. G. K. York, A. J. Gabor, and P. M. Dreyfus, "Paroxysmal Genital Pain," *Neurology* 29 (1979), 516–519.

31. R. L. Ruff, "Orgasmic Epilepsy," *Neurology* 30 (1980), 1252–1253.

32. Joseph, 1988, 1990.

33. R. Joseph, "Confabulation and Delusional Denial: Frontal Lobe and Lateralized Influences," *Journal of Clinical Psychology* 42 (1986), 507–518.

34. York *et al.*, 1979; Ruff, 1980; S. S. Spencer, D. D. Spencer, P. D. Williamson, and R. H. Mattson, "Sexual Automatisms in Complex Partial Seizures," *Neurology* 33 (1983), 527–533.

35. M. R. Cohen, R. C. Rosen, and I. Goldstein, "EEG Laterality Changes during Human Sexual Orgasm," *Archives of Sexual Behavior* 5 (1976), 189–200.

Chapter 4

1. R. Joseph, "The Neuropsychology of Development: Hemispheric Laterality, Limbic Language, and the Origin of Thought," *Journal of Clinical Psychology* 38 (1982), 4–33.
2. Joseph, 1982; R. Joseph, *Neuropsychology, Neuropsychiatry, and Behavioral Neurology* (New York: Plenum Press, 1990); R. Joseph, "The Right Cerebral Hemisphere: Emotion, Music, Visual-Spatial Skills, Body Image, Dreams, and Awareness," *Journal of Clinical Psychology* 44 (1988), 630–673; R. Joseph, "The Limbic System: Emotion, Id, Unconscious Mind," *The Psychoanalytic Review* 79 (1992).
3. Joseph, 1982, 1990, 1992.
4. Joseph, 1990.
5. R. Joseph, "Confabulation and Delusional Denial: Frontal Lobe and Lateralized Influences," *Journal of Clinical Psychology* 42 (1986), 507–518.
6. E. Ross, "The Aprosodias: Functional-Anatomic Organization of the Affective Components of Language in the Right Hemisphere," *Archives of Neurology* 38 (1981), 561–589; Joseph, 1988, 1990.
7. Ross, 1981; Joseph, 1988, 1990.
8. R. Joseph, "Awareness and the Role of Conscious Self-Deception in Resistance and Repression," *Psychological Reports* 46 (1980), 767–781; Joseph, 1982, 1988, 1990.
9. R. W. Doty and W. H. Overman, "Mnemonic Role of Forebrain Commissures in Macaques." In *Lateralization in the Nervous System*, S. Harnad et al., Eds. (pp. 75–88) (Orlando: Academic Press, 1977).
10. G. L. Risse and M. S. Gazzaniga, "Well Kept Secrets of the Right Hemisphere," *Neurology* 28 (1979), 950–953.
11. M. S. Gazzaniga and J. E. LeDoux, *The Integrated Mind* (New York: Plenum Press, 1978); R. Sperry, E. Zaidel, and D. Zaidel, "Self-Recognition and Social Awareness in the Deconnected Minor Hemisphere," *Neuropsychologia* 17 (1979), 153–166.
12. Joseph, 1988, 1990.
13. R. Joseph and R. E. Gallagher, "Gender and Early Environmental Influences on Activity, Overresponsiveness and Exploration," *Developmental Psychobiology* 13 (1980), 527–544.
14. V. H. Dennenberg, "Hemispheric Laterality in Animals and the Effects of Early Experience," *Behavioral Brain Science* 4 (1981), 1–49.
15. Joseph, 1980, 1982, 1988, 1990; R. E. Gallagher and R. Joseph, "Non-Linguistic Knowledge, Hemispheric Laterality, and the Conservation of Inequivalence," *Journal of General Psychology* 107 (1982), 31–40; R. Joseph and R. E. Gallagher, "Interhemispheric Transfer and the Completion of Reversible Operations in Non-Conserving Children," *Journal of Clinical Psychology*

41 (1985), 796–800; R. Joseph, R. E. Gallagher, W. Holloway, and J. Kahn, "Two Brains—One Child: Interhemispheric Transfer Deficits and Confabulation in Children Aged 3, 7, 10," *Cortex* 20 (1984), 317–331.

16. P. I. Yakovlev and A. Lecours, "The Myelogenetic Cycles of Regional Maturation of the Brain." In *Regional Development of the Brain in Early Life*, A. Minkowski, Ed. (pp. 404–491) (London: Blackwell, 1967).

17. Joseph and Gallagher, 1985; Joseph *et al.*, 1984; Gallagher and Joseph, 1982; M. A. J. Finlayson, "A Behavioral Manifestation of the Development of Interhemispheric Transfer of Learning in Children," *Cortex* 12 (1975), 290–295; D. Galin, R. Diamond, and J. Herron, "Development of Cross and Uncrossed Tactile Localization on the Fingers," *Brain and Language* 4 (1979), 588–590; D. Galin, J. Johnstone, L. Nakell, and J. Herron, "Development of the Capacity for Tactile Information Transfer between Hemispheres in Normal Children," *Science* 204 (1979), 1330–1332; R. H. Kraft, O. R. Mitchell, M. L. Languis, and G. H. Wheatley, "Hemispheric Asymmetries during Six-to-Eight Year-Olds' Performance on Piagetian Conservation and Reading Tasks," *Neuropsychologia* 18 (1980), 637–644; D. S. O'Leary, "A Developmental Study of Interhemispheric Transfer in Children Aged 5 to 10," *Child Development* 51 (1980), 743–750; A. Salamy, "Commissural Transmission: Maturational Changes in Humans," *Science* 200 (1978), 1409–1411.

18. Joseph *et al.*, 1984.

Chapter 5

1. R. Joseph, "The Neuropsychology of Development: Hemispheric Laterality, Limbic Language and the Origin of Thought," *Journal of Clinical Psychology* 38 (1982), 4–33; R. Joseph, *Neuropsychology, Neuropsychiatry, and Behavioral Neurology* (New York: Plenum Press, 1990); R. Joseph, "Awareness and the Role of Conscious Self-Deception in Resistance and Repression," *Psychological Reports* 46 (1980), 767–781.

2. J. Piaget, *The Origins of Intelligence in Children* (New York: Norton, 1952); J. Piaget, *Play, Dreams and Imitation in Childhood* (New York: Norton, 1962); J. Piaget, *The Child and Reality* (New York: Viking Press, 1974).

3. L. S. Vygotsky, *Thought and Language* (Cambridge, MIT Press, 1962).

4. Joseph, 1982, 1990.

5. Piaget, 1952, 1962, 1974; Vygotsky, 1962.

6. Ibid.

7. Vygotsky, 1962.

8. Joseph, 1982, 1990.

9. Ibid., 1982, 1990.

10. R. Ornstein, *The Psychology of Consciousness* (San Francisco: Freeman, 1972); J. E. Bogen, "The Other Side of the Brain," *Bulletin of the Los Angeles Neurological Societies* 34 (1969), 135–162.

11. J. E. Bogen, "The Other Side of the Brain," *Bulletin of the Los Angeles Neurological Societies* 34 (1969), 135–162; J. E. Bogen, "The Callosal Syndrome." In *Clinical Neuropsychology*, K. M. Heilman and E. Valenstein, Eds. (pp. 308–358) (New York: Oxford University Press, 1979).

12. J. Levy, "Language, Cognition, and the Right Hemisphere," *American Psychologist* 38 (1983), 538–541.

13. J. Levy and C. Trevarthen, "Metacontrol of Hemispheric Function in Human Split-Brain Patients," *Journal of Experimental Psychology: Human Perception and Performance* 2 (1976), 299–312.

14. J. Levy, C. Trevarthen, and R. W. Sperry, "Perception of Bilateral Chimeric Figures Following Hemispheric Deconnection," *Brain* 95 (1972), 61–78.

15. R. Sperry, "Lateral Specialization in the Surgically Separated Hemispheres." In *The Neurosciences: Third Study Program*, F. O. Schmitt and F. G. Worlden, Eds. (pp. 1–12) (Cambridge: MIT Press, 1974).

16. R. Sperry, "Some Effects of Disconnecting the Cerebral Hemispheres," *Science* 217 (1982), 1223–1226.

17. M. S. Gazzaniga, *The Bisected Brain* (New York: Appleton, 1970).

18. M. S. Gazzaniga and J. E. LeDoux, *The Integrated Mind* (New York: Plenum Press, 1978).

19. R. Sperry, "Brain Bisection and the Neurology of Consciousness." In *Brain and Conscious Experience*, J. C. Eccles, Ed. (pp. 298–313) (New York: Springer Verlag, 1978).

20. R. Joseph, "Dual Mental Functioning in a Split-Brain Patient," *Journal of Clinical Psychology* 44 (1988), 770–779.

21. Gazzaniga and LeDoux, 1978.

22. R. W. Sperry, E. Zaidel, and D. Zaidel, "Self-Recognition and Social Awareness in the Deconnected Minor Hemisphere," *Neuropsychologia* 17 (1979), 153–166.

23. A. J. Akelaitis, "Studies on the Corpus Callosum: 4. Diagnostic Dyspraxia in Epileptics Following Partial and Complete Section of the Corpus Callosum," *American Journal of Psychiatry* 101 (1945), 594–599.

24. Bogen, 1979.

25. Joseph, 1988.

26. Gazzaniga and LeDoux, 1978.

27. J. Levy, "Psychobiological Implications of Bilateral Asymmetry." In *Hemispheric Function in the Human Brain*, S. J. Dimond and J. G. Beaumont, Eds. (New York: Wiley, 1974).

Chapter 6

1. W. H. Sweet, F. Ervin, and V. Mark, "The Relationship of Violent Behavior in Focal Cerebral Disease." In *Aggressive Behavior*, S. Garattini and E. Sigg, Eds. (New York: Wiley, 1969).

2. P. Gloor, "Amygdala." In *Handbook of Physiology*, J. Field, Ed. (pp. 300–370) (Washington, DC: American Physiological Society, 1960); A. Kling, "Effects of Amygdalectomy on Social-Affective Behavior in Non-Human Primates." In *The Neurobiology of the Amygdala*, B. E. Eleftheriou, Ed. (pp. 127–170) (New York: Plenum Press, 1972); H. Ursin and B. R. Kaada, "Functional Localization within the Amygdaloid Complex," *EEG and Clinical Neurophysiology* 12 (1960), 1–20.

3. R. Joseph, *Neuropsychology, Neuropsychiatry, and Behavioral Neurology* (New York: Plenum Press, 1990); R. Joseph, "The Limbic System, Emotion, Id, Unconscious Mind," *The Psychoanalytic Review* 79 (1992).

4. E. C. Crosby, "Evidence for Some of the Trends in the Phylogenetic Development of the Vertebrate Telencephalon." In *Evolution of the Forebrain*, R. Hassler and H. Stephan, Eds. (pp. 333–371) (Stuttgart: Verlag, 1966).

5. W. G. Lisk, "Neural Localization for Androgen Activation of Copulatory Behavior," *Endocrinology* 80 (1967), 754–780.

6. P. D. Maclean, "New Findings of Brain Function and Sociosexual Behavior." In *Contemporary Sexual Behavior*, J. Zubin and J. Money, Eds. (pp. 90–117) (Baltimore: Johns Hopkins University Press, 1973).

7. R. Bleier, "Cytoarchitectonic Sexual Dimorphisms of the Medial Preoptic and Anterior Hypothalamic Area," *Journal of Comparative Neurology* 66 (1982), 603–605.

8. T. C. Rainbow, "Sex Differences in Brain Receptors," *Nature* 300 (1982), 648–649.

9. G. Raisman and P. Field, "Sexual Dimorphism in the Preoptic Area of the Rat," *Science* 173 (1971), 731–733.

10. G. Morgenson, "Septal-Hypothalamic Relationships." In *The Septal Nuclei*, J. F. DeFrance, Ed. (New York: Plenum Press, 1976); R. G. Health, "Brain Function in Epilepsy," *Journal of Neurology, Neurosurgery, and Psychiatry* 39 (1976), 1037–1051.

11. Joseph, 1990, 1992.

12. J. O'Keefe and H. Bouma, "Complex Sensory Properties of Certain Amygdala Units in the Freely Moving Cat," *Experimental Neurology* 23 (1969), 384–398.

13. Ursin and Kaada, 1960; P. Gloor, "Electrophysiological Studies on the Connections of the Amygdaloid Nucleus of the Cat. I & II," *Electroencephalography and Clinical Neurophysiology* 7 (1955), 223–262.

14. M. Fukuda, T. Ono, and K. Nakamura, "Functional Relation among Infero-temporal Cortex, Amygdala and Lateral Hypothalamus," *Journal of Neurophysiology* 57 (1987), 1060–1077.

15. P. Fedio and J. Van Buren, "Memory Deficits during Electrical Stimulation of the Speech Cortex in Conscious Man," *Brain and Language* 1 (1974), 29–42.

16. B. Milner, "Amnesia Following Operations on the Temporal Lobe." In *Amnesia*, K. Pribram and E. D. Broadbent, Eds. (pp. 75–89) (London: Butterworth, 1970).

17. Joseph, 1990.

18. R. B. Kesner and R. G. Andrus, "Amygdala Stimulation Disrupts the Magnitude of Reinforcement Contribution to Long-Term Memory," *Physiological Psychology* 10 (1982), 55–59.

19. C. Blakemore, *Mechanics of the Mind* (New York: Cambridge University Press, 1977).

20. R. G. Heath, "Physiological Basis of Emotional Expression," *Biological Psychiatry* 5 (1972), 172–184; P. D. Maclean, "The Hypothalamus and Emotional Behavior." In *The Hypothalamus*, W. Haymaker, Ed. (pp. 127–167) (Springfield, IL: Thomas, 1969); M. E. Olds and J. L. Forbes, "The Central Basis of Motivation: Intracranial Self Stimulation Studies," *Annual Review of Psychology* 32 (1981), 523–574; Gloor, 1960.

21. Ibid.

22. Joseph, 1990.

23. Olds and Forbes, 1981.

24. *Hallucinations and the Primary Process.* The amygdala and the hippocampus, particularly those of the right hemisphere, are very important in the production and recollection of visual, nonlinguistic images associated with past experience. Direct electrical stimulation of this region within the temporal lobes results not only in the recollection of visual images, but in the creation of fully formed visual and auditory hallucinations, as well as feelings of familiarity (e.g., déjà vu). The most complex forms of hallucination are associated with tumors in the most anterior portion of the temporal lobe, that is, the region containing the amygdala and the anterior hippocampus. Similarly, electrical stimulation of the anterior-lateral temporal cortical surface, particularly of the right temporal lobe, results in visual hallucinations of people, objects, faces, and various sounds. Stimulation of the right amygdala produces visual hallucinations, body sensations, déjà vus, illusions, and gustatory and alimentary experiences, and the surgical removal of the right amygdala abolishes hallucinations. Stimulation of the right hippocampus has also been associated with the production of memory- and dreamlike hallucinations; in fact, hallucinations seem to occur most frequently following hippocampal activation. Overall, it appears that interactions of the amygdala, the hippocampus, and the neocortex of the

temporal lobe are highly involved in the production of hallucinatory experiences. Presumably, it is the neocortex of the temporal lobe that interprets this material as perceptual phenomena.

Dreaming. The right hippocampus—and the right hemisphere in general—also appears to be involved (at least in part) in the production of dream imagery as well as rapid eye movement (REM) during sleep. Presumably, during REM sleep, the hippocampus acts as a reservoir from which various images, words, and ideas are drawn and incorporated into the matrix of dreamlike activity being woven by the right hemisphere. The hippocampus is probably just as likely the source of the material in a daydream.

Dreams and Infancy. In the newborn, and up to six to nine months, there are two distinct stages of sleep, which correspond to REM and N-REM periods in adults. Among infants, however, rapid eye movements occur during wakefulness as well as during sleep. In fact, REMs can be observed when the eyes are open, when the infant is crying, fussing, eating, or sucking. Moreover, REMs are also observed or occur within a few moments after an infant begins to engage in nutritional sucking, and they appear identical to those that occur during sleep.

The production of REMs during waking seems paradoxical in some respects, and a number of different mechanisms are probably responsible. Nevertheless, it is safe to assume that, like an adult, when the infant is showing REMs, he or she is dreaming or is, at least, in a dreamlike state. Possibly, this state corresponds to what Freud described as the primary process. That is, when the infant cries or fusses, it is *dreaming* of whatever relief it seeks. Correspondingly, REMs that occur while eating or sucking may be produced by the amygdala and the hippocampus, which are involved not only in the production of dreamlike activity, but in the identification, learning, and retention of motivationally significant information.

The Primary Process. The hypothalamus, our exceedingly ancient and primitive id, sees only inward. It can tell if the body needs nourishment but cannot determine what might be good to eat. It can feel thirst but has no way of slaking this thirst. The hypothalamus can only say, "I want" and "I need," and can only signal pleasure and displeasure. However, being the seat of pleasure, the hypothalamus may be exceedingly gracious in rewarding the organism when its needs are met. Conversely, when its needs go unmet, the hypothalamus may respond not only with displeasure and feelings of aversion, but with undirected fury and rage, the organism may cry out. The cry itself does not produce the immediately desired relief or reduction in tension. There is therefore a *pressure* on the limbic system and the organism to engage in environmental surveillance so as to meet the needs monitored by the hypothalamus.

Over the course of the first months of life, as the amygdala and then the hippocampus develop, the organism begins not only to see outward but to register and recall the events, objects, people, and so on associated with tension reduction, pleasure, and the fulfillment of the infant's internal needs (e.g., the taste, smell, and feeling of mother's breast and milk, and the experience of sucking and relief). This is called *learning*. With the maturation of the amygdala and the hippocampus, the infant is increasingly able to differentiate what occurs in the external environment based on needs monitored by the hypothalamus and the emotional and motivational significance of what is experienced. The infant can now orient, selectively attend, determine what brings satisfaction, and store this information in memory.

When the amygdala and the hippocampus are stimulated by a hungry hypothalamus, the events and images associated with past experiences of pleasure can be not only searched out externally, but recalled in imagination. For example, as an infant experiences hunger and stomach contractions as well as its own cries of displeasure, these states become associated with the sound, smell, and taste of the mother and her movements and other stimuli associated with being fed (cf. Piaget, 1952, pp. 37, 407–408). Repetitively experienced, the sequence from hunger to satiety evokes and becomes associated with the activation of certain neural pathways. Eventually, when the infant becomes hungry, if the hunger is prolonged the entire neural sequence associated with hunger and feeding (i.e., hunger, mother, food, and satiety) may become involuntarily triggered and activated (via association), so that an "image" of being fed is experienced. The activation of these rudimentary and infantile memory images is probably what constitutes, at least in part, the primary process. Behavioral, this process is manifested by REMs and by sucking and tongue movements as if eating when there is no food present. That is, when hungry, the infant begins to cry, rapid eye movements may be observed, and then the infant stops crying, smacks its lips, and makes sucking movements (mediated by the amygdala) *as if* it were being fed. The infant experiences being fed in the form of a dream or hallucination. Given the limited amount of reality contact infants are able to achieve, these rudimentary memories and images (even when they occur during waking) are probably indistinguishable from actual experience simply because they are experience. The infant presumably reexperiences, to some degree, the sensations and emotions originally linked to tension reduction. Thus, the young infant, as yet unable to distinguish between representation and reality, responds to the hallucination as reality, even while awake. When hunger is prolonged, the associations linked to feeding are triggered, and for a brief time, the infant behaves as if its hunger has been sated. Reality is replaced by an image, or rather, a "dream." This is the primary process.

25. D. Dicks, R. E. Myers, and A. Kling, "Uncus and Amygdaloid Lesions on Social Behavior in the Free Ranging Monkey," *Science* 160 (1969), 69–71.

26. K. R. Johanson and L. J. Enloe, "Alterations in Social Behavior Following Septal and Amygdaloid Lesions in the Rat," *Journal of Comparative and Physiological Psychology* 75 (1972), 280–301.

27. A. Kling, "Effects of Amygdalectomy on Social-Affective Behavior in Non-Human Primates." In *The Neurobiology of the Amygdala*, B. E. Elefitherious, Ed. (pp. 127–170) (New York: Plenum Press, 1972).

28. H. F. Harlow, "The Heterosexual Affectional System in Monkeys," *American Psychologist* 17 (1962), 1–9.

29. Johanson and Enloe, 1972.

30. J. Langmeier and Z. Matejcek, *Psychological Deprivation in Childhood* (New York: Wiley, 1975); W. Dennis, "Causes of Retardation among Institutionalized Children," *Journal of Genetic Psychology*, 96 (1975), 47–59; R. A. Spitz, "Hospitalism: An Inquiry into the Genesis of Psychiatric Conditions in Early Childhood," *Psychoanalytical Study of the Child* 1 (1945), 53–74.

31. Joseph, 1990; Heath, 1972.

32. M. C. Diamond, "Rat Forebrain Morphology: Right-Left, Male-Female, Young-Old, Enriched-Impoverished." In *Cerebral Lateralization in Non-Human Primates*, S. D. Glick, Ed. (pp. 181–201) (Orlando: Academic Press, 1985).

Chapter 7

1. S. de Beauvoir, *The Second Sex* (New York: Bantam Books, 1961); D. Tannen, *You Just Don't Understand* (New York: Ballantine Books, 1990).

2. J. Croates, *Women, Men and Language* (London: Longman, 1986); B. Emil and R. Stutman, "Sex Role Differences in the Relational Control of Dyadic Interaction," *Women's Studies in Communication* 6 (1983), 96–103; W. Farrell, "The Politics of Vulnerability." In *The Forty-Nine Percent Majority*, D. S. David and R. Brannon, Eds. (pp. 51–54) (Menlo Park, CA: Addison-Wesley, 1976); J. O. Balswick and C. W. Peek, "The Inexpressive Male." In *The Forty-Nine Percent Majority*, D. S. David and R. Brannon, Eds. (Menlo Park, CA: Addison-Wesley, 1976); Tannen, 1990.

3. J. Brooks-Gunn and W. S. Matthews, *He and She: How Children Develop Their Sex Role Identity* (Englewood Cliffs, NJ: Prentice-Hall, 1979); D. Eder and M. Hallinan, "Sex Differences in Children's Friendships," *American Sociological Review* 43 (1978), 237–250; R. C. Savin-Williams, "Dominance and Submission among Adolescent Boys." In *Dominance Relations*, D. R. Omark *et al.*, Eds. (New York: Garland Press, 1980); W. C. McGrew, *An Ethological Study of*

Children's Behavior (New York: Academic Press, 1979); J. Lever, "Sex Differences in the Games Children Play," *Social Problems* 23 (1976), 478–483.

4. R. Joseph, "Competition between Women," *Psychology* 22 (1985), 1–11; Brooks-Gunn and Matthews, 1979; Eder and Hallinan, 1978; N. Friday, *My Mother My Self* (New York: Dell, 1977). de Beauvoir, 1961; Lever, 1976.

5. Eder and Hallinan, 1978; P. Ekert, "Cooperative Competition in Adolescent Girl Talk," *Discourse Processes* 13 (1990), 1.

6. Tannen, 1990; Croates, 1986; Emil and Stutman, 1983.

7. Joseph, 1985; W. Wickler, *The Sexual Code* (Garden City, NY: Anchor Books, 1973); A. Jolly, *The Evolution of Primate Behavior* (New York: Macmillan, 1972) Among other primates, males attempt to maintain their status through similar sexual posturing. However, rather than make sexual comments, they actually mount other males so as to establish and demonstrate their dominance. In fact, lesser ranking males sometimes assume the female position and present themselves to higher status males for the purposes of being mounted, and so as to show respect and lessen tension.

8. L. A. Hughes, "But That's Not Really Mean: Competing in a Cooperative Mode," *Sex Roles* (1988), 669–687.

9. Tannen, 1990; Croates, 1986.

10. Farrell, 1976; Balswick and Peek, 1976.

11. Ibid.

12. Farrell, 1976.

13. R. Gould, "Measuring Masculinity by the Size of a Paycheck." In *The Forty-Nine Percent Majority*, D. S. David and R. Brannon, Eds. (pp. 113–118) (Menlo Park, CA: Addison-Wesley, 1976).

14. R. Joseph and R. E. Gallagher, "Gender and Early Environmental Influences on Activity, Overresponsiveness, and Exploration," *Developmental Psychobiology* 13 (1980), 527–544; R. Joseph, S. Hess, and E. Birecree, "Effects of Sex Hormones Manipulations and Exploration on Sex Differences in Learning," *Behavioral Biology* 24 (1978), 364–377.

15. A. Montague, *The Natural Superiority of Women* (New York: Collier, 1970).

16. L. J. Harris, "Sex Differences in Spatial Ability: Possible Environmental, Genetic, and Neurological Factors." In *Asymmetrical Function of the Brain*, M. Kinsbourne, Ed. (pp. 405–522) (New York: Cambridge University Press, 1978); D. M. Broverman, E. L. Klaiber, Y. Kobayashi, and W. Vogel, "Roles of Activation and Inhibition in Sex Differences in Cognitive Abilities," *Psychological Review* 76 (1968), 328–331; J. McGlone, "Sex Differences in Human Hemispheric Laterality," *Behavioral Brain Sciences* 4 (1982), 3–33; J. Durden-Smith and O. DeSimone, *Sex and the Brain* (New York: Warner Books, 1980); Joseph and Gallagher, 1980; Joseph *et al.*, 1978.

17. Harris, 1978; Durden-Smith and DeSimone, 1980.

18. R. Joseph, *Neuropsychology, Neuropsychiatry, and Behavioral Neurology* (New

York: Plenum Press, 1990); Harris, 1978; McGlone, 1982; Broverman *et al.*, 1968; Durden-Smith and DeSimone, 1980.

19. Broverman *et al.*, 1968.

20. Differences in visual-spatial capabilities exist not only among humans, but in animals as well. Although current environmental pressures no doubt contribute to these differences, it has been shown experimentally by my colleagues and me that, when animals are exposed to similar environments, these gender-based differences continue to be demonstrated. Roberta Gallagher and I have shown that, when animals are reared in an enriched as opposed to an impoverished environment, males continue to outperform females across tasks of visual-spatial ability. However, although males reared in an enriched environment are superior on such tasks to females, females reared in an enriched environment are superior on such tasks to males raised in a deprived environment. Hence, environment influences cognitive capabilities. Moreover, my colleagues and I have also demonstrated that these differences are affected by one's hormonal environment. For example, if male and female animals are exposed to testosterone (a male hormone) before the brain and the genitals become sexually differentiated (i.e., during the early stages of development), on reaching adulthood they perform similarly on tests of visual-spatial ability. If during this same critical period males are castrated or are administered a drug that blocks the reception of testosterone in the brain, they perform similarly to females. Eliminating or administering testosterone in adults has minimal effects on these abilities, however. These results indicate that males are biologically predisposed to outperform females on tests of visual-spatial capability and that environmental pressures play a significant role only when females are provided opportunities that males are denied, in which case, the sex-based differences are reversed.

21. R. J. Blumenschine, "Characteristics of Early Hominid Scavenging Niche," *Current Anthropology* 28 (1987), 383–407; R. Lee and I. DeVore (Eds.), *Man the Hunter* (Chicago: Aldine, 1968); R. S. Harding and G. Teleki (Eds.), *Omnivorous Primates* (New York: Columbia University Press, 1980); K. Hill, "Hunting and Hominid Evolution," *Journal of Human Evolution* 11 (1982), 521–544; R. A. Dart, "The Predatory Transition from Ape to Man," *International Anthropological and Linguistics Review* 1 (1953), 201–217; P. S. Martin, "Pleistocene Overkill." In *Pleistocene Extinctions*, P. S. Martin and H. E. Wright, Eds. (pp. 75–120) (New Haven: Yale University Press, 1969); L. R. Binsford, *In Pursuit of the Past: Decoding the Archaeological Records* (London: Thames & Hudson, 1983); L. R. Binsford, "Faunal Remains from Klaisis River Mouth," *Journal of Anthropology* 4 (1984), 292–327; L. R. Binsford, "Comments: Rethinking the Middle/Upper Paleolithic Transition," *Current Anthropology* 23 (1982), 177–179; R. Leakey and R. Lewin, *Origins* (New York:

Dutton, 1977); S. J. Washburn and C. S. Lancaster, "The Evolution of Hunting." In *Man the Hunter*, R. Lee and I. DeVore, Eds. (pp. 293–303) (Chicago: Aldine, 1968); R. Ardrey, *The Hunting Hypothesis* (New York: Bantam, 1977).

22. G. Isaac and D. C. Crader, "To What Extent Were Early Hominids Carnivorous?" In *Omnivorous Primates*, R. S. O. Harding and G. Telecki, Eds. (pp. 37–103) (New York: Columbia University Press, 1981); R. Klein, "The Ecology of Early Man in Southern Africa," *Science* 197 (1977), 115–126; H. Kaplan and K. Hill, "Hunting Ability and Reproductive Success among Male Ache Foragers," *Current Anthropology* 26 (1985), 131–133; G. P. Murdock and C. Provost, "Factors in the Division of Labor by Sex," *Ethnology* 12 (1967), 203–235.

23. J. Goodale, *Tiwi Wives* (Seattle: University of Washington Press, 1971); G. P. Murdock, "Comparative Data on the Division of Labor by Sex," *Social Forces* 16 (1937), 551–553; M. K. Martin and B. Voorhies, *Females of the Species* (New York: Columbia University Press, 1975); S. Slocum, "Women in Cross Cultural Perspective." In *Toward an Anthropology of Women*, R. R. Reiter, Ed. (pp. 36–50) (New York: Monthly Review Press, 1975); L. Fedigan, *Primates Paradigms: Sex Roles and Social Bonds* (Montreal: Eden Press, 1982); J. K. Brown, "A Note on the Division of Labor by Sex," *American Anthropologist* 72 (1970), 1073–1078; Leakey and Lewin, 1977.

24. J. K. Wolpoff, "Modern Homo Sapiens Origins." In *The Origins of Modern Humans*, F. Smith and F. Spender, Eds. (pp. 411–483) (New York: Liss, 1980).

25. G. P. Rightmire, *The Evolution of Homo Erectus* (New York: Cambridge University Press, 1990); P. V. Tobias, *The Brain in Hominid Evolution* (New York: Columbia University Press, 1971).

26. Ibid. *Homo erectus* was a purported evolutionary ancestor who lived from approximately 1.9 million years ago to about 300,000 years ago. This species of humans were also the first individuals who appear to have developed crude shelters and home bases, and who perhaps used various earth pigments (ocher) for cosmetic or artistic purposes. During the time of *Homo erectus* the human brain appears to have become significantly enlarged, big-game hunting began, and the females' pelvis enlarged.

27. A. L. Zilman, "Women as Shapers of the Human Adaptation." In *Woman the Gatherer*, F. Dahlberg, Ed. (pp. 75–120) (New Haven: Yale University Press, 1981).

28. Dart, 1953.

29. Zilman, 1981.

30. G. P. Murdock, "Comparative Data on the Division of Labor by Sex," *Social Forces* 16 (1937), 551–553; Goodale, 1971; Martin and Voorhies, 1975; Slocum, 1975.

31. Ibid.

32. Murdock; Martin and Voorhies, 1975.

33. H. G. Bandi, *Art of the Stone Age* (New York: Crown, 1961); T. Prideaux, *Cro-Magnon Man* (New York: Time-Life, 1973).

34. Martin and Voorhies, 1975.

35. C. W. M. Hart and A. Pilling, *The Tiwi of North Australia* (New York: Holt, 1960); Kaplan and Hill, 1985.

36. Joseph, 1990.

37. Binsford, 1982, 1983; Prideaux, 1973.

38. A. Jolly, *The Evolution of Primate Behavior* (New York: Macmillan, 1972); J. Goodall, *In the Shadow of Man* (New York: Houghton Mifflin, 1971).

39. V. Geist, "Comment: Characteristics of an Early Hominid Scavenging Niche," *Current Anthropology* 28 (1987), 396–387.

40. Goodall, 1971; Jolly, 1972.

41. T. Gibson, "Meat Sharing as a Political Ritual." In *Hunters and Gatherers*, T. Ingold *et al.*, Eds. (pp. 165–180) (New York: Berg, 1988); Kaplan and Hill, 1985.

42. Zilman, 1981; Kaplan and Hill, 1985.

43. Kaplan and Hill, 1985.

44. Zilman, 1981; Kaplan and Hill, 1985.

45. Jolly, 1972; Wickler, 1973.

46. D. Falk, "Brain Lateralization in Primates and Its Evolution in Hominids," *Yearbook of Physical Anthropology* 30 (1987a), 107–125; D. Falk, "Hominid Paleoneurology," *Annual Review of Anthropology* 30 (1987b), 107–125.

47. W. H. Oswalt, *An Anthropological Analysis of Food Getting Technology* (New York: Wiley, 1976).

48. F. Dahlberg (Ed.), *Woman the Gatherer* (New Haven: Yale University Press, 1981).

49. Brown, 1970.

50. Prideaux, 1973; Bandi, 1961.

51. For those who are bothered by discussions of evolution and who are firm believers in the biblical notion of God as Creator, it is noteworthy that a careful reading of Genesis indicates that God created man and woman on the seventh day and stated (Chapter 1): Let us make man in our image and after our likeness: and let *them* have dominion . . . and . . . male and female created he them." However, what was the length of a day in the mind of God? In any case, it was only after creating man and woman and after a great mist inundated the earth that God then created Adam and Eve (in Chapter 2). Moreover, if we consider the complaint of Cain after God banished him for killing Abel, it is also clear that a considerable number of people were alive and thriving at the time, people who were not related to Cain or Adam or Eve. That is, either God had pulled off another act of

creation in the next county, or these people were the descendants of those whom he first fashioned.

Chapter 8

1. J. Langmeier and Z. Matejcek, *Psychological Deprivation in Childhood* (New York: Wiley, 1975); W. Dennis, "Causes of Retardation among Institutionalized Children," *Journal of Genetic Psychology* 96 (1975), 47–59; R. A. Spitz, "Hospitalism: An Inquiry into the Genesis of Psychiatric Conditions in Early Childhood," *Psychoanalytical Study of the Child* 1 (1945), 53–74.
2. Langmeier and Matejcek, 1975.
3. R. Joseph, *Neuropsychology, Neuropsychiatry, and Behavioral Neurology* (New York: Plenum Press, 1990).
4. H. F. Harlow, "The Heterosexual Affectional System in Monkeys," *American Psychologist* 17 (1962), 1–9.
5. Although I have been tremendously influenced by the work of Eric Berne, Fred Harris, and W. H. Missildine, it is important for the reader to note that my conceptions and arguments regarding the unconscious Child and Parent also differ markedly from those of these authors.

Chapter 12

1. R. F. Schmidt, *Fundamentals of Neurophysiology* (New York: Springer-Verlag, 1978); J. R. Cooper, F. E. Bloom, and R. H. Roth, *The Biochemical Basis of Neuropharmacology* (New York: Oxford University Press, 1974); G. Lynch, *Synapses, Circuits, and the Beginnings of Memory* (Cambridge: MIT Press, 1986).
2. R. Joseph, *Neuropsychology, Neuropsychiatry, and Behavioral Neurology* (New York: Plenum Press, 1990); R. Joseph, "The Right Cerebral Hemisphere: Emotion, Music, Visual-Spatial Skills, Body Image, Dreams and Awareness," *Journal of Clinical Psychology* 44 (1988), 630–673.
3. Joseph, 1990.
4. H. Hecaen and M. L. Albert, *Human Neuropsychology* (New York: Wiley, 1978).
5. D. F. Kripke and D. Sonnenschein, "A 90 Minute Daydream Cycle," *Sleep Research* 2 (1973), 187–188.

6. W. D. Foulkes, "Dream Reports from Different Stages of Sleep," *Journal of Abnormal and Social Psychology* 65 (1962), 14–25.

7. B. Monroe, A. Rechtschaffen, D. Foulkes, and J. Jensen, "Discriminability of REM and NREM Reports," *Personality and Social Psychology* 2 (1965), 456–460.

8. Foulkes, 1962.

9. D. R. Goodenough, A. Shapiro, M. Holden, and R. Steinschriber, "Comparison of 'Dreamers' and 'Non-Dreamers,'" *Journal of Nervous and Mental Disease* 59 (1959), 295–302.

10. Monroe *et al.*, 1965.

11. J. Kamiya, "Behavioral, Subjective and Physiological Aspects of Drowsiness and Sleep." In *Function of Varied Experience*, D. W. Fiske and S. R. Maddi, Eds. (pp. 145–174) (Homewood, IL: Dorsey, 1961).

12. L. Goldstein, N. W. Stolzfus, and J. F. Gardocki, "Changes in Inter-hemispheric Amplitude Relationships in the EEG during Sleep," *Physiology and Behavior* 8 (1972), 811–815.

13. D. Hodoba, "Paradoxic Sleep Facilitation by Interictal Epileptic Activity of Right Temporal Origin," *Biological Psychiatry* 21 (1986), 1267–1278.

14. J. S. Meyer, "Cerebral Blood Flow in Normal and Abnormal Sleep and Dreaming," *Brain and Cognition* 6 (1987), 266–294.

15. Joseph, 1988, 1990.

16. M. Bertini, C. Violani, P. Zoccolotti, A. Antonelli, and L. DiStephano, "Performance on a Unilateral Tactile Test during Waking and upon Awakenings from REM and NREM." In *Sleep*, P. Koella, Ed. (pp. 122–155) (Basel: Karger, 1983).

17. R. Broughton, "Human Consciousness and Sleep/Waking Rhythms: A Review and Some Neuropsychological Considerations," *Journal of Clinical Neuropsychology* 4 (1982), 193–218.

18. J. P. Banquet, "Inter- and Intrahemispheric Relationships of the EGG Activity during Sleep in Man," *Electroencephalography and Clinical Neurophysiology* 55 (1983), 51–59.

19. Kripke and Sonnenschein, 1973.

20. R. Klein and R. Armitage, "Rhythms in Human Performance: 1½ Hour Oscillations in Cognitive Style," *Science* 204 (1979), 1326–1328.

21. Bertini *et al.*, 1983.

22. Hodoba, 1986.

23. Joseph, 1988, 1990.

24. R. D. Cartwright, L. W. Tipton, and J. Wicklund, "Focusing on Dreams," *Archives of General Psychiatry* 37 (1980), 275–288.

Chapter 14

1. Some religious groups, including some Muslim and Christian sects, have believed that the left hand (which is controlled by the right brain) is "evil" or uncouth and should be used (in place of the right hand) for functions considered "unclean," such as wiping after defecating. In contrast, the right hand has long been considered good and just. Hence, for example, the origin of the term *righteous* and the need to "sit at the right hand of God," concepts that are also associated with strength and, thus, the right hand. It is for such reasons that, in former years, left-handed children in Catholic and some public schools were forced to write with their right hand. In contrast, some Eastern religions have regarded the left hand as being associated with wisdom.

Chapter 16

1. N. Geschwind, F. A. Quadfasel, and J. M. Segarra, "Isolation of the Speech Area," *Neuropsychologia* 6 (1968), 327–340.
2. J.-P. Sartre, *Being and Nothingness* (New York: Philosophical Library, 1956).

Chapter 17

1. R. M. Rilke, *The Notebooks of Malte Laurids Brigge* (New York: Norton, 1949).

Index

Abuse. *See also* Child abuse
  child/parent ego interactions and,
    211–212, 227, 236–237
  control and, 283–286
  love as reward for, 293–294
  projection and, 361–363
  unconscious compatibility in, 289–292
Acalculia, 51
The Adjusted American (Putney), 326
Adult ego personality, 177–180, 187, 259
  activation of, 192
  child ego personality vs., 198–200
  ego vs., 191
  as mediator, 183
Agnosia, 51
AIDS, 113
Akelaitis, A. J., 103, 104
Alcohol abuse, 258
  child/parent ego interactions and,
    211–212, 217, 219, 222, 224, 227,
    238, 292
  misinterpreted needs and, 322
All-or-none principle, in axon firing,
  263

Amnesia, 114, 330–331
Amygdala, 80, 108–109, 110, 113, 114,
  118, 119, 120, 134, 137, 175
  attachment and, 128–129, 131–132, 133
  pleasure principle and, 126
  psychiatric disturbances and, 121,
    122, 124
Angular gyrus, 75, 80, 93, 190, 329
  functions of, 32–33
Anomia, 51
Anorexia, 72
Anterior commissure, 57
Apathy, 71, 73
Aphasia, 79
  dreams and, 273–274
  expressive (Broca's), 30
  gender differences in recovery from,
    147
  receptive (Wernicke's), 30–32, 73
Apraxia, 51
Archetypes, 18
Arcuate fasciculus, 32, 80
Association needs, 130–136, 167–168
  misinterpretation of, 134–135

Atman, 12
Attachment, 128–130, 131–133
Australopithecines, 151
Axons, 188, 263–266
Aztecs, 18

Ba, 12
Bad needs, 289–292, 304
Bed-wetting, 361
Behaviorism, 1–2
Being and Nothingness (Sartre), 2
Bennett, F. L., 86
Berne, Eric, 170
Blame
  misinterpreted needs and, 325–
    327
  of self, 254–255
Body image, 37–38, 68, 92
  disturbances of, 40–43, 72–73
  ego personalities and, 188–189
  projection and, 344–345
Body language, 33, 60, 159, 248, 249,
    256, 338
Bogen, J. E., 99, 104
Bonding, 151, 155, 158, 160
Bradshaw, J. L., 50
Brady, James, 46
Brahma, 12
Brain, 21. *See also* Brain damage; Left
    brain; Right brain
  evolution of, 22–28, 48–49
  gender differences in, 145–152, 156,
    158
  misinterpreted needs and, 327–328
  repetition and, 262–267
  self-deception and, 335
  as source of mind, 12, 13–14, 16
Brain damage, 14, 43, 329–332, 333. *See
    also* Left-brain damage; Right-brain
    damage
Brain stem, 57, 272
Brion, 104
Broca, Paul, 30
Broca's (expressive) aphasia, 30
Broca's expressive speech area, 30, 31,
    32–33, 75, 79, 93, 329
  gender differences in, 148, 158

Central core, 165–166, 193, 196, 267, 270
Child abuse, 206, 216–218, 262
  anger caused by, 289
  as legacy, 260–261
  love needs and, 168–169
  projection and, 361–362, 363
  scapegoating and, 256, 257
Child ego personality, 166–167, 177,
    178, 180, 185–200, 201–226, 227–
    249, 285
  activation of, 191, 196–197
  adult ego personality vs., 198–200
  complexes and, 270
  control by, 183, 259–260
  denial and, 334
  dreams and, 279–280
  failure to become conscious of, 225
  interaction with other child ego, 292–
    293
  manipulation by, 287
  meeting expectations of, 230–231
  misinterpreted needs and, 318, 326,
    327
  neglect and, 219–221
  positive aspects of, 197–198
  projection and, 352, 362, 363
  rejection and, 219, 255, 289
  repetition compulsions and. *See*
    Repetition compulsions
  scapegoating and, 257
  seeking of in partner, 231–235
  self-deception and, 336
  taking responsibility for, 298, 373,
    376
Child molestation, 213–214, 217, 260,
    262
  inability to recall, 61–62, 85, 89
Children, 115, 160, 167–173. *See also*
    Child abuse; Child ego
    personality; Child molestation;
    Infants; Parents
  complexes and, 269–270
  corpus callosum in, 85, 87–88, 96
  death and, 169–170
  divorce and, 169–170
  egocentric speech in, 94–96
  memory in, 85–90

Children (*Cont.*)
  modeling and mimicry by, 287–288
  names and, 251–252
  projection and, 360–361
  rejection of, 169, 254, 255, 256–257, 289
  repetition and, 261–262
  scapegoating of, 255–259
  television and, 204–206
  unconscious self-image of, 6–7
  unimportance felt by, 172
Chromosomes, 146
Cognition, gender differences in, 146–148
Collective unconscious, 16, 17–19, 119
Communication
  gender differences in, 139–145
  left/right-brain, 58–67, 87
  in repetition compulsions, 239–242
Compensation, 183, 216, 307, 334, 351, 352, 356
  reaction formation and, 308–310
Complexes, 267–270
Conditioning, 1
Confabulation, 43, 71, 72, 178, 297, 302, 305, 335, 341, 358, 376
  explained, 329–333
Conscience, critical, 204–206
Conscious mind, 91, 304. *See also* Left brain
  abuse justified by, 285
  competition for control by, 182–183
  confabulation and, 331, 333
  denial and, 333–334
  ego and, 188
  ego personalities and, 178–179, 192
  evolution of, 48–49
  Freud's theory of, 19–20
  language and, 29–30
  limbic system and, 115–118, 120–121
  misinterpreted needs and, 314–315
  parent ego personality and, 208–209
Conscious self-image, 304
  confabulation and, 332
  denial and, 334
  ego personalities and, 180–182
  misinterpreted needs and, 317

Conscious self-image (*Cont.*)
  projection and, 348, 349, 350, 351, 358
  reaction formation and, 309
  rejection and, 254
  self-deception and, 339, 340
Control, in relationships, 283–286
Corpus callostomy. *See* Split-brain surgery
Corpus callosum, 57, 66, 94, 267
  egocentric speech and, 95–96
  immature, 85, 87–88, 96
Counting, 50–52
Creativity, 201–203
Critical conscience, 204–206
Critical parent, 173–176, 183, 206, 292
Criticism, 281–283
Cro-Magnons, 24–25, 54–55, 77, 150, 153, 154, 157, 202
Crying, 30

Dancing, 33
Danly, M., 79
Daydreams, 69, 96, 271
  hemispheric oscillation in, 273–275
Death, children's perception of, 169–170
Defense mechanisms, 183, 301–312, 373, 376. *See also* specific types
  desires increased by, 307–308
  reasons for, 303–305
Delusions, 72, 79
  of grandeur, 47, 71
  of persecution, 345
Dendrites, 263–266
Denial, 71, 183, 304, 306, 307, 308, 310, 329–342, 360. *See also* Self-deception
  in body image, 40–43, 73
Dennenberg, V. H., 86
Dennis, M., 26
Depression, 71, 73, 216, 218, 362
  dreams as therapy for, 279
  misinterpreted needs and, 327
Diamond, M. C., 86
Dimond, S. J., 104
Disinhibition, 71
Displacement, 329

Displacement (*Cont.*)
   explained, 311–312
   misinterpreted needs and, 325–327
Divorce, children's perception of, 169–
      170
Doty, R. W, 84, 88
Dreams, 33, 67, 68, 113, 201–203, 270–
      280
   archetypes in, 18
   backward, 276
   collective unconscious and, 19
   forgetting of, 272–273
   hemispheric oscillation in, 271–272,
      273–275
   limbic system and, 119
   patterns in, 277
   problem solving in, 275
   repetitive, 277–280
   stimulants of, 275–277
Drug abuse
   axons and, 264–265
   child/parent ego interactions and,
      212–213, 215–216, 217, 219, 227

Eating disorders
   limbic system and, 124, 134–135, 136
   misinterpreted needs in, 321
Ego, 187–188, 267
   activation of, 192
   adult ego personality vs., 191
   dreams and, 274–275
   Freud's theory of, 20–21
   self-deception and, 335
Egocentric speech, 93–96
Ego personalities, 187–192. *See also*
      Adult ego personality; Child ego
      personality; IDfant ego
      personality; Lesser Ego; Parent ego
      personality
   activation of, 191–192
   competition for control among, 182–
      183
   conscious recognition of, 178–179
   self-image and, 180–182, 189–190
Egyptians, 12, 13, 14, 18
Emotional abnormalities. *See* Psychiatric
      disturbances

Emotional blunting, 73
Emotional lability, 71
Emotional memory, 47, 82
Emotions, 33, 43–44, 46–47, 71–74, 86
   gaining control over, 373–374
   in language, 79–80
   limbic system and, 109–110
Epilepsy, 14, 73–74, 98, 113. *See also*
      Seizure disorders
Estrus, 152
Euphoria, 71, 73
Evolution
   of gender differences. *See* Gender
      differences
   of handedness, 49, 51
   of mind-brain, 22–28, 48–49
Expressive (Broca's) aphasia, 30

Familiarity, seeking to maintain, 173,
      227–230, 243–245
Fathers, scapegoating by, 257
Fighting
   love and, 294–295
   sex and, 295–296
Finger agnosia, 51
Finger recognition, 50–52
Food, providing of, 151, 152, 153–156.
      See also Gathering; Hunting
Free association, 267, 280
Freud, Sigmund, 13, 16–17, 18, 19–21,
      51, 125, 187, 204, 274, 277, 316
Frontal lobes, 66–67, 75, 113, 126–127,
      133, 154
   denial and, 334
   dreams and, 274–275
   ego and, 187–188
   evolution of, 24
   in limbic system control, 114–115
   projection and, 344
   repetition and, 266, 267
   self-deception and, 335

Galin, David, 21, 87
Gallagher, R. E., 86
Gambling, 294
Gap filling. *See* Confabulation

Gathering, 148–152, 153, 160–161. *See also* Hunting
  language and, 156–158
Gazzaniga, M. S., 84, 88, 99
Gender differences, 139–161
  in brain function, 145–152, 156, 158
  in language, 139–145, 147–148, 156–158
  limbic system and, 111–113, 152
Genesis, book of, 18, 155, 158
Gerstmann's syndrome, 51
Geschwind, Norman, 104, 331
Gestalt, ability to perceive, 35, 80, 82
Gott, P. S., 26
Gray matter. *See* Neocortex
Greeks, 13, 14
Greenough, W. T., 86
Gur, Reuben, 102

Hallucinations, 14, 271
Hand, the, 50–56
Handedness, 3, 33, 50–52, 115, 148
  dreams and, 273
  evolution of, 49, 51
  gathering and, 149, 151, 158
  hunting and, 149, 151
  language and, 50
  tool making and, 52
Harlow, Harry, 129, 132, 168
Harris, Fred, 170
Hate, projection of, 356–358
Heilman, Kenneth, 79
Hippocampus, 110, 119, 122, 137, 175, 268, 279
  functions of, 113–114
  pleasure principle and, 126
Hippocrates, 13
Homeostasis, 120, 125, 179, 182, 316
Homo erectus, 149–150, 151
Homo habilis, 52
Homophobia, 310
Homo sapiens, 152
Homosexuality, 111–113, 228
Hoppe, K. D., 21
Human face, ability to perceive, 35–37
Humor, sense of, 63, 79

Hunting, 48–49, 148–152, 153, 154–156. *See also* Gathering
  visual-spatial capabilities and, 159–161
Hypergraphia, 14
Hypothalamus, 114–115, 118, 122, 131
  attachment and, 133
  conscious awareness and, 120–121
  ego personalities and, 179
  functions of, 110–113
  misinterpreted needs and, 316, 324, 326
  pleasure principle and, 125, 126
Hypoxia, 114
Hysteria, 47, 71, 72–73

Id, 20–21
IDfant ego personality, 179–180, 187, 192, 199, 259, 297
  activation of, 191
  competition for control by, 183
  misinterpreted needs and, 325
Imagination, 96
Indus Valley people, 12
Infants, 126, 160. *See also* Children
  association needs of, 130–131
  attachment needs of, 132, 133
  brain function in, 150
  language development in, 76–79
  socialization skill abnormalities in, 133–134
  trauma and, 136–138
  unconscious self-concept and, 7
Inference, 43–45
Inferiority complexes, 307
Ink blot tests, 267
Insight, 96
Interhemispheric fissure, 57
Intonation, 43–44
Intuition, 63, 96, 249

James, William, 13
Jaynes, J., 21
Jedynak, 104
Jung, Carl Gustav, 13, 16–17, 18, 19, 21, 181, 202, 268, 277, 352, 375
Justification, 311, 327, 334, 337, 358

Ka, 12
Kant, Immanuel, 12
Kimura, D., 50
Kinsbourne, M., 50
Kraft, R. H., 87
Krashen, S., 26

Language, 22, 30–31, 33, 59–60, 67,
    75–76, 115, 117, 167, 202. *See also*
    Language axis; Linguistic thinking
  angular gyrus and, 32–33
  conscious mind and, 29–30
  in dreams, 274
  egocentric. *See* Egocentric speech
  ego personalities and, 177
  emotional content in, 79–80
  evolution of, 23–24, 25–27
  gender differences in, 139–145, 147–
    148, 156–158
  handedness and, 50
  limbic system and, 76–79
  memory and, 85, 86
  reality confused with, 375–376
  repetition and, 262
  reprogramming via, 373–374
  tool making and, 52–56, 156–158
  unconscious self-image and, 7
Language axis, 31–32, 76, 79, 285, 334,
    375–376
  confabulation and, 329, 330, 331–333,
    376
  projection and, 358
Lansdell, H., 26
Lateralized memory, 81–85, 88
Lateralized specialization, 66, 97, 102
Laughter, uncontrolled, 122–124
Learning, repetition and, 262–267
Left brain, 3, 21, 34, 37, 38, 43, 44, 115–
    118, 119–120, 375. *See also*
    Conscious mind; Left-brain
    damage
  abuse justified by, 285, 291
  association needs and, 134
  body image and, 41–43
  child ego personality and, 186, 187,
    230
  competition for control by, 182

Left brain (*Cont.*)
  complexes and, 270
  confabulation and, 331, 332, 333, 335,
    341
  creativity and, 201–202
  critical parent and, 175
  defense mechanisms and, 304
  dreams and, 67, 201–202, 271–275,
    277
  ego and, 188, 192
  egocentric speech and, 95–96
  ego personalities and, 177, 178
  evolution of, 24, 25–26, 48–49, 51
  functions of, 22, 29–33, 45, 75–76
  gender differences in, 147
  limbic system and, 76–79, 80
  limitations of, 47–48
  manipulation and, 286
  memory and, 42, 81–82, 83–85, 86,
    87, 114, 117
  misinterpreted needs and, 313, 314–
    315, 321, 327–328
  overreliance on, 63, 301–302
  parent ego personality and, 176,
    230
  projection and, 343–344, 345, 348,
    358
  repetition and, 267
  repetition compulsions and, 230,
    240–241, 246
  right-brain communication with, 58–
    67, 87
  right-brain conflicts with, 103–105
  right-brain cooperation with, 67
  right-brain damage and, 27
  self-deception and, 338, 340–341, 342
  self-image and, 6, 180
  split-brain surgery and, 102–103
  supporting role of, 68, 69, 70–71, 96–
    97
  thinking and, 93
  tool making and, 52, 53, 56
  unconscious mind and, 91–92
Left-brain damage, 14, 26–27, 30–32,
    47, 73
Lesser ego, 180, 187, 210, 284, 292
Levy, J., 99

Limbic system, 13, 14, 21, 49, 57, 76–80, 107–138, 167, 168, 169, 185, 248, 256–257, 279, 297, 326, 376
    association needs and, 130–136
    child ego personality and, 186, 221
    choice and, 368, 370
    competition for control by, 182
    complexes and, 268, 270
    confabulation and, 333
    critical parent and, 175
    distrust of functions mediated by, 301
    ego and, 188
    ego personalities and, 177, 179–180, 191
    emotions and, 109–110
    evolution of, 51
    gender differences and, 111–113, 152
    language development and, 76–79
    limitations of, 314–315
    misinterpreted needs and, 127–130, 134–135, 314–315, 319, 328
    nuclei of, 110–114, 118–119, 133, 295
    pleasure principle and, 125–127
    projection and, 344, 348, 349
    self-deception and, 339, 340, 342
    sexual disturbances and, 73, 121–122
    television and, 206
    trauma and, 136–138
    unconscious mind and, 115–120, 127, 136–138
Linguistic thinking, 7, 29, 30, 56, 92-93, 136, 301, 329
Loneliness, 134–135, 136
Love
    abnormalities in, 133–134
    abuse rewarded with, 293–294
    child abuse and, 168–169
    child/parent ego interactions and, 218–219
    fighting and, 294–295
    limbic system and, 109, 110, 113, 120, 128–130, 133–134
    misinterpretation of need for, 317, 319, 320
    projection and, 349–352, 353–356
    sex confused with, 202, 296
LSD, 113

Mania, 47, 71
Manipulation, 286–287
Marriage
    parent ego personality activated in, 238–239
    projection and, 353–356
Masturbation, 122, 128, 137, 252
Mathematics, 22, 25, 29, 48, 114
Meaning, 43–44
Melody, 34, 59–60, 79–80, 201. *See also* Music
    vocal, 43–44, 46, 60, 68, 79
Memory, 97, 178
    in children, 85–90
    collective unconscious and, 19
    confabulation and, 330
    emotional, 47, 82
    lateralized, 81–85, 88
    left-brain, 42, 81–82, 83–85, 86, 87, 114, 117
    limbic system and, 113–114, 119, 138
    repetition and, 263–266
    right-brain, 61–63, 81–82, 83–85, 86, 87, 113
    split-brain surgery and, 84, 86, 87, 106
    unconscious, 81–85
    unconscious self-image and, 6
    verbal, 6, 82, 114, 268
    visual-spatial, 82
Menstrual cycle, 111
Miller, Laurence, 21
Milner, Brenda, 114
Mimicry, 287–288
Mind, 21
    abode of, 14–16
    duality of, 11–14, 16, 335
    evolution of, 22–28, 48–49
    self-deception and, 335
    structure of, 185–187
Misinterpreted needs, 127–130, 134–135, 313–328
    brain in, 327–328
    inappropriate discharge of tension and, 324–327
    projection, displacement, and rationalization in, 325–327

Misinterpreted needs (*Cont.*)
  secondary/primary need confusion
    in, 322–323, 325
  seeking to fulfill, 323–324
Missildine, W. H., 170
Modeling, 287–288
  of abuse, 361–363
Monroe, Marilyn, 217–218
Moses, 14, 155
Mothers
  limbic system in behavior of, 129–130
  scapegoating by, 256–257
Music, 34, 46, 67, 79–80, 201. *See also*
    Melody; Singing
Mythology, collective unconscious and,
    18, 19

Names, power of, 251–253
Naming, 50–52
Neanderthals, 12, 23–24, 52–53, 56, 150
Needs
  bad, 289–292, 304
  defense mechanisms and, 303–304
  good, 304
  misinterpretation of. *See*
    Misinterpreted needs
  primary, 322–323, 325
  secondary, 322–323, 324, 325
Neglect, 40–43, 71, 73, 256
  child ego personality affected by,
    219–221
Neocortex, 86, 110, 119, 126, 180, 316,
    327–328
Nettleton, N., 50
Neurons, 263. *See also* Axons; Dendrites
Neurotransmitters, 263, 265
Nietzsche, Friedrich, 20, 28
Nisbett, R. E., 2
Non-rapid eye movement sleep (N-
    REM), 271–272, 273
"Not OK" feelings, 167–173, 225–226,
    286, 318
Novelly, Robert, 26
N-REM. *See* Non-rapid eye movement
    sleep

Occipital lobes, 32, 272

Old cortex, 109, 110
O'Leary, D. S., 87
Orgasm, 73–74, 133
Original Mind, 17
Ornstein, Robert, 21
Overman, W. H., 84, 88

Paradoxical sleep, 271
Paranoia, 71
Parent ego personality, 175–176, 177,
    178, 179–180, 186, 187, 196, 199–
    200, 201–226, 227–249, 260–261,
    291, 292–293
  activating in one's spouse, 238–
    239
  activation of, 191, 192
  becoming conscious of, 208–209
  contradictory messages of, 207–208
  control by, 183, 259–260
  critical, 173–176, 183, 206, 292
  dreams and, 280
  examining and confronting, 209
  failure to become conscious of, 225
  manipulation by, 286, 287
  meeting expectations of, 230–231
  misinterpreted needs and, 315, 318,
    319
  modeling and mimicry in, 287
  positive aspects of, 208
  projection and, 348, 352, 353, 363
  reaction formation and, 306
  rejection and, 289
  repetition compulsions and. *See*
    Repetition compulsions
  seeking of in partner, 231–235
  self-deception and, 336
  taking responsibility for, 298, 373,
    376
Parents, 5. *See also* Children; Infants;
    Parent ego personality
  busy, 220–221
  confronting of, 374–375
  limbic system in behavior of, 129–130
  "not OK" characteristics of, 225–226
  "not OK" messages sent by, 171–172
  rejection by, 254, 255, 256–257
  unconscious self-image and, 6

Parietal lobes, 32, 37, 38, 39, 40, 51, 72, 73, 75, 79, 148, 188
Pavlov, Ivan, 137, 316
Personal unconscious, 17
Phantom limbs, 38
Piaget, Jean, 94
Pineal gland, 13, 16
Plato, 12, 13
Pleasure, 315, 316
Pleasure principle, 20, 179, 316
    limbic system and, 125–127
Potentiation, of axons, 264
Praying, 22, 30, 331, 332
Preconscious mind, 16, 17, 19–20
Pregnancy, 111
Primary auditory area, 80
Primary needs, 322–323, 325
Primary process, 126
Primary receiving area, 188
Primary unconscious, 118, 119, 268
"Project for a Scientific Psychology" (Freud), 21
Projection, 183, 306, 329, 335, 341–342, 343–363
    in abuse modeling, 361–363
    in body image, 344–345
    children and, 360–361
    explained, 310–311
    hate in, 356–358
    instant dislike as, 347
    love and, 349–352, 353–356
    marriage and, 353–356
    misinterpreted needs and, 325–327
    scapegoating in, 345–346
    sexual, 347–349
    of tension, 358–360
Prosopagnosia, 36
Psychiatric disturbances, 71–72, 73. See also specific disorders
    limbic system and, 121–124
Psychoanalytic theory, 21. See also Freud, Sigmund
Psychotherapy, 266–267
Puerility, 71
Putney, 326

Quadfasel, F. A., 331

Rage, limbic system and, 121, 133
Rape, self-deception and, 336–338
Rapid eye movements (REMs), 173, 271–272, 277
Rationalization, 183, 305, 329, 334, 337
    explained, 310
    misinterpreted needs and, 325–327
Reaction formation, 183, 216, 319, 352, 356
    compensation and, 308–310
    explained, 305–307
Reading, 25, 67
Reagan, Ronald, 46
Receptive (Wernicke's) aphasia, 30–32, 73
Reinforcement, 1, 2
    of abuse, 293–294
    of fighting, 296
Rejection, 219, 254–261, 289
    anticipation of, 245–249
    children and, 169, 254, 255, 256–257, 289
    criticism and, 281–283
    need for, 242–243
    self-blame and, 254–255
    self-image and, 254
Relationship addictions, 297–298
Religion. See Spiritual beliefs
REMs. See Rapid eye movements
Repetition, 261–267
    axons and dendrites in, 263–266
    dreams and, 277–280
Repetition compulsions, 221–225, 227–249
    attraction of opposites in, 236
    in choice of partner, 231–235
    communication breakdown in, 239–242
    meeting expectations in, 230–231
    rejection anticipation in, 245–249
    rejection need in, 242–243
    rewriting of family script in, 238
Reptilian brain, 110
Rescuing, 288, 292
Resting potential, of axons, 263, 264
Right brain, 3, 21, 30, 91–92, 115–118, 119–120, 319. See also Right-brain damage; Unconscious mind

Right brain (*Cont.*)
  child ego personality and, 186, 187, 221
  in children, 169, 171
  competition for control by, 182
  complexes and, 268, 270
  creativity and, 201–203
  critical parent in, 175
  distrust of, 301–302
  dreams and, 33, 67, 68, 201–203, 270–276, 278–279, 280
  ego and, 188
  ego personalities and, 177, 178
  evolution of, 24, 26–27, 33, 48–49, 51
  functions of, 22, 33–47
  gender differences in, 159
  left-brain communication with, 58–67, 87
  left-brain conflicts with, 103–105
  left-brain cooperation with, 67
  left-brain damage and, 26–27
  memory and, 61–63, 81–82, 83–85, 86, 87, 113
  misinterpreted needs and, 327–328
  parent ego personality and, 176
  projection and, 341, 342, 344, 348
  repetition compulsions and, 222, 230, 231, 236, 240–241, 246, 249
  scapegoating and, 256
  self-deception and, 335, 338, 340
  self-image and, 6, 180
  in social-emotional intelligence, 63–66
  split-brain surgery and, 102–103
  in thinking, 96–97
Right-brain damage, 27, 34, 46, 73, 114
  body image and, 40
  confabulation and, 329–330, 331–332, 333
  emotions and, 47
  inference drawing and, 44, 45
  memory and, 82
  projection and, 344–345
  psychiatric disturbances and, 71–72
  sexual disturbances and, 72, 73–74
Risse, G. L., 84, 88
Rosenzweig, M. R., 86
Ross, Eliot, 79

Salamy, A., 87
Sartre, Jean-Paul, 2, 340, 341
Scapegoating, 255–259
  in projection, 345–346
Schizophrenia, 15, 73
Scripts, familial, 238
Secondary needs, 322–323, 324, 325
Secondary unconscious, 119
Segarra, J. M., 331
Seizure disorders, 14, 72, 266. *See also* Epilepsy
Self-blame, rejection and, 254–255
Self-concept. *See* Self-image
Self-control, 298, 373–374
Self-deception, 183, 308, 334, 335–341, 342. *See also* Denial
Self-destructive behavior, 5, 97
  in child/parent ego interactions, 216–218, 219, 221
  projection and, 362
Self-fulfilling prophecies, 196, 197, 326
Self-image, 6–7, 37–38, 304
  abuse and, 285
  adult. *See* Adult ego personality
  confabulation and, 332
  conscious. *See* Conscious self-image
  denial and, 334
  dreams and, 274
  ego personalities and, 180–182, 189–190
  inaccurate, 253
  misinterpreted needs and, 317, 318
  projection and, 346, 348, 349, 350, 351, 358, 362–363
  reaction formation and, 306, 309
  rejection and, 254
  self-deception and, 337, 339, 340
  unconscious. *See* Unconscious self-image
Septal nuclei, 113, 120, 131–132
Sex, 156
  fighting and, 295–296
  hunting/gathering contribution to, 151–152, 153
  limbic system and, 109, 111, 125, 127, 136, 137–138, 295

Sex (*Cont.*)
love confused with, 202, 296
Sex differences. *See* Gender differences
Sexual disturbances, 72, 73–74, 121–122
Sexual molestation. *See* Child
molestation
Sexual projection, 347–349
Shapiro, B. E., 79
Shopping, 160
Sign language, 24
Singing, 22, 30, 33, 48, 331, 332. *See
also* Music
Skinner, B. F., 1
Sleeptalking, 272
Social-emotional disinhibition, 71
Social-emotional intelligence, 63–66,
368–373
Socialization skills, 133–134
Sodium amytal procedures, 46, 84
Speech. *See* Language
Speech release, 79
Sperry, R. W., 99
Spiritual beliefs, 11–12, 13–15, 18
Split-brain surgery, 98–103, 106, 179
behavior following, 100–101
left/right-brain conflicts and, 103–105
left/right-brain cooperation and, 102–
103
memory and, 84, 86, 87, 106
Stereopsis, 68
Stickiness, 132
Stimuli, 1
Strokes, 30, 40, 71–72, 114, 147, 344
Subconscious mind, 16, 68
Suicide, 215, 218
Sumerians, 11–12, 13, 78
Superego, 20–21, 204, 209
Suppression, 304, 307, 334
Swearing, 22, 30
Synapses, 263, 265

Television, 204–206
Temporal lobes, 14, 30, 32, 46, 76, 79–
80, 82, 113, 114, 147, 268
Temporal-sequential functions, 29, 53,
54, 56, 58, 167, 202
dreams and, 276

Temporal-sequential functions (*Cont.*)
ego personalities and, 177–178
evolution of, 48
gender differences in, 147, 148, 156,
158
handedness and, 50, 51, 52, 149, 151
limbic system in, 76, 80
Tension, 315, 317, 333, 342
inappropriate discharge of, 324–327
projection of, 358–360
reduction of, 315–317
Testosterone, 111, 148
Thinking, 92–97
internalization of, 93–95
linguistic. *See* Linguistic thinking
repetition and, 266–267
Tool making, 51–56, 156–158
Touch, sense of, 37
Touching, 256–257
Trauma
complexes and, 270
dreams as therapy for, 279–280
limbic system and, 136–138
repetition in, 262, 266–267
Tucker, Don, 79
Tumors, limbic system, 108, 109, 121
Type A personality, 216

Unconscious child within. *See* Child
ego personality
Unconscious compatibility, 289–292
Unconscious memories, 81–85
Unconscious mind, 1–7, 16–17, 68, 91–
92. *See also* Right brain
competition for control by, 182–183
ego and, 188
ego personalities and, 192
evolution of, 48–49
Freud's theory of, 17, 19–20
limbic system and, 115–120, 127, 136–
138
origins of, 3–4
in repetition compulsions, 236–237,
241–245
spiritual beliefs concerning, 13
Unconscious parent within. *See* Parent
ego personality

Unconscious self-image, 6–7, 304
    denial and, 334
    ego personalities and, 180–182
    misinterpreted needs and, 317
    projection and, 343–344, 362–363
    rejection and, 254
Unimportance, children's feeling of, 172

Vedic literature, 12
Verbal memory, 6, 82, 114, 268
Verbal thinking. *See* Linguistic thinking
Violence, limbic system and, 121, 133
Visual closure, 34–35, 36
Visualization, 266–267
Visual-spatial capabilities, 34, 58, 202
    gender differences in, 146, 148, 159–
        160

Visual-spatial memory, 82
Vocal melody, 43–44, 46, 60, 68, 79
Vygotsky, L. S., 94

Watson, John B., 1
Wernicke, Carl, 31
Wernicke's (receptive) aphasia, 30–32,
        73
Wernicke's receptive language area, 30,
        31–33, 75–76, 93, 329
Whitaker, H. A., 26
Whitman, Charles, 107–108, 121
Wilson, T. D., 2
Wish fulfillment, 126, 274
Word association tests, 267
Writing, 22, 25, 33